U0141242

深智數位
股份有限公司

深智數位
股份有限公司

前言

隨著科技的高速發展，邊緣計算和人工智慧正成為推動社會生產和日常生活變革的引擎。近年來，人工智慧、物聯網、行動網際網路、巨量資料和邊緣計算等技術的迅猛發展，深刻地改變了社會的生產方式，極大地提高了生產效率和社會生產力。邊緣計算和人工智慧這兩個領域的交匯，帶來了前所未有的機遇和挑戰。身為分散式運算模型，邊緣計算使資料的處理不再侷限於中心化的雲端，可以在資料來源附近進行資料處理，從而降低了延遲、提高了效率，更進一步地適應大規模物聯網時代的現實需求。

本書詳細闡述邊緣計算和人工智慧的基礎知識和開發技術，採用專案式開發的學習方法，旨在推動人工智慧人才的培養。全書共 5 章：

第 1 章為邊緣計算與人工智慧概述，主要內容包括邊緣計算概述、人工智慧概述、邊緣計算與人工智慧的結合、邊緣計算與人工智慧的發展歷程、邊緣計算與人工智慧的應用領域。

第 2 章為邊緣計算與人工智慧基本開發方法，主要內容包括邊緣計算與人工智慧框架、邊緣計算的演算法開發、邊緣計算的硬體設計、邊緣計算的應用程式開發。

第 3 章為邊緣計算與人工智慧模型開發，主要內容包括資料獲取與標註、YOLOv3 模型的訓練與驗證、YOLOv5 模型的訓練與驗證、YOLOv3 模型的推理與驗證、YOLOv5 模型的推理與驗證、YOLOv3 模型的介面應用、YOLOv5 模型的介面應用、YOLOv3 模型的演算法設計、YOLOv5 模型的演算法設計。

第 4 章為邊緣計算與人工智慧基礎應用程式開發，主要內容包括人臉門禁應用程式開發、人體入侵監測應用程式開發、手勢開關風扇應用程式開發、視覺火情監測應用程式開發、視覺車牌辨識應用程式開發、視覺智慧抄表應用程式開發、語音窗簾控制應用程式開發、語音環境播報應用程式開發。

第 5 章為邊緣計算與人工智慧綜合應用程式開發，主要內容包括智慧家居系統設計與開發、輔助駕駛系統設計與開發。

本書既可作為高等學校相關專業的教材或教學參考書，也可供相關領域的工程技術人員參考。對於邊緣計算與人工智慧的開發同好，本書也是一本深入淺出、貼近社會應用的技術讀物。

在撰寫本書的過程中，作者借鑑和參考了專家、學者、技術人員的相關研究成果，在此表示感謝。我們盡可能按學術規範予以說明，但難免會有疏漏之處，如有疏漏，請及時透過出版社與作者聯繫。

感謝中智訊（武漢）科技有限公司在本書撰寫過程中提供的幫助，特別感謝電子工業出版社的編輯在本書出版過程中給予的大力支持。

由於本書涉及的知識面廣、撰寫時間倉促，加之作者的水準和經驗有限，疏漏之處在所難免，懇請讀者們和專家批評指正。

作者

2024 年 2 月

繁體中文版說明

本書作者為中國大陸人士，使用軟體為中國國內之雲端平台，因此本書部分圖例使用簡體中文介面，在此特別說明。

目錄

第 1 章　邊緣計算與人工智慧概述

第 2 章　邊緣計算與人工智慧基本開發方法

第 3 章 邊緣計算與人工智慧模型開發

第 4 章 邊緣計算與人工智慧基礎應用程式開發

第 5 章 邊緣計算與人工智慧綜合應用程式開發

1 邊緣計算與人工智慧概述

邊緣計算（Edge Computing）是一種新興的計算模式，是指將計算、儲存和網路等資源盡可能地靠近資料來源和終端使用者，使資料能夠在本地進行處理和分析，從而減少資料傳輸延遲和網路壅塞，提高應用的回應速度和效率。邊緣計算是一種分散式的計算模式，可以將計算和儲存功能從傳統的雲端資料中心遷移到資料生成的邊緣裝置，如手機、平板電腦、智慧穿戴裝置、智慧家居裝置、無人駕駛車輛、工業機器人等。邊緣計算的目標是提供低延遲、高頻寬、高可靠性、高安全性的計算服務，滿足人工智慧、物聯網、5G、工業自動化等應用場景的需求。邊緣計算如圖 1.1 所示。

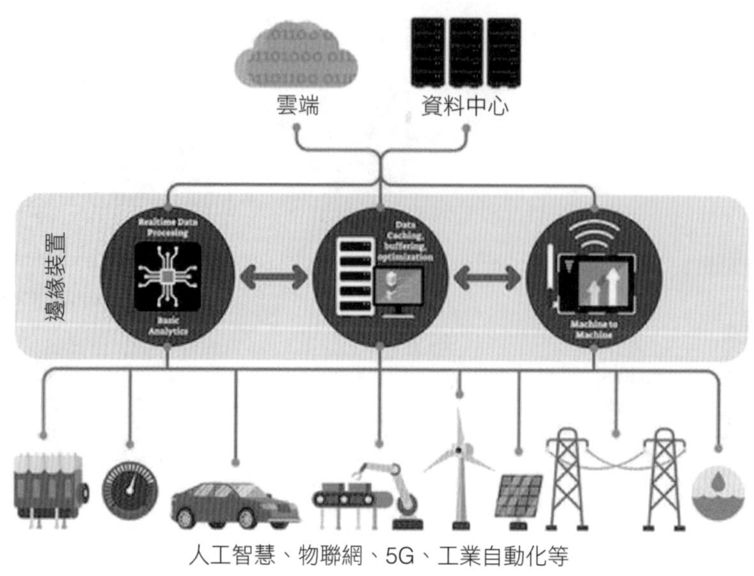

▲ 圖 1.1 邊緣計算

邊緣計算和人工智慧（Artificial Intelligence，AI）是相互連結的，它們可以相互促進和增強。

1.1 邊緣計算概述

與雲端運算相比，邊緣計算強調在離終端使用者和資料來源更近的位置進行資料處理，以提供更低的延遲、更高的頻寬和更好的使用者體驗。

邊緣計算的目標是透過在邊緣裝置上進行即時的資料處理、分析和決策，使算力更接近資料來源，從而減少對雲端運算的依賴。邊緣裝置可以是智慧型手機、物聯網裝置、感測器、邊緣伺服器等。邊緣計算使得資料可以在邊緣裝置上得到處理，只需要將必要的結果傳輸到雲端，從而降低資料傳輸延遲和網路壅塞，減少對頻寬和儲存資源的需求。

邊緣計算的核心思想是將資料和計算任務分配到不同的節點上，以實現更快速、更高效的資料處理。邊緣計算可以將計算任務分配到更靠近資料來源的裝置上，減少了資料傳輸的時間和頻寬消耗，能夠更進一步地保護資料的安全性和隱私性。另外，邊緣計算還可以分離網路和應用程式，提高應用程式的回應速度和可靠性。

邊緣計算的主要組成部分包括邊緣裝置、邊緣閘道和邊緣服務。邊緣裝置是指能夠執行應用程式和進行資料處理的裝置，如智慧型手機、平板電腦、感測器、攝影機等。邊緣閘道則是指連接邊緣裝置和雲端伺服器的網路裝置，可以提供網路傳輸、儲存和計算等服務。邊緣服務是一種透過雲端和邊緣裝置之間協作方式，提供給使用者服務和應用程式的方法。

邊緣計算可以應用於多個領域，如智慧製造、智慧城市、智慧醫療、智慧交通等。在智慧製造領域，邊緣計算可以用於工業物聯網，以便對裝置狀態和生產過程進行監控和最佳化；在智慧城市領域，邊緣計算可以用於智慧交通、智慧能源管理和智慧環境監測等。

1.2 人工智慧概述

人工智慧（Artificial Intelligence，AI）是一門研究如何使電腦模仿人類智慧的學科，涵蓋了多個技術領域，如機器學習、深度學習、自然語言處理、電腦視覺和專家系統等。

人工智慧的目標是使電腦具備像人類一樣的智慧和學習能力，如理解、推理、決策能力。透過學習和分析大量的資料，人工智慧可以發現模式、趨勢和規律等資訊，並根據這些資訊做出預測和決策。人工智慧還可以透過自然語言處理技術，使電腦理解和處理人類的語言。

人工智慧的應用非常廣泛，涵蓋了許多領域。在醫療領域，人工智慧可以幫助醫生進行診斷和治療，提高醫療效果和準確性；在交通領域，人工智慧可以用於智慧交通和自動駕駛，提高交通安全和效率；在金融領域，人工智慧可以用於風險評估、詐騙檢測和智慧投資等。

人工智慧的發展受益於巨量資料技術的發展、運算能力的提升，以及演算法的進步。隨著雲端運算和邊緣計算的發展，人工智慧的應用變得更加普遍和即時；但人工智慧也帶來了許多挑戰，如資料隱私、倫理道德和社會影響等方面的問題。

人工智慧正在改變我們的工作方式和生活方式。隨著技術的不斷進步和創新，人工智慧將在更多領域得到應用和發展，為人類帶來更多的便利。

1.3 邊緣計算和人工智慧的結合

邊緣計算提供了一種將運算資源和資料處理能力遷移到資料來源或終端使用者的計算模式，人工智慧則是一種模擬和實現人類智慧的技術和方法。當二者結合在一起時，可以實現更高效、更強大的智慧應用和服務。邊緣計算與人工智慧的結合可以帶來以下優勢：

（1）降低延遲：人工智慧應用通常需要大量的運算資源和較高的資料處理能力，將人工智慧模型和演算法部署在邊緣裝置或邊緣伺服器上，可以減少資料傳輸的距離、降低延遲，使即時回應和決策成為可能。

（2）保護隱私：某些人工智慧應用需要處理敏感性資料，如人臉辨識、語音辨識等，將人工智慧應用遷移到邊緣裝置，可以在本地進行資料處理和分析，減少了將敏感性資料傳輸到雲端的風險，增強了隱私的保護。

（3）支援離線執行：邊緣計算使人工智慧應用可以在離線環境中執行，邊緣裝置可以儲存和執行人工智慧模型，在網路連接不穩定或無法連接雲端時，仍然可以進行智慧決策和任務執行。

（4）節約頻寬：人工智慧應用通常需要傳輸和處理大量的資料，將人工智慧應用遷移到邊緣裝置，可以減少對網路頻寬的依賴，降低傳輸成本、減緩網路壅塞。

（5）實現分散式學習：邊緣計算可以支援分散式的人工智慧模型訓練和學習，多個邊緣裝置可以協作工作，共用本地的資料和模型，進行聯合學習和模型更新，提高整體智慧水準。

邊緣計算為人工智慧應用提供了更強大、更智慧的計算和決策能力，使人工智慧應用更接近資料來源和終端使用者，降低了延遲、提高了隱私保護能力、支援離線執行，並促進了分散式學習和合作的發展。邊緣計算和人工智慧的結合將為各行各業帶來更多智慧化的創新和應用。

1.4 邊緣計算與人工智慧的發展歷程

邊緣計算和人工智慧的發展歷程如下：

1）早期階段

邊緣計算和人工智慧是在不同時期被提出的。邊緣計算是在 2009 年由 IBM 提出的，旨在將計算和儲存資源盡可能地遷移到資料來源和終端使用者。人工智慧作為一個學科和研究領域，始於 20 世紀 50 年代。在邊緣計算的早期，邊緣計算和人工智慧並沒有明確的聯繫。

2）人工智慧在雲端運算的推動下得到快速發展

在人工智慧的發展過程中，雲端運算有著至關重要的作用。隨著雲端運算技術的成熟和普及，人工智慧應用可以將資料傳輸到雲端進行大規模的計算和訓練。在這種模式下，雲端運算提供了強大的計算和儲存能力，但也存在延遲和資料隱私等問題。

3）邊緣計算的興起

隨著物聯網、智慧裝置和感測器的快速發展，相關的應用對即時性和低延遲的需求越來越高。邊緣計算應運而生，它將運算資源遷移到資料來源和終端使用者的邊緣位置，實現了快速回應和即時決策。邊緣計算的興起為人工智慧應用提供了更好的計算和處理環境。

4）邊緣計算與人工智慧的結合

近年來，邊緣計算和人工智慧開始相互融合。邊緣裝置上的小型化、低功耗的運算資源使得其可以承載一部分人工智慧應用的任務，如影像辨識、語音辨識、自然語言處理等。在這種模式下，人工智慧應用中的推理和決策過程可以在邊緣裝置上進行，減少了人工智慧應用與雲端之間傳輸的資料量和延遲。

5）分散式智慧和邊緣智慧

邊緣計算和人工智慧的結合還催生了分散式智慧和邊緣智慧的概念。邊緣裝置可以共用本地的資料和模型，進行聯合學習和模型更新，實現智慧的協作工作；同時，邊緣裝置上的智慧模型可以不斷學習和最佳化，提高自身的智慧水準。

邊緣計算提供了更接近資料來源和終端使用者的計算和處理能力，為人工智慧應用帶來了更低的延遲、更高的即時性、更好的隱私保護。隨著分散式智慧和邊緣智慧的興起，邊緣計算和人工智慧將進一步融合，為智慧化的應用場景帶來更多創新和發展。

1.5 邊緣計算與人工智慧的應用領域

邊緣計算和人工智慧在許多領域中都有廣泛的應用，以下是邊緣計算和人工智慧在一些主要領域的應用範例。

1）智慧交通

邊緣計算和人工智慧可用於交通管理和智慧交通系統。透過在邊緣裝置上部署人工智慧演算法，可以即時地進行交通監測、車輛辨識、交通流量最佳化和事故預測等。邊緣計算和人工智慧在智慧交通的應用如圖 1.2 所示。

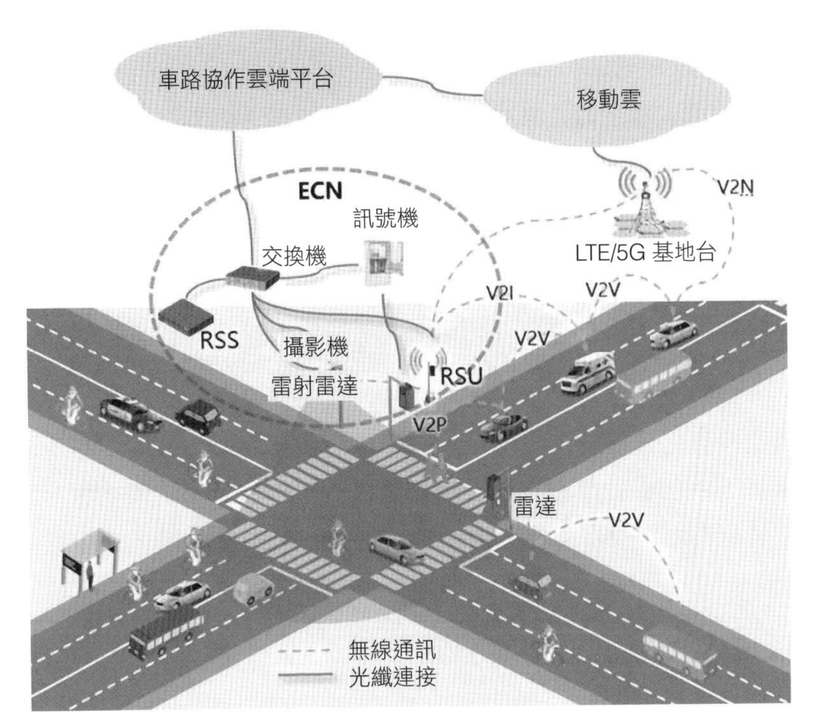

▲ 圖 1.2 邊緣計算和人工智慧在智慧交通的應用

以下是一些在智慧交通系統中應用邊緣計算和人工智慧的範例：

（1）即時交通監測和預測：透過攝影機、感測器等邊緣裝置，可以即時地收集交通資料，如車流量、車速、擁堵情況等，這些資料可以在邊緣裝置上進行即時處理和分析，以實現交通狀態的即時監測和預測。

（2）車輛辨識和行為分析：透過在邊緣裝置上進行影像辨識和行為分析，可以自動辨識車輛，收集車型分類、車道偏移監測等資料，這些資料可以用於交通流量統計、違章檢測和交通事故預測等應用。

（3）交通訊號最佳化：利用邊緣裝置上的人工智慧演算法，可以對交通訊號進行即時最佳化排程。邊緣裝置可以收集交通資料並進行即時的交通流分析，以確定最佳的交通訊號控制策略，從而最佳化交通流量、減少擁堵和提高交通效率。

（4）事故預測和智慧導航：透過在邊緣裝置上進行資料分析和模式辨識，可以實現交通事故的預測和智慧導航系統。邊緣裝置可以收集交通、天氣等資料，並利用人工智慧演算法來預測潛在的交通事故，為駕駛員提供即時的導航建議。

（5）自動駕駛和車聯網：邊緣計算和人工智慧在自動駕駛與車聯網領域中也起著重要的作用。透過在邊緣裝置上部署人工智慧模型和演算法，可以實現自動駕駛車輛的感知、決策和控制；同時，邊緣計算也可以實現車輛之間的通訊和協作，提高交通安全和效率。

上述這些應用可使智慧交通系統即時監測交通狀態、最佳化交通流量、預測交通事故，並為駕駛員和交通管理部門提供即時資訊和決策支援。邊緣計算的優勢在於能夠將智慧計算和資料處理能力遷移到交通源頭，減少資料傳輸延遲，並提供即時的回應和決策能力。

2）工業自動化

邊緣計算和人工智慧在工業自動化領域中有重要應用。透過將人工智慧模型和演算法部署在邊緣裝置上，可以實現即時的裝置監測、故障診斷和預測性維護，從而提高生產效率和降低故障風險。以下是一些在工業自動化中應用邊緣計算和人工智慧的範例：

（1）即時裝置監測：邊緣裝置可以搭載感測器和監測裝置，用於即時監測工業裝置的狀態和性能參數，如溫度、壓強、震動等。邊緣計算將資料的收集和處理遷移到裝置端，減少了資料傳輸延遲，實現了即時監測和控制。

（2）故障診斷和預測性維護：透過在邊緣裝置上部署人工智慧模型和演算法，可以對工業裝置進行即時的故障診斷和預測性維護。邊緣裝置可以即時分析感測器擷取的資料，並與預先訓練好的模型進行比對，以檢測裝置故障的跡象，提前預測裝置的維護需求。

（3）生產過程的最佳化：邊緣計算和人工智慧可用於生產過程的最佳化，提高生產效率和產品品質。透過在邊緣裝置上進行即時的資料分析和演算法推理，可以對生產過程進行監測和最佳化，實現即時的排程和控制。

（4）品質控制和缺陷檢測：邊緣計算和人工智慧可用於品質控制和缺陷檢測。透過在邊緣裝置上進行影像處理和模式辨識，可以對產品品質進行檢測和分類，對生產線上的缺陷進行即時辨識和警告。

（5）工人安全和人機協作：邊緣計算和人工智慧在提升工人安全和人機協作方面也能發揮重要的作用。邊緣裝置可以即時監測工作環境中的安全風險，並透過人工智慧演算法進行預警和控制；此外，邊緣計算還可以實現機器人和工人之間的即時協作和互動。

上述這些應用可使工業自動化系統更加智慧化、高效化和可靠化。邊緣計算將人工智慧模型和演算法遷移到裝置端，實現了即時的資料分析和決策，降低了對雲端運算的依賴，提供了更快的回應速度和更強的隱私保護。

3）智慧城市

邊緣計算和人工智慧在智慧城市領域中發揮著重要的作用。透過在邊緣裝置上進行資料分析和決策，可以實現智慧路燈控制、垃圾管理、環境監測、智慧保全等功能，提升城市的可持續性和居民的生活品質。邊緣計算和人工智慧在智慧城市中的應用如圖 1.3 所示。

以下是一些在智慧城市中應用邊緣計算和人工智慧的範例：

（1）智慧路燈控制：邊緣裝置透過可以感知環境和交通狀況的感測器，即時控制路燈的亮度和開關；透過使用人工智慧演算法和資料分析，根據即時需求和節能目標來最佳化路燈的控制策略，提高能源利用效率。

▲ 圖 1.3 邊緣計算和人工智慧在智慧城市中的應用

（2）垃圾管理：邊緣計算和人工智慧可用於垃圾管理系統的最佳化。透過在垃圾桶或垃圾箱上安裝感測器並在感測器上執行人工智慧演算法，可以即時監測垃圾量，最佳化垃圾收集的路線和時間，減少垃圾收集車輛的行駛距離和成本。

（3）環境監測：透過配備各種感測器，邊緣裝置可以即時監測環境參數，如空氣品質、雜訊、溫度等。透過在邊緣裝置上部署人工智慧模型和演算法，可以對環境資料進行即時分析和預警，為城市居民提供健康舒適的生活環境。

（4）智慧保全：邊緣計算和人工智慧可用於智慧保全系統。透過在邊緣裝置上部署視訊監控和影像辨識演算法，可以實現即時的視訊監控和異常行為檢測。邊緣裝置可以自動辨識異常時間並及時警告，提高城市的安全性。

（5）公共服務最佳化：邊緣計算和人工智慧可用於公共服務的最佳化，如智慧公共汽車網站、智慧停車管理和智慧公共設施管理。透過在邊緣裝置上進行資料分析和決策，可以提供個性化的公共服務，提升城市居民的生活品質。

上述這些應用可使智慧城市更加智慧化、高效化和可持續化。邊緣計算將人工智慧模型、演算法和決策遷移到裝置端，實現了即時的資料處理和決策能力，降低了對雲端運算的依賴，並提供了更快的回應速度和更強的資料隱私保護。

4）醫療保健

邊緣計算和人工智慧在醫療保健領域有廣泛應用。透過在邊緣裝置上進行即時的生物訊號監測、健康資料分析和遠端醫療，可以實現個性化的醫療診斷、疾病預測和健康管理。邊緣計算和人工智慧在醫療保健領域中的應用如圖 1.4 所示。

▲ 圖 1.4 邊緣計算和人工智慧在醫療保健領域中的應用

以下是一些在醫療保健中應用邊緣計算和人工智慧的範例：

（1）遠端醫療：邊緣計算和人工智慧可以用於遠端醫療服務。透過在邊緣裝置上部署視訊通訊和醫學影像分析演算法，可以實現醫生和患者之間的遠端即時交流和診斷。邊緣裝置可以提供高品質的視訊傳輸和影像分析服務，減少對網路頻寬的需求，提供即時的醫療服務。

（2）健康監測與預警：透過在邊緣裝置上配備感測器和監測裝置，可以即時監測患者的健康指標，如心跳、血壓、血糖等；透過在邊緣裝置上部署人工智慧模型和演算法，可以對患者的健康資料進行即時分析和預警，及時發現異常情況並提供相應的處理建議。

（3）醫療影像分析：邊緣計算和人工智慧在醫學影像分析方面有廣泛應用。透過在邊緣裝置上部署醫學影像處理和辨識演算法，可以實現對 X 射線、CT 掃描、MRI 等醫學影像的自動分析和診斷。邊緣裝置可以減少影像資料傳輸延遲，提供即時的影像分析結果，有助醫生快速做出準確的診斷。

（4）智慧藥物管理：邊緣計算和人工智慧可用於智慧藥物管理系統。透過在邊緣裝置上部署人工智慧演算法和感測器，可以對藥物的儲存、配送和用量進行即時監測和管理。邊緣計算和人工智慧的應用，可以提供準確的用藥提醒和用量控制，減少藥物的錯誤使用和不良反應的發生。

（5）疾病預測和預防：邊緣計算和人工智慧可以用於疾病預測和預防。透過在邊緣裝置上進行資料分析和模式辨識，可以利用患者的健康資料和生活習慣來預測潛在的疾病風險，並提供個性化的預防措施和建議。

上述這些應用可使醫療保健更加智慧化、個性化和可及性。邊緣計算將人工智慧模型和演算法遷移到醫療裝置端，實現了即時的資料處理和決策能力，提供了更快速、準確和個性化的醫療服務。此外，邊緣計算還降低了對網路頻寬和雲端運算的依賴，增加了資料隱私的保護。

5）零售業

邊緣計算和人工智慧可用於零售業的個性化行銷和供應鏈管理。透過在邊緣裝置上部署即時的使用者行為分析和推薦演算法，可以為消費者提供個性化的產品推薦和購物體驗。同時，邊緣計算還可以在即時庫存管理、物流最佳化和預測需求等方面發揮作用。以下是一些在零售業中應用邊緣計算和人工智慧的範例：

（1）個性化推薦：透過在邊緣裝置上部署人工智慧模型和演算法，可以即時分析顧客的購買歷史、偏好和行為，並根據這些資訊提供個性化的產品推薦。邊緣計算能夠處理巨量的資料，快速生成推薦結果，提升顧客的購物體驗和銷售轉換率。

（2）庫存管理：邊緣裝置可以即時監測零售店鋪的庫存情況，並透過人工智慧演算法進行預測和最佳化。基於銷售資料和供應鏈資訊，邊緣計算可以幫助零售商進行精準的庫存管理，避免庫存過剩或缺貨，提高營運效率和客戶滿意度。

（3）智慧支付：邊緣計算和人工智慧可用於智慧支付。透過在邊緣裝置上部署人臉辨識、指紋辨識和聲紋辨識等技術，可以實現安全、快速和無接觸的支付。邊緣計算能夠處理本地支付交易，減少對雲端運算的依賴和支付延遲。

（4）即時分析和預測：邊緣裝置可以收集和分析即時的銷售資料、顧客行為和市場趨勢等資訊。透過在邊緣裝置上部署人工智慧模型和演算法，可以進行即時的資料分析和預測，快速做出決策和調整銷售策略。

（5）智慧保全和防詐騙：邊緣計算和人工智慧可用於零售店鋪的保全和詐騙檢測。透過在邊緣裝置上部署視訊監控和影像辨識演算法，可以即時監測零售店鋪內的安全情況和異常行為。邊緣計算可以自動辨識潛在的詐騙行為並及時警告，提高安全性，保護零售商的利益。

上述這些應用可使零售業更加智慧化、高效化和個性化。邊緣計算將人工智慧演算法和決策遷移到裝置端，實現了即時的資料處理和決策能力，降低了對雲端運算的依賴，提供了更快的回應速度和更強的資料隱私保護。

6）農業

邊緣計算和人工智慧在農業領域中有廣泛的應用。透過在農田、溫室等邊緣裝置上部署感測器和人工智慧演算法，可以實現即時的土壤監測、作物生長預測、灌溉控制和病蟲害預警，提高農業生產的效率和可持續性。邊緣計算和人工智慧在智慧農業中的應用如圖 1.5 所示。

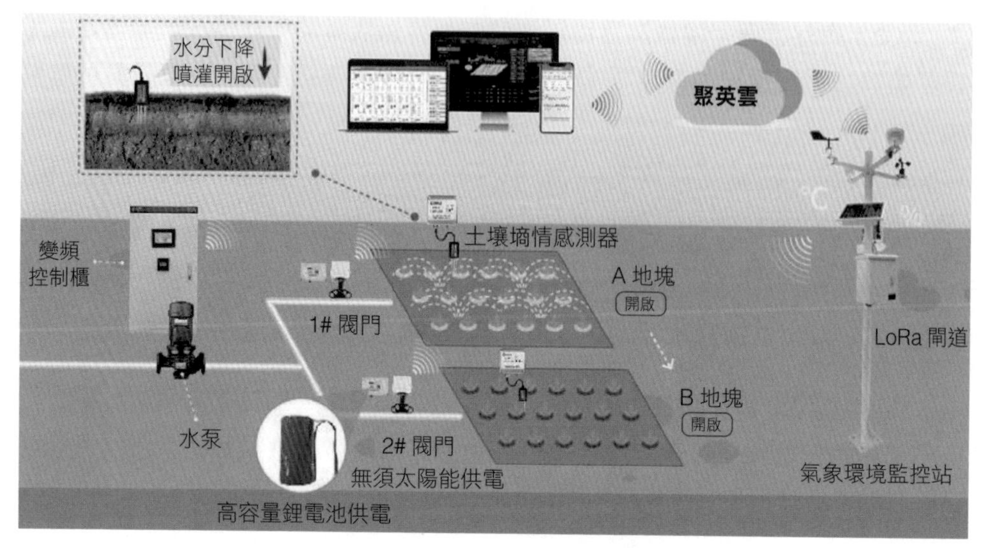

▲ 圖 1.5 邊緣計算和人工智慧在智慧農業中的應用

以下是一些在農業中應用邊緣計算和人工智慧的範例：

（1）作物監測和管理：透過在農田中部署感測器和邊緣裝置，可以即時監測土壤濕度、溫度、光照等環境指標，並利用人工智慧演算法分析和預測作物的生長情況。邊緣計算可以提供作物生長模型和決策支援，最佳化灌溉、施肥和病蟲害管理，提高作物的產量和品質。

（2）智慧灌溉系統：透過在邊緣裝置上部署感知技術和人工智慧演算法，可以實現智慧灌溉系統。邊緣裝置可以即時監測土壤濕度、氣象條件和作物需水量，並根據資料進行智慧決策和控制灌溉裝置。這樣可以實現精確的灌溉，避免浪費水資源，提高水資源利用效率。

（3）無人機農業：邊緣計算和人工智慧在無人機農業中有重要的應用。透過在邊緣裝置上部署影像辨識和人工智慧演算法，無人機可以即時擷取農田的影像資料，並對作物生長、病蟲害和營養狀態進行分析和辨識，及時發現問題並採取措施，提高農田管理的效果。

（4）農產品品質檢測：邊緣裝置可以對農產品進行品質檢測。透過在邊緣裝置上部署影像處理和人工智慧演算法，可以即時分析農產品的外觀、大小、成熟度等特徵，進行品質評估和分級，提高農產品的市場競爭力和溯源能力。

（5）預測和決策支援：邊緣計算和人工智慧可以利用農業資料進行預測和決策支援。透過在邊緣裝置上部署決策演算法和預測模型，可以分析氣象資料、市場需求和供應鏈資訊，預測作物產量、市場價格和最佳銷售策略，幫助農民制訂合理的種植計畫和銷售策略，提高農業經濟效益。

上述這些應用可使農業更加智慧化、高效化和可持續發展。邊緣計算和人工智慧將決策和分析能力遷移到農田現場，實現了即時的資料處理和決策能力，減少了對雲端運算的依賴，提供了更快速的回應和更好的資料隱私保護。

此外，邊緣計算和人工智慧還可以用於能源管理、金融服務、環境保護等多個領域。隨著技術的不斷發展和創新，邊緣計算和人工智慧的應用領域將繼續擴大，並為各行各業帶來更多的智慧化解決方案。

1.6 本章小結

邊緣計算和人工智慧是兩個相互連結且互相促進的技術領域。邊緣計算強調將資料處理和決策遷移到離資料來源更近的邊緣裝置，以實現即時回應、降低延遲和減少對雲端運算的依賴。人工智慧則涉及機器學習、深度學習和自然語言處理等技術，使電腦能夠模仿人類智慧，進行自主學習和自主決策。

邊緣計算和人工智慧的結合為許多領域帶來了巨大的創新和改進，如智慧物聯網、智慧城市、智慧工廠、智慧交通和智慧農業等。邊緣計算和人工智慧的應用使得裝置能夠即時處理大量的資料，並做出智慧決策，從而提供更快速、高效和個性化的服務。

透過將人工智慧模型和演算法部署在邊緣裝置上，邊緣計算使得資料可以在本地進行處理和分析，減少了資料傳輸的銷耗和延遲，增強了資料隱私和安全性。同時，人工智慧為邊緣計算提供了強大的分析能力和智慧決策支援，使得邊緣裝置能夠更進一步地理解和應對不同的場景和需求。

邊緣計算和人工智慧的結合為多個領域帶來了許多創新和改進的機會，推動了智慧化和自動化的發展。隨著邊緣計算和人工智慧技術的不斷演進和成熟，將使更多的應用場景和領域從中受益，為人們的生活和工作帶來更多的便利。

邊緣計算與人工智慧基本開發方法

本章結合 AiCam 平臺學習邊緣計算與人工智慧基本開發方法，本章節內容包括：

（1）邊緣計算與人工智慧框架：主要介紹邊緣計算的參考框架，AiCam 平臺的執行環境、主要特性、開發流程、主程式 aicam 和啟動指令稿，AiCam 平臺的組成、演算法、部分案例，邊緣計算的硬體設計平臺，以及相應的開發步驟和驗證。

（2）邊緣計算的演算法開發：主要介紹機器視覺應用導向的邊緣計算（邊緣視覺）框架、演算法介面和演算法設計，以及相應的開發步驟和驗證。

（3）邊緣計算的硬體設計：主要介紹邊緣計算的虛擬模擬平臺、硬體偵錯和應用程式開發，以及相應的開發步驟和驗證。

（4）邊緣計算的應用程式開發：主要介紹邊緣視覺的應用程式開發邏輯、開發框架、開發介面和開發工具，以及相應的開發步驟和驗證。

2.1 邊緣計算與人工智慧框架

邊緣計算在物端或資料端附近的網路邊緣側，融合了網路、計算、儲存、應用等核心能力，能夠就近提供邊緣智慧服務，滿足敏捷連接、即時業務、資料最佳化、應用智慧、安全與隱私保護等方面的關鍵需求。作為連接物理世界和數位世界的橋樑，邊緣計算能夠實現智慧資產、智慧閘道、智慧系統和智慧服務。

本節的基礎知識如下：

（1）掌握邊緣計算的參考框架。

（2）了解 AiCam 平臺的執行環境、開發流程及典型應用案例。

（3）結合邊緣計算框架和 AiCam 平臺，了解邊緣計算閘道和邊緣計算平臺。

（4）結合人臉辨識案例，掌握在 AiCam 平臺上邊緣視覺應用程式開發的全流程。

2.1.1 原理分析與開發設計

2.1.1.1 邊緣計算的參考框架

邊緣計算的參考框架是基於模型驅動工程（Model-Driven Engineering，MDE）方法設計的，它對物理世界和數位世界的知識進行模型化，從而實現了物理世界和數位世界的協作、跨產業的生態協作。邊緣計算的參考框架如圖 2.1 所示，該框架減少了系統的異質性，簡化了跨平臺的移植，能夠有效支撐系統的全生命週期活動。

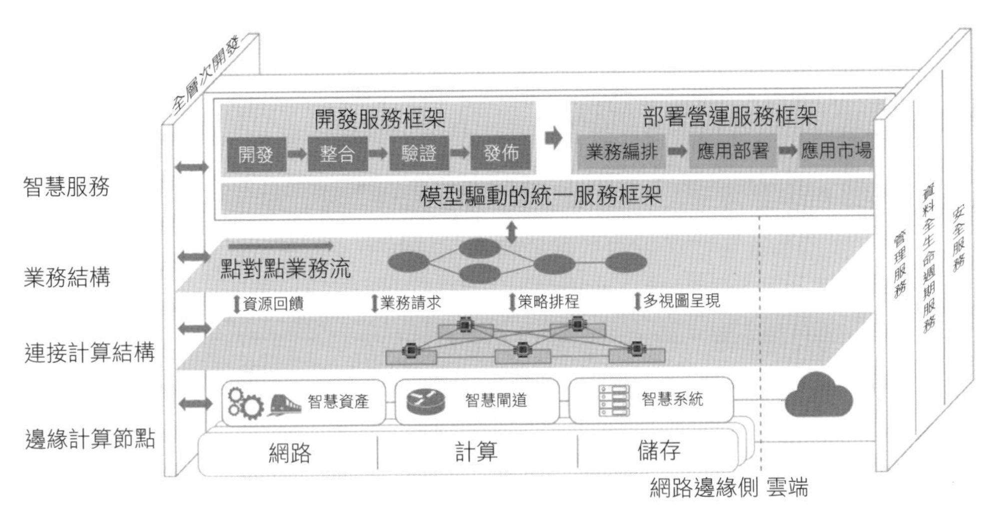

▲ 圖 2.1 邊緣計算的參考框架

在邊緣計算的參考框架中，每層都提供了模型化的開放介面，實現了框架的全層次開放。從橫向看，智慧服務基於模型驅動的統一服務框架，透過開發服務框架和部署營運服務框架實現了開發與部署的智慧協作、軟體開發介面的一致性和部署營運的自動化。智慧服務透過模型化的工作流〔即業務結構（Fabric）〕定義了點對點業務流，實現了敏捷業務；連接計算結構針對業務遮蔽了邊緣智慧分散式架構的複雜性；邊緣計算節點可相容多種異質連接，支援即時處理與回應，提供了軟硬一體化的安全服務。從縱向看，透過管理服務、資料全生命週期服務和安全服務實現了業務的全流程、全生命週期的智慧服務。

人工智慧開發平臺 AiCam 可用於開發部署與影像辨識、影像分析和電腦視覺相關的人工智慧應用的工具和框架。AiCam 平臺提供了豐富的功能和函式庫，使開發者能夠建構高性能的機器學習和深度學習模型，從而實現自動化的影像處理和視覺分析任務。人工智慧開發平臺一般具有以下特點：

（1）資料管理和前置處理：提供用於處理和管理影像資料的工具，可進行資料前置處理（如影像標準化、尺度調整、增強，以及資料清洗），以確保資料品質和一致性。

（2）模型訓練和調優：提供強大的機器學習和深度學習框架，如 TensorFlow、PyTorch 和 Keras 等，以支持影像分類、物件辨識、語義分割等任務的模型訓練；提供預訓練的模型和經過驗證的網路框架，以便開發者在此基礎上進行遷移學習和微調，從而加快模型開發和訓練的過程。

（3）模型部署和推理：提供用於將訓練好的模型部署到生產環境中的工具和介面，這些工具和介面可以將模型部署為 API 或整合到現有的應用程式中，提供高性能的推理引擎，以便即時處理和分析影像資料。

（4）輔助工具和函式庫：提供各種輔助工具和函式庫，以簡化開發過程並提高開發效率。輔助工具和函式庫提供了影像註釋和標註工具，用於生成訓練集；還提供了模型評估和驗證工具，以衡量模型的性能和準確性。

（5）可擴充性和靈活性：通常具有良好的可擴充性和靈活性，以適應不同規模和要求的專案。人工智慧開發平臺可以在本地電腦或雲端環境中執行，支援平行計算和分散式訓練，以處理大規模的影像資料和複雜的計算任務。

2.1.1.2 AiCam 平臺

AiCam 平臺是人工智慧開發導向的一套開發系統，其主介面如圖 2.2 所示，可以實現數位影像處理、機器視覺、邊緣計算等應用，內建的 AiCam 核心引擎整合了演算法、模型、硬體、應用輕量級開發框架，能夠快速整合和開發更多的案例。

▲ 圖 2.2 AiCam 平臺的主介面

1）執行環境

AiCam 平臺採用 B/S 架構，其組成如圖 2.3 所示，使用者透過瀏覽器即可執行專案。人工智慧演算法模型和演算法透過邊緣本地雲端服務的方式為應用提供互動介面，軟體平臺可部署到各種邊緣端裝置執行，包括 GPU 伺服器、CPU 伺服器、ARM 開發板、百度 EdgeBorad 開發板（FZ3/FZ5/FZ9）、英偉達 Jetson 開發板等。

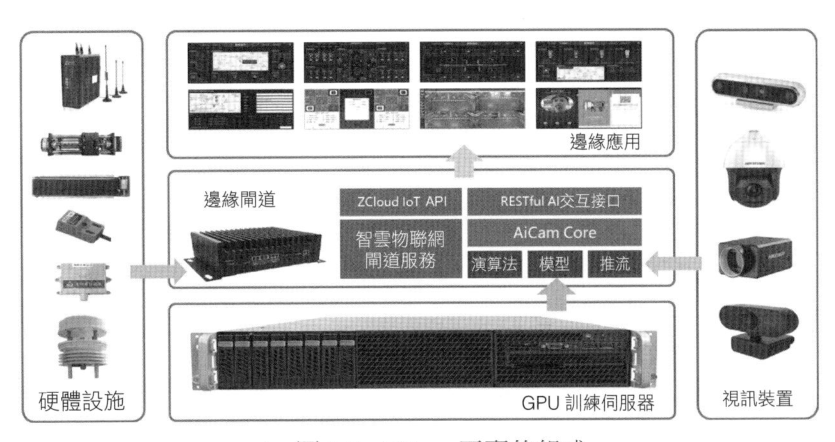

▲ 圖 2.3 AiCam 平臺的組成

2）主要特性

AiCam 平臺的主要特性如下：

（1）可實現多平臺邊緣端部署：AiCam 平臺支援 x86、ARM、GPU、FPGA、MLU 等異質計算環境的部署和離線計算的推理，可滿足多樣化的邊緣專案應用需求。

（2）即時視訊推送分析：支援本地攝影機、網路攝影機的連線，提供即時的視訊推串流服務，透過 Web HTTP 介面可實現快速的預覽和存取。

（3）統一模型呼叫介面：不同演算法框架採用統一的模型呼叫介面，開發者可以輕鬆切換不同的演算法模型，進行模型驗證。

（4）統一硬體控制介面：AiCam 平臺連線了物聯網雲端平台，不同的硬體資源採用統一的硬體控制介面，遮敝了底層硬體的差異，方便開發者連線不同的控制裝置。

（5）清晰簡明應用介面：採用了基於 Web 的 RESTful 介面，可快速地進行模型的呼叫，並即時傳回視訊分析的計算結果影像和計算結果資料。

3）開發流程

AiCam 平臺整合了演算法、模型、硬體、應用輕量級開發框架，其開發框架如圖 2.4 所示。

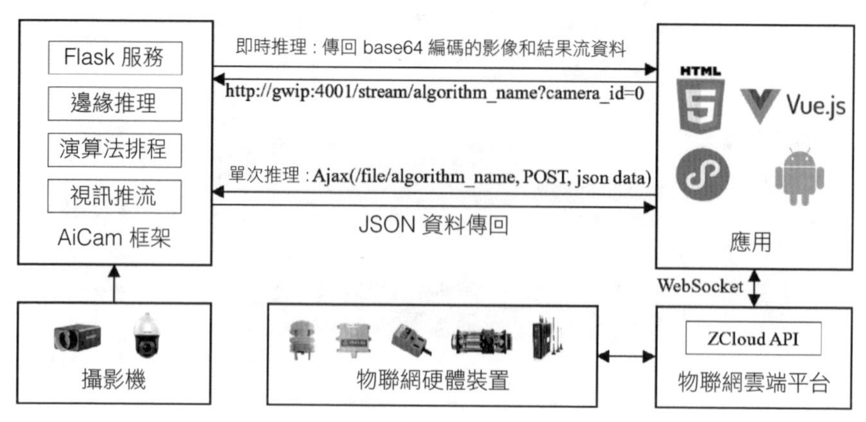

▲ 圖 2.4 AiCam 平臺的開發框架

AiCam 平臺的功能框架如圖 2.5 所示。

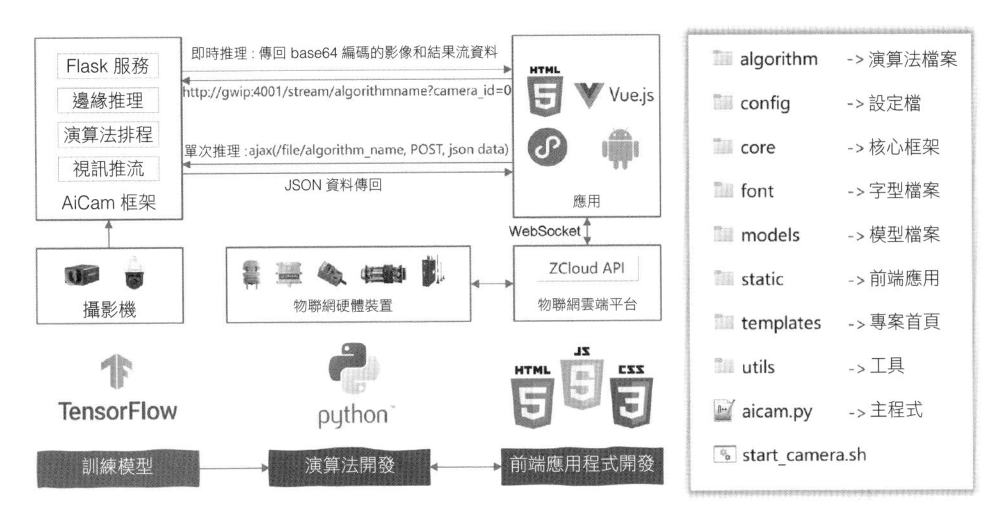

▲ 圖 2.5 AiCam 平臺的功能框架

4）主程式 aicam

主程式 aicam.py 核心程式如下：

```
#獲取當前專案根目錄
basedir = os.path.abspath(os.path.dirname(__file__))
#全域參數
__app = Flask(__name__, static_folder="static", template_folder='templates')
#cross-domain
CORS(__app, supports_credentials=True)

#進入首頁路由
@__app.route('/')
def index():
    return render_template('index.html')

#設定 icon 圖示
@__app.route('/favicon.ico')
def favicon():
    return send_from_directory(os.path.join(__app.root_path, 'static'), 'favicon.ico',
                                                mimetype='image/vnd. microsoft.icon')
```

```
class Stream:
    def __init__(self, cd):
        print("INFO: Stream create.")
        self.cd = cd
    def __iter__(self):
        return self

    def __next__(self):
        return self.cd()

    def __del__(self):
        print("INFO: Stream delete.")

@__app.route('/ptz/preset', methods=["POST"])
def ptzPreset():
    if request.method == 'OPTIONS':
        res = Response()
        res.headers['Access-Control-Allow-Origin'] = '*'
        res.headers['Access-Control-Allow-Method'] = '*'
        res.headers['Access-Control-Allow-Headers'] = '*'
        return res
    dat = request.stream.read()
    cmd = 39
    param = 1
    if len(dat) > 0:
        jo = json.loads(dat)
        cmd = jo['cmd']
        param = jo['param']
    camera = None
    camera_id = request.values.get("camera_id")
    if camera_id != None:
        camera_id = camera_id.strip()
        camera = cam.getCamera(camera_id)
    else:
        camera_url = request.values.get("camera_url")
        if camera_url != None:
            camera = cam.loadCamera(camera_url)
    if camera != None:
        presetPtz = getattr(camera, "presetPtz", None)
```

```
        if presetPtz is not None:
            presetPtz(cmd, param)

    res = Response()
    res.headers['Access-Control-Allow-Origin'] = '*'
    res.headers['Access-Control-Allow-Method'] = '*'
    res.headers['Access-Control-Allow-Headers'] = '*'
    return res

@__app.route('/ptz/relativemove', methods=["POST"])
def ptz():
    if request.method == 'OPTIONS':
        res = Response()
        res.headers['Access-Control-Allow-Origin'] = '*'
        res.headers['Access-Control-Allow-Method'] = '*'
        res.headers['Access-Control-Allow-Headers'] = '*'
        return res
    # 獲取攝影機編號、基礎應用編號、基礎應用中子應用的編號
    dat = request.stream.read()
    jo = json.loads(dat)
    x = 0
    y = 0
    z = 0
    if 'x' in jo:
        x = jo['x']
    if 'y' in jo:
        y = jo['y']
    if 'z' in jo:
        z = jo['z']
    camera = None
    camera_id = request.values.get("camera_id")
    if camera_id != None:
        camera_id = camera_id.strip()
        camera = cam.getCamera(camera_id)
    else:
        camera_url = request.values.get("camera_url")
        if camera_url != None:
            camera = cam.loadCamera(camera_url)
    if camera != None:
```

```
        runPtz = getattr(camera, "runPtz", None)
        if runPtz is not None:
            runPtz(x,y,z)
    res = Response()
    res.headers['Access-Control-Allow-Origin'] = '*'
    res.headers['Access-Control-Allow-Method'] = '*'
    res.headers['Access-Control-Allow-Headers'] = '*'
    return res
# 即時視訊應用路由
@__app.route('/stream/<action>')
def video_stream(action):
    if request.method == 'OPTIONS':
        res = Response()
        res.headers['Access-Control-Allow-Origin'] = '*'
        res.headers['Access-Control-Allow-Method'] = '*'
        res.headers['Access-Control-Allow-Headers'] = '*'
        return res

    # 獲取攝影機編號、基礎應用編號、基礎應用中子應用的編號
    camera_id = request.values.get("camera_id")
    camera = None
    if camera_id != None:
        camera_id = camera_id.strip()
        camera = cam.getCamera(camera_id)
    else:
        camera_url = request.values.get("camera_url")
        if camera_url != None:
            camera = cam.loadCamera(camera_url)
    if camera != None:
        def cam_read():
            return camera.read()
    else:
        def cam_read():
            return False, None
    mimetype = 'text/event-stream'
    boundary = '\r\nContent-Type: text/event-stream\r\n\r\ndata:'
    if type == 'image':
        mimetype = 'multipart/x-mixed-replace; boundary=frame'
        boundary = b"\r\n--frame\r\nContent-Type: image/jpeg\r\n\r\n"
```

```
    def gen():
        while True:
            ret, img = cam.read(camera_id)
            if ret:
                result = alg.request(img, action)
                if type == 'image':
                    img = base64.b64decode(result['result_image'].encode('utf-8'))
                    yield boundary+img
                else:
                    yield boundary+json.dumps(result)
            else:
                time.sleep(1)
    res = Response(gen(), mimetype=mimetype)
    res.headers['Access-Control-Allow-Origin'] = '*'
    res.headers['Access-Control-Allow-Method'] = '*'
    res.headers['Access-Control-Allow-Headers'] = '*'
    return res

# 非即時視訊處理（影像、視訊、音訊檔案）
@__app.route('/file/<action>', methods=["POST"])
def file_handle(action):
    result = {}
    param_data = request.form.get("param_data")              # 參數 JSON 字串
    file_name = request.files.get("file_name")
    file_data = None
    if file_name != '' and file_name is not None :
        file_data = file_name.read()                         # 檔案資料

    param_json = {}
    # 將參數字典設定到共用陣列
    if param_data is not None and file_util.is_json(param_data):
        param_json = json.loads(param_data)
    result = alg.request(file_data, action, param_json)
    res = Response(json.dumps(result))
    res.headers['Access-Control-Allow-Origin'] = '*'
    res.headers['Access-Control-Allow-Method'] = '*'
    res.headers['Access-Control-Allow-Headers'] = '*'

    return res
```

```
#即時視訊應用路由
@__app.route('/image/stream/<action>')
def video_stream_image(action):
    if request.method == 'OPTIONS':
        res = Response()
        res.headers['Access-Control-Allow-Origin'] = '*'
        res.headers['Access-Control-Allow-Method'] = '*'
        res.headers['Access-Control-Allow-Headers'] = '*'
        return res
    #獲取攝影機編號、基礎應用編號、基礎應用中子應用的編號
    camera_id = request.values.get("camera_id")
    camera = None
    if camera_id != None:
        camera_id = camera_id.strip()
        camera = cam.getCamera(camera_id)
    else:
        camera_url = request.values.get("camera_url")
        if camera_url != None:
            camera = cam.loadCamera(camera_url)
    if camera != None:
        def cam_read():
            return camera.read()
    else:
        def cam_read():
            return False, None
    mimetype = 'multipart/x-mixed-replace; boundary=frame'
    boundary = b"\r\n--frame\r\nContent-Type: image/jpeg\r\n\r\n"
    def gen():
        i=0
        while i<30:
            ret, img = cam_read()
            if ret:
                result = alg.request(img, action)
                img = base64.b64decode(result['result_image'].encode('utf-8'))
                return boundary+img
            else:
                i += 1
                time.sleep(1)
        raise StopIteration
```

```
    res = Response(Stream(gen), mimetype=mimetype)
    res.headers['Access-Control-Allow-Origin'] = '*'
    res.headers['Access-Control-Allow-Method'] = '*'
    res.headers['Access-Control-Allow-Headers'] = '*'
    return res

if __name__ == '__main__':
    __app.run(host='0.0.0.0', port=4001, debug=False)
```

5）啟動指令稿

啟動指令稿 start_aicam.sh 主要用於建構執行環境、啟動主程式 aicam.py，程式如下：

```
#!/bin/bash
echo " 開始執行指令稿 "
ps -aux | grep "aicam.py"|awk '{print $2}'|xargs kill -9

cd `dirname $0`
PWD=`pwd`
export LD_LIBRARY_PATH=$PWD/core/pyHCNetSDK/HCNetSDK_linux64:$LD_LIBRARY_PATH

#>>> conda initialize >>>
#!! Contents within this block are managed by 'conda init' !!
__conda_setup="$('/home/zonesion/miniconda3/bin/conda' 'shell.bash' 'hook' 2> /dev/
null)"
if [ $? -eq 0 ]; then
    eval "$__conda_setup"
else
    if [ -f "/home/zonesion/miniconda3/etc/profile.d/conda.sh" ]; then
        . "/home/zonesion/miniconda3/etc/profile.d/conda.sh"
    else
        export PATH="/home/zonesion/miniconda3/bin:$PATH"
    fi
fi
unset __conda_setup
#<<< conda initialize <<<
```

```
conda activate py36_tf25_torch110_cuda113_cv345
python3 aicam.py
echo " 指令稿啟動完成 "
```

2.1.1.3 開發資源

1）AiCam 平臺的組成

利用 AiCam 平臺，使用者能夠方便快捷地開展深度學習的教學、競賽和科學研究等工作。從最基礎的 OpenCV、模型訓練到邊緣裝置的部署，AiCam 平臺進行了全端式的封裝，降低了開發難度。AiCam 平臺的組成如圖 2.6 所示。

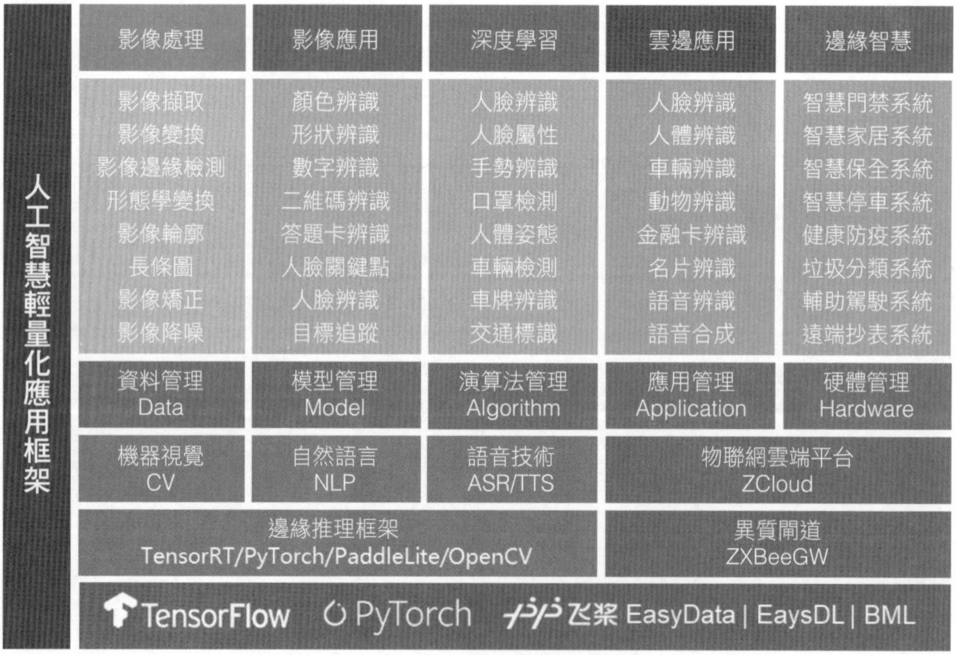

人工智慧輕量化應用框架

影像處理	影像應用	深度學習	雲邊應用	邊緣智慧
影像擷取 影像變換 影像邊緣檢測 形態學變換 影像輪廓 長條圖 影像矯正 影像降噪	顏色辨識 形狀辨識 數字辨識 二維碼辨識 答題卡辨識 人臉關鍵點 人臉辨識 目標追蹤	人臉辨識 人臉屬性 手勢辨識 口罩檢測 人體姿態 車輛檢測 車牌辨識 交通標識	人臉辨識 人體辨識 車輛辨識 動物辨識 金融卡辨識 名片辨識 語音辨識 語音合成	智慧門禁系統 智慧家居系統 智慧保全系統 智慧停車系統 健康防疫系統 垃圾分類系統 輔助駕駛系統 遠端抄表系統
資料管理 Data	模型管理 Model	演算法管理 Algorithm	應用管理 Application	硬體管理 Hardware
機器視覺 CV	自然語言 NLP	語音技術 ASR/TTS	物聯網雲端平台 ZCloud	
邊緣推理框架 TensorRT/PyTorch/PaddleLite/OpenCV			異質閘道 ZXBeeGW	
TensorFlow	PyTorch	飞桨 EasyData | EaysDL | BML		

▲ 圖 2.6 AiCam 平臺的組成

AiCam 平臺支援以下應用：

- 影像處理：基於 OpenCV 開發的數位影像處理演算法。

- 影像應用：基於 OpenCV 開發的影像應用。
- 深度學習：基於深度學習技術開發的影像辨識、影像檢測等應用。
- 視覺雲端應用：基於百度雲介面開發的影像辨識、影像檢測、語音辨識、語音合成等應用。
- 邊緣智慧：結合硬體場景的邊緣應用。
- 綜合案例：結合行業軟 / 硬體應用場景的邊緣計算。

2）平臺的演算法清單

透過實驗常式的方式，AiCam 平臺為機器視覺演算法提供了單元測試，並開放了程式。影像基礎演算法、影像基礎應用、深度學習應用和百度雲邊應用的介面及其描述如表 2.1 到表 2.4 所示。

▼ 表 2.1 影像基礎演算法

類別	介面名稱	介面描述
影像擷取	image_capture	即時視訊流擷取和輸出
影像標註	image_lines_and_rectangles	繪製直線與矩形
	image_circle_and_ellipse	繪製圓和橢圓
	image_polygon	繪製多邊形
	image_display_text	顯示文字
影像轉換	image_gray	灰度實驗
	image_simple_binary	二值化
	image_adaptive_binary	自我調整設定值二值化實驗
影像變換	image_rotation	影像旋轉
	image_mirroring	影像鏡像旋轉實驗
	image_resize	影像縮放實驗
	image_perspective_transform	影像透視變換

（續表）

類 別	介 面 名 稱	介 面 描 述
影像邊緣檢測	image_edge_detection	影像邊緣檢測實驗
形態學變換	image_eroch	腐蝕
	image_dilate	膨脹
	image_opening	開運算
	image_closing	閉運算
影像輪廓	image_contour_experiment	影像輪廓實驗
	image_contour_search_rectangle	透過影像輪廓特徵查詢外接矩形
	image_contour_search_minrectangle	透過影像輪廓特徵查詢最小外接矩形
	image_contour_search_mincircle	透過影像輪廓特徵查詢最小外接圓
長條圖	image_simple_histogram	原始影像 + 長條圖資料
	image_equalization_histogram	長條圖 + 均衡化長條圖資料
	image_self_adaption_equalization_histogram	自我調整均衡化長條圖資料
範本匹配	image_template_matching	影像的範本匹配
霍夫變換	image_standard_hough_transform	霍夫變換檢測直線
	image_asymptotic_probabilistic_hough_transform	漸進機率式霍夫變換檢測直線
	image_hough_transform_circular	影像的霍夫變換檢測圓
梯度變換	image_sobel	Sobel 運算元
	image_scharr	Scharr 運算元
	image_laplacian	Laplacian 運算元

（續表）

類 別	介 面 名 稱	介 面 描 述
影像矯正	image_correction	影像矯正
影像增加浮水印	image_watermark	影像增加浮水印
影像雜訊消除	image_noise	雜訊影像
	image_box_filter	方框濾波
	image_blur_filter	均值濾波
	image_gaussian_filter	高斯濾波
	image_bilateral_filter	高斯雙邊濾波
	image_medianblur	中值濾波

▼ 表 2.2 影像基礎應用

類 別	介 面 名 稱	介 面 描 述
顏色辨識	image_color_recognition	辨識目標的顏色
形狀辨識	image_shape_recognition	辨識目標的形狀
數字辨識	image_mnist_recognition	辨識手寫數字
二維碼辨識	image_qrcode_recognition	辨識二維碼內容
人臉檢測	image_face_detection	利用 Dlib 函式庫的人臉檢測演算法
人臉關鍵點	image_key_detection	利用 Dlib 函式庫的人臉關鍵點辨識演算法
人臉辨識	image_face_recognition	基於 HAAR 人臉特徵分類器進行人臉辨識
目標追蹤	image_motion_tracking	對移動目標進行追蹤標註

▼ 表 2.3 深度學習應用

類 別	介 面 名 稱	介 面 描 述
人臉檢測	face_detection	人臉檢測模型及演算法
人臉辨識	face_recognition	人臉辨識模型及演算法
人臉屬性	face_attr	多種人臉屬性資訊（如年齡、性別、表情等）
手勢辨識	handpose_detection	辨識人體手部的主要關鍵點
口罩檢測	mask_detection	檢測是否佩戴口罩
人體姿態	personpose_detection	辨識人體的 21 個主要關鍵點
車輛檢測	car_detection	辨識 ROS 智慧小車
車牌辨識	plate_recognition	辨識車牌號碼
行人檢測	person_detection	辨識行人並進行標註
交通號誌	traffic_detection	辨識各種交通號誌

▼ 表 2.4 百度雲邊應用

類 別	介 面 名 稱	介 面 描 述
人臉辨識	baidu_face_recognition	人臉註冊及辨識
人體辨識	baidu_body_attr	人體檢測與屬性辨識演算法
車輛檢測	baidu_vehicle_detect	車輛屬性及檢測演算法
手勢辨識	baidu_gesture_recognition	手勢辨識演算法
數字辨識	baidu_numbers_detect	數字辨識演算法
文字辨識	baidu_general_characters_recognition	通用文字辨識演算法
語音辨識	baidu_speech_recognition	百度語音辨識（標準版）應用
語音合成	baidu_speech_synthesis	百度語音合成服務應用

3）AiCam 平臺的部分案例截圖

AiCam 平臺的部分案例截圖如圖 2.7 到圖 2.11 所示。

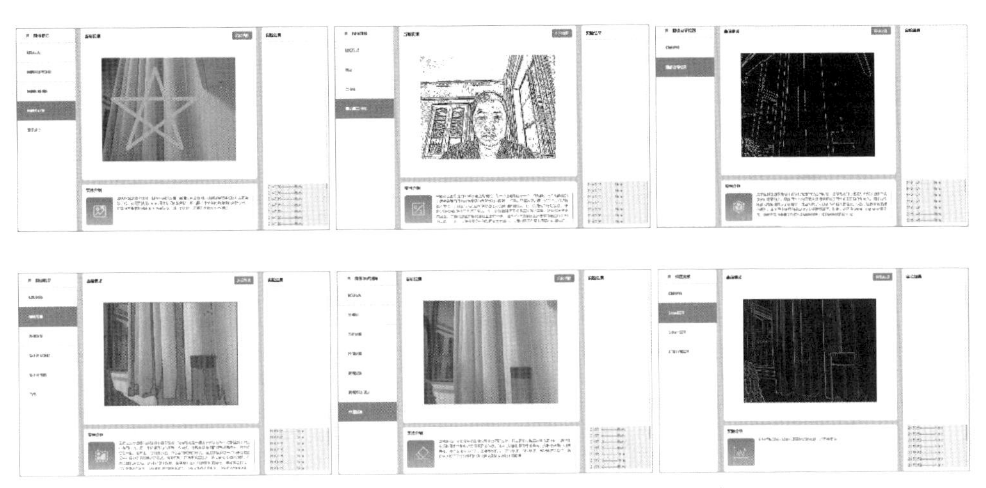

▲ 圖 2.7 基礎演算法案例截圖

▲ 圖 2.8 基礎應用案例截圖

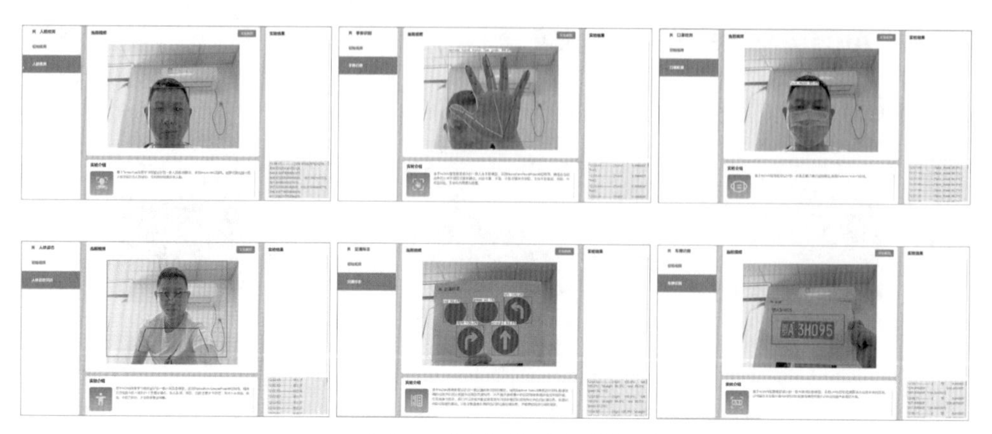

▲ 圖 2.9 深度學習案例截圖

圖 2.10 雲邊應用案例截圖

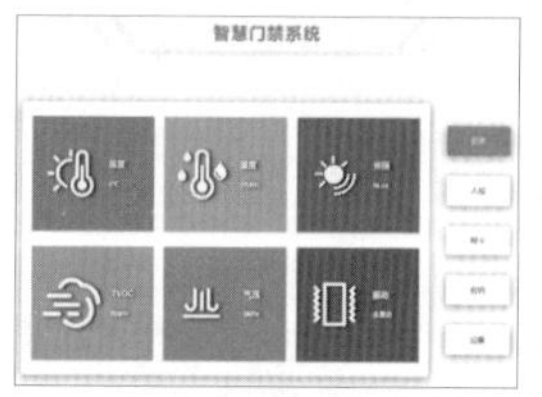

▲ 圖 2.11 邊緣智慧案例截圖

2.1.1.4 邊緣計算的硬體設計平臺

本書採用的是 GW3588 邊緣計算閘道。為了深化無線節點在無線感測網路中的使用，本書中的部分專案需要使用感測器和控制裝置，涉及 xLab 開發平臺，該平臺設計了豐富的硬體裝置，包括擷取類、控制類、保全類、顯示類、辨識類等開發平臺。

1）GW3588 邊緣計算閘道

GW3588 邊緣計算閘道採用全新的商業產品級一體機外觀設計，以及搭載 AI 嵌入式邊緣計算處理器的 RK3399 開發板。RK3399 採用了 4 核心 Cortex-A76 和 4 核心 Cortex-A55 的微處理器，4 核心的 Mali-G610 的 GPU、6 TOP 算力的神經網路處理器（Neural Processing Unit，NPU）、16 GB 的 RAM 和 128 GB 的 ROM，15.6″的高畫質電容螢幕（LCD），可執行 Ubuntu、Android 等作業系統。

GW3588 邊緣計算閘道不僅提供了豐富的外接裝置介面（如圖 2.12 所示），可方便使用者開發偵錯；還提供了豐富的擴充模組，可完成人工智慧機器視覺、語音語言、邊緣計算、綜合專案等的開發任務。

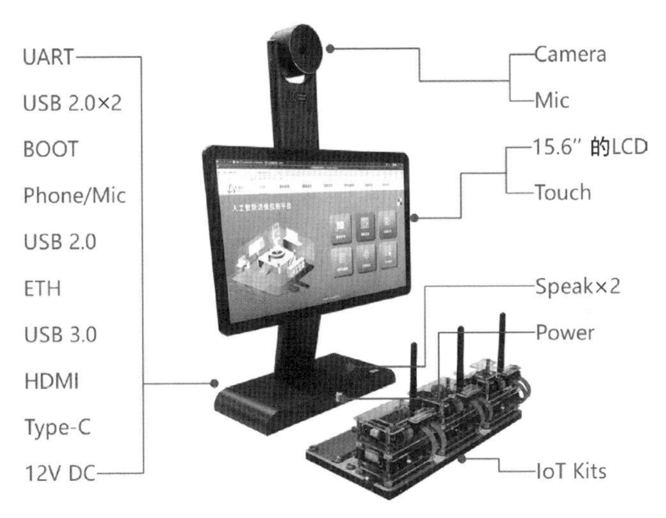

▲ 圖 2.12　GW3588 邊緣計算閘道的外接裝置介面

2）擷取類開發平臺（Sensor-A）

擷取類開發平臺包括溫濕度感測器、光照度感測器、空氣品質感測器、氣壓海拔感測器、三軸加速度感測器、距離感測器、繼電器、語音辨識感測器等，如圖 2.13 所示。

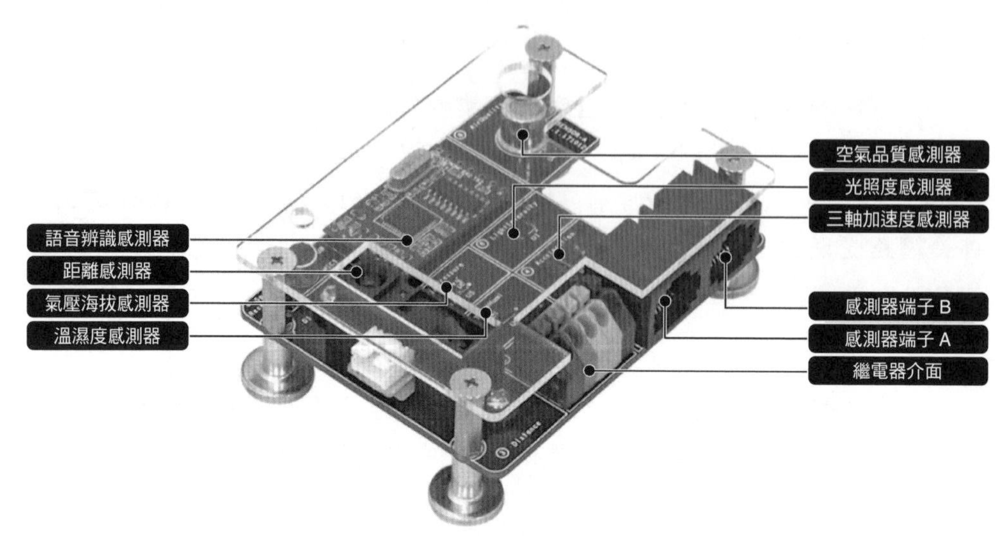

▲ 圖 2.13 擷取類開發平臺

- 兩路 RJ45 工業介面，包含 I/O、DC 3.3 V、DC 5 V、UART、RS-485、兩路繼電器輸出等功能，提供兩路 3.3 V、5 V 電源輸出。
- 溫濕度感測器的型號為 HTU21D，採用數位訊號輸出和 I2C 通訊介面，測量範圍為 -40 ～ 125℃，以及 5%RH ～ 95%RH。
- 光照度感測器的型號為 BH1750，採用數位訊號輸出和 I2C 通訊介面，對應廣泛的輸入光範圍，相當於 1 ～ 65535 lx。
- 空氣品質感測器的型號為 MP503，採用類比訊號輸出，可以檢測氣體酒精、煙霧、異丁烷、甲醛，檢測濃度為 10 ～ 1000 ppm（酒精）。
- 氣壓海拔感測器的型號為 FBM320，採用數位訊號輸出和 I2C 通訊介面，測量範圍為 300 ～ 1100 hPa。
- 三軸加速度感測器的型號為 LIS3DH，採用數位訊號輸出和 I2C 通訊介面，量程可設定為 ±2*g*、±4*g*、±8*g*、±16*g*（*g* 為重力加速度），16 位元資料輸出。
- 距離感測器的型號為 GP2D12，採用類比訊號輸出，測量範圍為 10 ～ 80 cm，更新頻率為 40 ms。

- 採用繼電器控制，輸出節點有兩路繼電器介面，支援 5 V 電源開關控制。
- 語音辨識感測器的型號為 LD3320，支援非特定人辨識，具有 50 筆辨識容量，採用序列埠通訊。

3）控制類開發平臺（Sensor-B）

控制類開發平臺包括：風扇、步進馬達、蜂鳴器、LED、RGB 燈、繼電器介面，如圖 2.14 所示。

- 兩路 RJ45 工業介面，包含 IO、DC 3.3 V、DC 5 V、UART、RS-485、兩路繼電器輸出等功能，提供兩路 3.3 V、5 V 電源輸出。
- 風扇為小型風扇，採用低電位驅動。
- 步進馬達為小型 42 步進馬達，驅動晶片為 A3967SLB，邏輯電源的電壓範圍為 3.0 ～ 5.5 V。
- 使用小型蜂鳴器，採用低電位驅動。

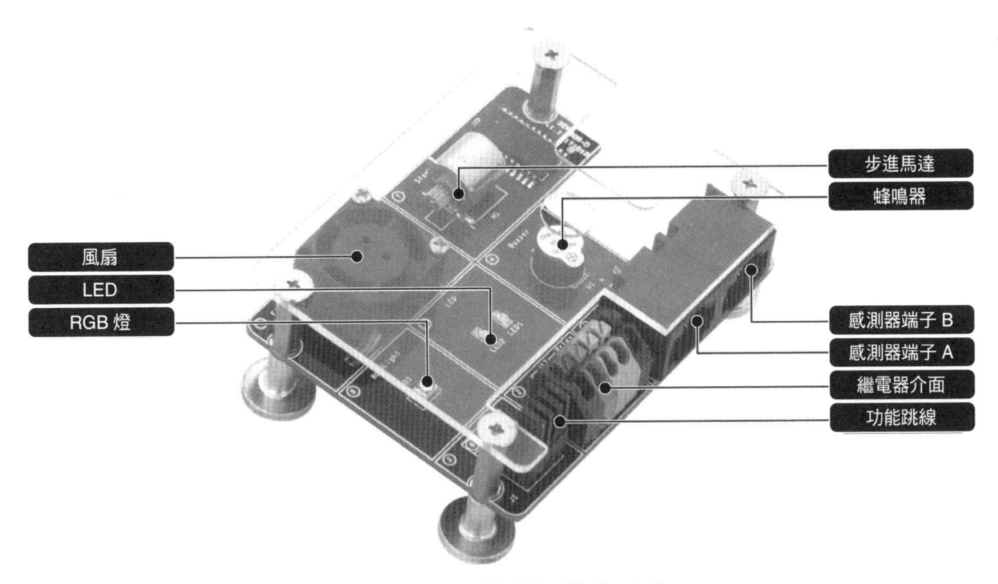

▲ 圖 2.14 控制類開發平臺

- 兩路反白 LED 燈，採用低電位驅動。
- RGB 燈採用低電位驅動，可組合出任何顏色。
- 採用繼電器控制，輸出節點有兩路繼電器介面，支援 5 V 電源開關控制。

4）保全類開發平臺（Sensor-C）

保全類開發平臺包括：火焰感測器、光柵感測器、人體紅外感測器、瓦斯感測器、觸控感測器、震動感測器、霍爾感測器、繼電器介面、語音合成感測器等，如圖 2.15 所示。

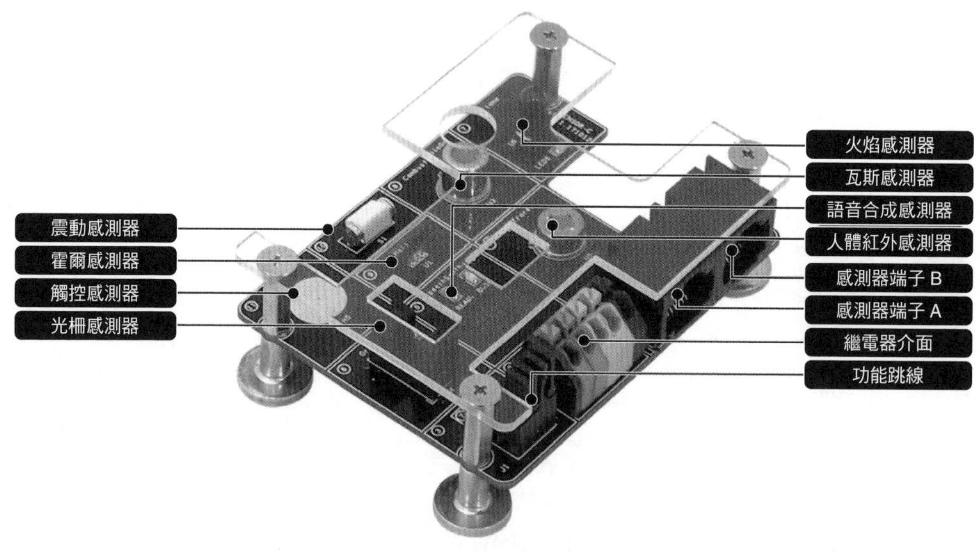

▲ 圖 2.15 保全類開發平臺

- 兩路 RJ45 工業介面，包含 IO、DC 3.3 V、DC 5 V、UART、RS-485、兩路繼電器輸出等功能，提供兩路 3.3 V、5 V 紅外線電源輸出。
- 火焰感測器採用 5 mm 的探測器，可檢測火焰或波長為 760 ～ 1100 nm 的紅外線，探測溫度為 600℃左右，採用數位開關量輸出。
- 光柵感測器的槽式光耦槽寬為 10 mm，工作電壓為 5 V，採用數位開關量訊號輸出。

- 人體紅外感測器的型號為 AS312，電源電壓為 3 V，感應距離為 12 m，採用數位開關量訊號輸出。
- 瓦斯感測器的型號為 MP-4，採用類比訊號輸出，感測器加熱電壓為 5 V，供電電壓為 5 V，可測量天然氣、甲烷、瓦斯氣、沼氣等。
- 觸控感測器的型號為 SOT23-6，採用數位開關量訊號輸出，檢測到觸控時，輸出電位翻轉。
- 震動感測器在低電位時有效，採用數位開關量訊號輸出。
- 霍爾感測器的型號為 AH3144，電源電壓為 5 V，採用數位開關量輸出，工作頻率寬（0 ～ 100 kHz）。
- 採用繼電器控制，輸出節點有兩路繼電器介面，支援 5 V 電源開關控制。
- 語音合成感測器的型號為 SYN6288，採用序列埠通訊，支援 GB2312、GBK、UNICODE 等編碼，可設定音量、背景音樂等。

5）顯示類開發平臺（Sensor-D）

顯示類開發平臺包括：OLED 顯示幕、數位管、五向開關和感測器端子，如圖 2.16 所示。

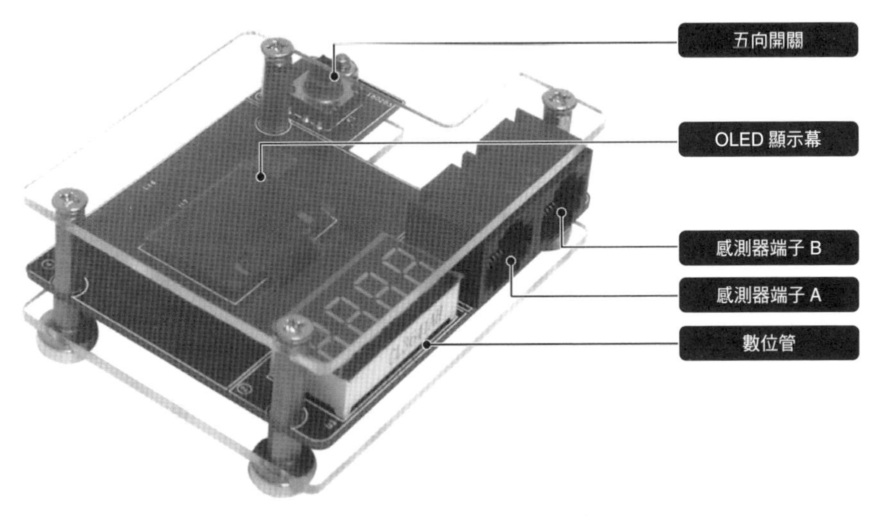

▲ 圖 2.16 顯示類開發平臺

- 兩路 RJ45 工業介面，包含 IO、DC 3.3 V、DC 5 V、UART、RS-485、兩路繼電器輸出等功能，提供兩路 3.3 V、5 V 電源輸出。
- 硬體採用分區設計，絲印方塊圖清晰易懂，包含感測器編號，模組採用壓克力防護。
- OLED 顯示幕：解析度為 128×64、尺寸為 0.96″，採用 I2C 通訊介面。
- 數位管：採用 4 位元共陰極數位管，驅動晶片的型號為 ZLG7290，採用 I2C 通訊介面。
- 五向開關：五向按鍵，驅動晶片的型號為 ZLG7290，採用 I2C 通訊介面。

6）125 kHz&13.56 MHz 二合一開發平臺（Sensor-EL）

125 kHz&13.56 MHz 二合一開發平臺包括：繼電器介面、蜂鳴器、OLED 顯示幕、RFID、感測器端子、USB 偵錯介面、功能跳線，如圖 2.17 所示。

- 兩路 RJ45 工業介面，包含 IO、DC 3.3 V、DC 5 V、UART、RS-485、兩路繼電器輸出等功能，提供兩路 3.3 V、5 V 電源輸出。
- 硬體採用分區設計，絲印方塊圖清晰易懂，包含感測器編號，模組採用壓克力防護。
- RFID：採用 125 kHz&13.56 MHz 的射頻感測器，UART 介面（TTL 電位），支援 ISO/IEC 14443 A/MIFARE、NTAG、MF1xxS20、MF1xxS70、MF1xxS50、EM4100、T5577 等射頻卡。
- OLED 顯示幕：解析度為 128×64、尺寸為 0.96″，採用 I2C 通訊介面。
- 支援 USB 供電、USB 轉序列埠，可連接電腦的 USB 介面。

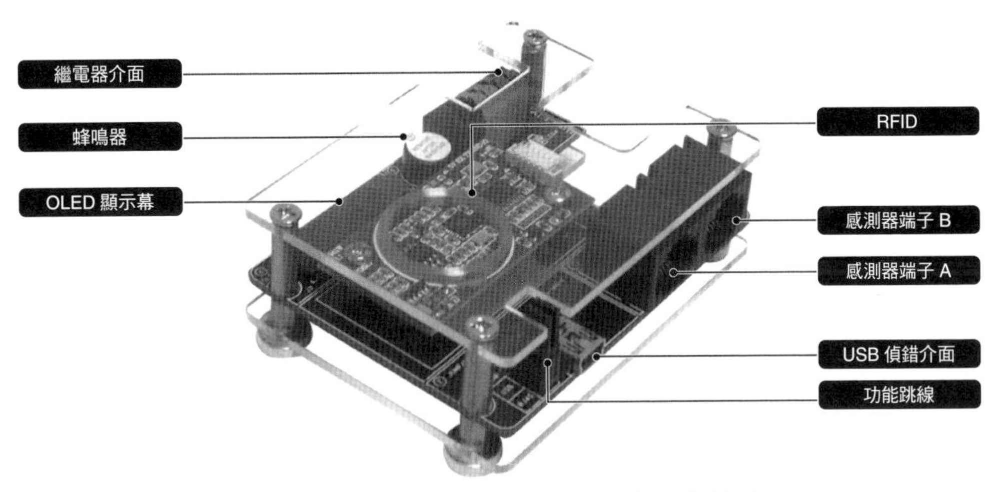

▲ 圖 2.17　125 kHz&13.56 MHz 二合一開發平臺

7）900 MHz&ETC 開發平臺（Sensor-EH）

900 MHz&ETC 開發平臺包括：RFID、感測器端子、ETC 欄桿、重置按鈕、USB 偵錯介面、功能跳線，如圖 2.18 所示。

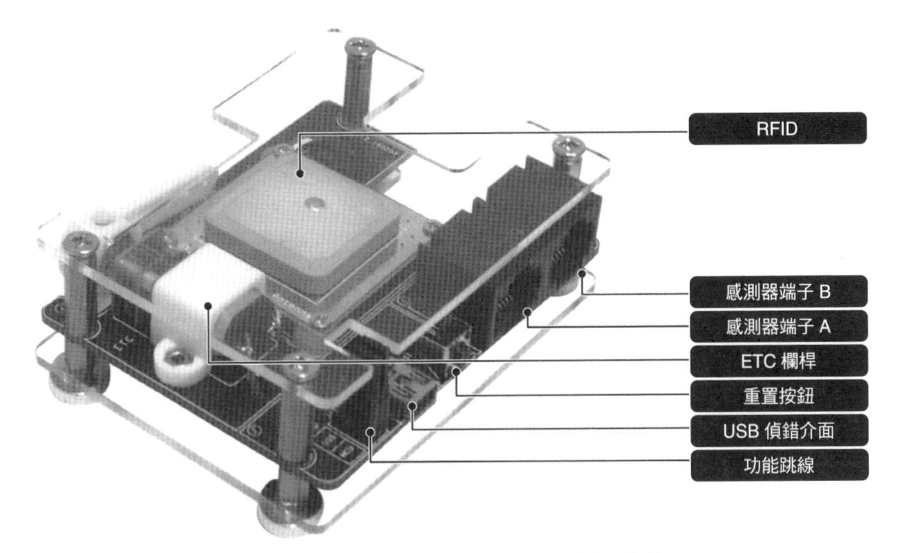

▲ 圖 2.18　900 MHz&ETC 開發平臺

- 兩路 RJ45 工業介面，包含 IO、DC 3.3 V、DC 5 V、UART、RS-485、兩路繼電器輸出等功能，提供兩路 3.3 V、5 V、12 V 電源輸出。
- 硬體採用分區設計，絲印方塊圖清晰易懂，包含感測器編號，模組採用壓克力防護。
- RFID：採用 900 MHz&ETC 的射頻感測器，支持 ISO 18000-6C 協定，工作頻率為 ISM 頻段的 902 ～ 928 MHz，工作模式為跳頻工作、定頻工作或軟體可調，功率在 0 dBm ～ 27 dBm 內可調，支持的讀取標籤協定有 EPC C1 Co Gen2 和 ISO-18000-6C，讀取距離為 1 ～ 5 cm，整合了 25×25 高性能微小型陶瓷天線（板載天線）和 ETC 欄桿。

2.1.1.5 功能與核心程式設計案例

AiCam 平臺能夠完成基於邊緣應用的演算法開發、模型開發、硬體開發、應用程式開發，開發常式可透過使用者端的瀏覽器執行。範例如下：

```
##########################################################################################
# 檔案：image_face_recognition.py
##########################################################################################
import glob
import face_recognition
import os
import sys
import cv2 as cv
import numpy as np
import base64
import json

class ImageFaceRecognition(object):
    def __init__(self, dir_path="algorithm/image_face_recognition"):
        # 讀取註冊人臉特徵的 npy 檔案
        self.dir_path = dir_path
        feature_path = os.path.join(dir_path, "*.npy")
        feature_files = glob.glob(feature_path)
        # 解析檔案名稱，作為註冊人姓名
```

```
        self.feature_names = [item.split(os.sep)[-1].replace(".npy", "") for item in
feature_files]
        #print(feature_names)
        self.face_cascade = cv.CascadeClassifier(dir_path+"/haarcascade_frontalface_
alt.xml")

        self.features = []
        for f in feature_files:
            feature = np.load(f)
            self.features.append(feature)

    def image_to_base64(self, img):
        image = cv.imencode('.jpg', img, [cv.IMWRITE_JPEG_QUALITY, 60])[1]
        image_encode = base64.b64encode(image).decode()
        return image_encode

    def base64_to_image(self, b64):
        img = base64.b64decode(b64.encode('utf-8'))
        img = np.asarray(bytearray(img), dtype="uint8")
        img = cv.imdecode(img, cv.IMREAD_COLOR)
        return img

    def face_id(self, img, classifier):
        gray = cv.cvtColor(img, cv.COLOR_BGR2GRAY)
        faces = classifier.detectMultiScale(gray, 1.3, 5)
        return faces

    def inference(self, image, param_data):
        #code：辨識成功傳回 200
        #msg：相關提示訊息
        #origin_image：原始影像
        #result_image：處理之後的影像
        #result_data：結果資料
        return_result = {'code': 200, 'msg': None, 'origin_image': None, 'result_
image': None, 'result_data': None}

        # 應用請求介面：@__app.route('/file/<action>', methods=["POST"])
        #image：應用傳遞過來的資料（根據實際應用可能為影像、音訊、視訊、文字）
```

```
#param_data：應用傳遞過來的參數，不能為空
if param_data != None:
    #讀取應用傳遞過來的影像
    image = np.asarray(bytearray(image), dtype="uint8")
    image = cv.imdecode(image, cv.IMREAD_COLOR)
    #保留原始影像資料
    origin = image.copy()

    #type=0 表示註冊
    if param_data["type"] == 0:
        try:
            image_encoding = face_recognition.face_encodings(image)[0]
            if len(image_encoding) != 0:
                flag = True
            else:
             return_result["code"] = 404
             return_result["msg"] = "未檢測到人臉！"
        except:
             return_result["code"] = 500
             return_result["msg"] = "註冊失敗！"
        if flag:
             feature_name = param_data["reg_name"] + ".npy"
             feature_path = os.path.join(self.dir_path, feature_name)
             np.save(feature_path, image_encoding)
             print("已儲存人臉")
             return_result["code"] = 200
             return_result["msg"] = "註冊成功！"
    #type=1 表示辨識
    if param_data["type"] == 1:
        #呼叫介面進行人臉比對
        rects = self.face_id(image, self.face_cascade)
        for x, y, w, h in rects:
            crop = image[y: y + h, x: x + w]
            #視訊流中人臉特徵編碼
            img_encoding = face_recognition.face_encodings(crop)
            if len(img_encoding) != 0:
                #獲取人臉特徵編碼
                img_encoding = img_encoding[0]
                #與註冊的人臉特徵進行對比
```

```
                    result = face_recognition.compare_faces(self.features, img_
encoding, tolerance=0.4)
                    if True in result:
                        result = int(np.argmax(np.array(result, np.uint8)))
                        rec_result = self.feature_names[result]
                        cv.putText(image, rec_result, (x, y), cv.FONT_HERSHEY_
SIMPLEX, 1.2, (0, 255, 0), thickness=1)
                        cv.rectangle(image, (x, y), (x + w, y + h), (0, 255, 0), 1)
                        return_result["result_data"]=rec_result
                    else:
                        cv.putText(image, 'unknown', (x, y), cv.FONT_HERSHEY_
SIMPLEX, 1.2, (0, 255, 0), thickness=1)
                      cv.rectangle(image, (x, y), (x + w, y + h), (0, 255, 0), 1)
                      return_result["result_data"]='unknown'


                else:
                    cv.putText(image, 'unknown', (x, y), cv.FONT_HERSHEY_SIMPLEX,
                            1.2, (0, 255, 0), thickness=1)
                    cv.rectangle(image, (x, y), (x + w, y + h), (0, 255, 0), 1)
                    return_result["result_data"]='unknown'
            return_result["msg"] = "辨識成功！"
            return_result["origin_image"] = self.image_to_base64(origin)
            return_result["result_image"] = self.image_to_base64(image)
        else:
            #呼叫介面進行人臉比對
            rects = self.face_id(image, self.face_cascade)
            for x, y, w, h in rects:
                crop = image[y: y + h, x: x + w]
                #視訊流中人臉特徵編碼
                img_encoding = face_recognition.face_encodings(crop)
                if len(img_encoding) != 0:
                    #獲取人臉特徵編碼
                    img_encoding = img_encoding[0]
                    #與註冊的人臉特徵進行對比
                    result = face_recognition.compare_faces(self.features, img_
encoding, tolerance=0.4)
                    if True in result:
                        result = int(np.argmax(np.array(result, np.uint8)))
                        rec_result = self.feature_names[result]
```

```
                    cv.putText(image, rec_result, (x, y), cv.FONT_HERSHEY_SIMPLEX,
                            1.2, (0, 255, 0), thickness=1)
                    cv.rectangle(image, (x, y), (x + w, y + h), (0, 255, 0), 1)
                    return_result["result_data"]=rec_result
                else:
                  cv.putText(image, 'unknown', (x, y), cv.FONT_HERSHEY_SIMPLEX,
                          1.2, (0, 255, 0), thickness=1)
                  cv.rectangle(image, (x, y), (x + w, y + h), (0, 255, 0), 1)
                  return_result["result_data"]='unknown'

            else:
                cv.putText(image, 'unknown', (x, y), cv.FONT_HERSHEY_SIMPLEX, 1.2,
                        (0, 255, 0), thickness=1)
                cv.rectangle(image, (x, y), (x + w, y + h), (0, 255, 0), 1)
                return_result["result_data"]='unknown'
        return_result["result_image"] = self.image_to_base64(image)
    return return_result

# 單元測試，如果處理類別中引用了檔案，則在單元測試中要修改檔案路徑
if __name__=='__main__':
    mode = sys.argv[1]
    c_dir = os.path.split(os.path.realpath(__file__))[0]
    # 建立影像處理物件
    img_object = ImageFaceRecognition(c_dir)

    # 讀取測試影像
    image = cv.imread(c_dir+"/test.jpg")
    # 將影像編碼成資料流程
    img = cv.imencode('.jpg', image, [cv.IMWRITE_JPEG_QUALITY, 60])[1]

    # 設定參數
    addUser_data = {"type":0, "reg_name":"lilianjie"}

    # 呼叫介面進行人臉註冊
    if mode == '0':
        result = img_object.inference(img, addUser_data)
        print(result)
```

```
# 呼叫介面進行人臉辨識
elif mode == '1':
    result = img_object.inference(image, None)
    frame = img_object.base64_to_image(result["result_image"])
    print(result)

    # 影像顯示
    cv.imshow('frame',frame)
    while True:
        key=cv.waitKey(1)
        if key==ord('q'):
            break
    cv.destroyAllWindows()
else:
    print(" 參數錯誤！ ")
```

（1）在邊緣計算閘道上執行 PyCharm 開發環境（見圖 2.19），匯入開發專案，在編輯視窗可以查看演算法程式，在終端執行專案。

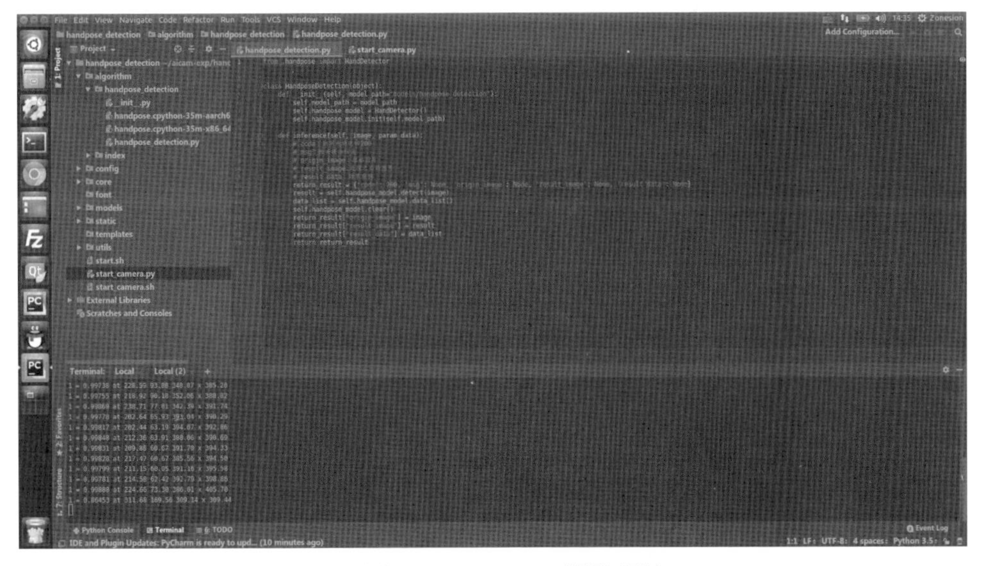

▲ 圖 2.19 PyCharm 開發環境

（2）透過使用者端瀏覽器存取專案的應用頁面，可以在前端的應用頁面中看到演算法即時處理視訊流後傳回的結果。

AiCam 平臺的應用頁面如圖 2.20 所示。

▲ 圖 2.20 AiCam 平臺的應用頁面

2.1.2 開發步驟與驗證

2.1.2.1 專案部署

1）硬體部署

（1）準備好 AiCam 平臺，並正確連接 Wi-Fi 天線、攝影機、電源。

（2）為 AiCam 平臺通電，啟動 Ubuntu 作業系統。

（3）連接區域網內的 Wi-Fi 網路，記錄邊緣計算閘道的 IP 位址，如 192.168.100.200。

2）專案部署

（1）邊緣計算閘道在出廠時已經預設部署了 aicam 專案套件，路徑為 /home/zonesion/aicam。

（2）如果需要對程式進行更新，則透過 SSH 將更新後的 aicam 專案套件上傳到 /home/ zonesion/ 目錄下。

（3）在 SSH 終端輸入以下命令解壓縮 aicam 開發專案：

```
$ cd ~/
$ unzip aicam.zip
```

2.1.2.2 專案執行

（1）在 SSH 終端輸入以下命令執行 aicam 開發專案：

```
$ cd ~/aicam
$ chmod 755 start_aicam.sh
$ conda activate py36_tf114_torch15_cpu_cv345    //Ubuntu 20.04 作業系統下需要切換環境
$ ./start_aicam.sh
// 開始執行指令稿
* Serving Flask app "start_aicam" (lazy loading)
* Environment: production
WARNING: Do not use the development server in a production environment.
Use a production WSGI server instead.
* Debug mode: off
* Running on http://0.0.0.0:4000/ (Press CTRL+C to quit)
```

（2）在使用者端或邊緣計算閘道端開啟 Chrome 瀏覽器，造訪 https://192.168.100.200:1443 即可查看專案內容。

2.1.2.3 專案體驗

（1）在 AiCam 平臺的主頁面可以看到六大版塊內容，如圖 2.2 所示。

（2）按一下任一應用即可查看專案的內容和結果，如圖 2.21 所示。

▲ 圖 2.21 查看專案內容和結果

（3）按一下「實驗截圖」按鈕可在「實驗結果」區域顯示當前視訊視窗中的影像，按一下「實驗結果」中的影像可進行預覽（見圖 2.22），或按右鍵該影像，透過右鍵選單可儲存截圖。

▲ 圖 2.22 預覽視訊視窗中的影像

第 3 章到第 5 章中的部分專案需要填寫一些配置資訊或連線硬體，相關的設定檔路徑為 aicam\static\js\config.js，讀者可根據實際的情況進行修改。

2.1.3 本節小結

本節主要介紹邊緣計算與人工智慧框架，首先介紹了邊緣計算的參考框架；然後介紹了 AiCam 平臺的執行環境、主要特性、開發流程、主程式 aicam、啟動指令稿，以及 AiCam 平臺的組成、演算法清單和部分案例的截圖；接著介紹了邊緣計算的硬體開發平臺；最後透過一個案例舉出了邊緣計算的開發步驟和核心程式。

2.1.4 思考與擴充

（1）當視訊視窗中無畫面時，請在 SSH 終端按下複合鍵「Ctrl+C」退出程式，檢查攝影機是否正確插入到 USB 介面，然後重新開機 AiCam 平臺進行測試。

（2）當視訊視窗中的畫面出現延遲，SSH 終端出現「select timeout」錯誤資訊時，請在 SSH 終端按下複合鍵「Ctrl+C」退出程式，重新將攝影機連接到 USB 介面，然後重新開機 AiCam 平臺進行測試。

2.2 邊緣計算的演算法開發

在邊緣計算中，在設計和執行演算法時通常要考慮資源受限的環境，如邊緣裝置的儲存容量、算力和能量是有限的，這些特點要求邊緣計算的演算法適應邊緣裝置的資源限制和即時需求。以下是邊緣計算演算法的主要特點：

（1）低計算複雜度：邊緣裝置的運算資源（如處理器性能、記憶體和儲存容量）是有限的，因此邊緣計算的演算法要具有較低的計算複雜度，以確保在資源受限的環境中能夠高效執行。

（2）輕量級和緊湊：由於邊緣裝置的資源受限，邊緣計算的演算法通常是輕量級和緊湊的，這就需要採用簡化的模型結構、特徵選擇或壓縮技術，以減小演算法的記憶體佔用和計算需求。

（3）即時性：邊緣計算需要快速處理資料並做出即時決策，因此邊緣計算的演算法通常需要在較短的時間內產生結果。

（4）本地決策：與傳統的雲端運算不同，邊緣計算更強調在本地裝置上進行決策，因此邊緣計算的演算法需要具備本地感知和決策的能力，減少對雲端的依賴。

（5）適應性：由於邊緣環境可能隨時變化，因此邊緣計算的演算法需要具備一定的適應性，能夠根據環境的變化動態地調整參數或選擇合適的處理方式。

本節的基礎知識如下：

- 掌握機器視覺應用導向的邊緣計算的框架。
- 結合 AiCam 平臺，掌握即時推理演算法介面和單次推理演算法介面的設計。
- 結合人臉辨識案例，掌握邊緣演算法的開發。

2.2.1 原理分析與開發設計

2.2.1.1 機器視覺應用導向的邊緣計算框架

身為新興的計算模式，邊緣計算可以有效滿足機器視覺應用領域的低延遲、高頻寬需求，其基本理念是在網路邊緣提供計算服務，把傳統雲端運算資源遷移到網路邊緣，更加貼近資料來源，從而擁有更快的回應速度和互動能力。邊緣計算具有協作、開放、彈性的特點，不僅可以實現與雲端運算的互相協作，也可以將計算和儲存等資源以服務的形式提供給使用者，還可以根據業務增加的規模和需求來靈活呼叫和配置邊緣節點，實現自動化的快速部署。

邊緣視覺處理的框架如圖 2.23 所示，主要承載影像處理功能（包括視訊編 / 解碼、視訊影像增強、視訊影像內容分析、視訊影像檢索、AI 推理等），可對視訊影像中人員、車輛、物體等物件的特徵、行為、數量、品質等進行檢測或辨識，提高視訊影像整體或特定部分的清晰度、對比度。

邊緣視覺處理平臺透過資料介面與雲端側資料進行互通，並對外提供應用和資料服務，包括裝置登入、登出、狀態、視訊流、特徵資料、結構化資料等資訊的上傳和下發；透過控制介面與端側、邊緣側、雲端側進行協作排程，包括配置下發、演算法模型下發、視訊查詢、任務下發、邊雲協作等；透過管理介面實現應用的下發部署及全生命週期管理等，並形成邊緣 AI 框架，包括深度學習演算法庫、模型庫、訓練、推理等，可完成輕量級、低延遲、高效的 AI 計算。

雲端側的主要功能包括但不限於視訊影像儲存管理、協作管理、演算法管理等，以及提供 AI 框架的上層能力，可實現視訊影像儲存、算力網路最佳化、端邊協作等，同時還可以透過 AI 框架（訓練、推理、深度學習演算法庫）實現 AI 模型的訓練和推理。

端側主要包括視訊影像資訊擷取裝置、智慧視訊影像處理裝置。根據不同的部署位置和應用場景，邊緣視覺硬體的形態有所不同，其主要功能是承載視訊影像處理、視訊影像辨識、目標比對、視訊影像檢索等一種或多種視訊影像處理功能，為邊緣視覺處理平臺提供計算、儲存、網路等資源，如智慧攝影機、智慧閘道、智慧視訊伺服器、AR/VR 裝置等。

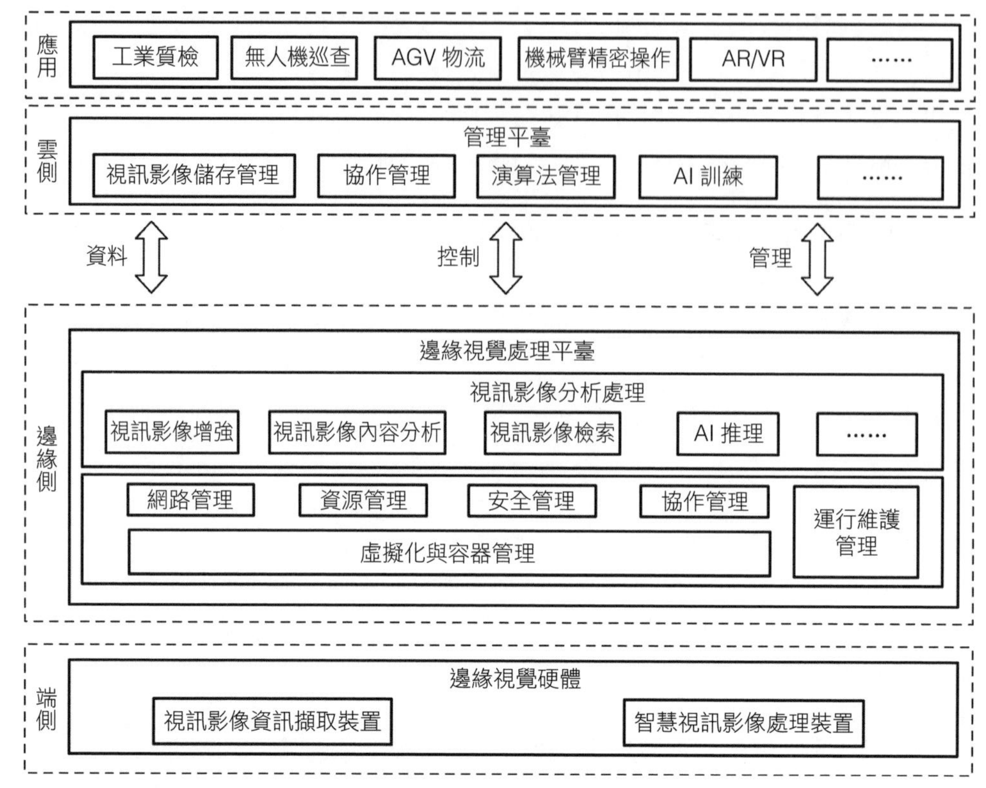

▲ 圖 2.23 邊緣視覺處理的框架

2.2.1.2 演算法介面

1）框架設計

AiCam 平臺採用統一模型呼叫、統一硬體介面、統一演算法封裝和統一應用範本的設計模式，可以在嵌入式邊緣計算環境下進行快速的應用程式開發和專案實施。AiCam 平臺透過 RESTful 介面呼叫模型的演算法，即時傳回視訊影像的分析結果，透過物聯網雲端平台的應用介面與硬體連接和互動，最終實現各種應用。AiCam 平臺的開發框架請參考圖 2.4。

2）即時推理

AiCam 平臺的即時推理介面主要實現了視訊流的即時 AI 計算，透過演算法即時計算攝影機擷取到的視訊影像，傳回計算結果影像（如框出目標位置和辨識內容的影像）。計算結果影像被即時推串流到應用端，並以視訊的方式進行顯示。應用端可將計算結果資料（如目標座標、目標關鍵點、目標名稱、推理時間、置信度等）用於業務的處理。即時推理的詳細邏輯如下：

（1）開啟邊緣計算閘道的攝影機，獲取即時的視訊影像。

（2）將即時視訊影像推送給演算法介面的 inference 方法。

（3）透過 inference 方法進行影像處理，或呼叫模型進行影像推理。

（4）透過 inference 方法傳回 base64 編碼的影像處理結果。

（5）AiCam 平臺將傳回的影像處理結果拼接為 text/event-stream 資料流程，供應用使用。

（6）應用層透過 EventSource 介面獲取即時推送的影像處理結果（計算結果影像和計算結果資料）。

（7）應用層解析資料流程，提取出計算結果影像和計算結果資料進行應用展示。

3）單次推理

AiCam 平臺的單次推理介面主要實現了應用層業務的單次推理計算請求，應用層將需要計算的影像及配置參數透過 Ajax 介面傳遞給演算法，演算法根據參數進行影像的推理計算並傳回計算結果影像（如框出目標位置和辨識內容的影像）和計算結果資料（如目標座標、目標關鍵點、目標名稱、推理時間、置信度等）。單次推理的詳細邏輯如下：

（1）應用層截取需要 AI 計算的影像，並將影像轉為 Blob 格式的資料。

（2）應用層以 JSON 格式封裝參數（如人臉註冊應用的人臉名稱、操作類型等）。

（3）以 formData 表單資料的形式透過 Ajax 介面將影像和參數傳遞到演算法。

（4）透過演算法的 inference 方法接收應用傳遞的影像和參數，呼叫模型進行影像推理。

（5）透過 inference 方法傳回 base64 編碼的計算結果影像（如框出目標位置和辨識內容的影像）和計算結果資料（如目標座標、目標關鍵點、目標名稱、推理時間、置信度等）。

（6）AiCam 平臺將演算法處理的計算結果影像和計算結果資料透過 Ajax 介面傳回。

（7）應用層解析傳回資料，提取出計算結果影像和計算結果資料進行應用展示。

2.2.1.3 邊緣計算的演算法設計

1）人臉註冊

人臉註冊透過單次推理介面呼叫演算法實現人臉註冊功能，應用層將需要註冊的人臉影像和參數〔所需要註冊的人名、處理類別（註冊）〕透過 Ajax 介面傳遞給演算法。演算法的 inference 方法透過傳遞過來的參數 param_data 是不是 None 來判斷是否單次推理，None 表示即時推理，非 None 表示單次推理。演算法檔案為 algorithm\image_face_recognition\image_face_ recognition.py，關鍵程式如下：

```
def inference(self, image, param_data):
    #code：辨識成功傳回 200
    #msg：相關提示訊息
    #origin_image：原始影像
```

```
    #result_image：處理之後的影像
    #result_data：結果資料
    return_result = {'code': 200, 'msg': None, 'origin_image': None, 'result_image':
None, 'result_data': None}

    # 應用請求介面：@__app.route('/file/<action>', methods=["POST"])
    #image：應用傳遞過來的資料（根據實際應用可能為影像、音訊、視訊、文字）
    #param_data：應用傳遞過來的參數，不能為空
    if param_data != None:
        # 讀取應用傳遞過來的影像
        image = np.asarray(bytearray(image), dtype="uint8")
        image = cv.imdecode(image, cv.IMREAD_COLOR)

        #type=0 表示註冊
        if param_data["type"] == 0:
            try:
                image_encoding = face_recognition.face_encodings(image)[0]
                if len(image_encoding) != 0:
                    flag = True
                else:
                return_result["code"] = 404
                return_result["msg"] = " 未檢測到人臉！ "
            except:
                return_result["code"] = 500
                return_result["msg"] = " 註冊失敗！ "
            if flag:
                feature_name = param_data["reg_name"] + ".npy"
                feature_path = os.path.join(self.dir_path, feature_name)
                np.save(feature_path, image_encoding)
                print(" 已儲存人臉 ")
                return_result["code"] = 200
                return_result["msg"] = " 註冊成功！ "
    return return_result
```

2）人臉比對

人臉比對透過即時推理介面呼叫演算法實現人臉比對功能，應用層透過 EventSource 介面呼叫演算法介面獲取資料流程，演算法檔案為 algorithm\image_face_recognition\image_face_ recognition.py，關鍵程式如下：

```
def inference(self, image, param_data):
    #code：辨識成功傳回 200
    #msg：相關提示訊息
    #origin_image：原始影像
    #result_image：處理之後的影像
    #result_data：結果資料
    return_result = {'code': 200, 'msg': None, 'origin_image': None, 'result_image':
None, 'result_data': None}

    # 應用請求介面：@__app.route('/file/<action>', methods=["POST"])
    #image：應用傳遞過來的資料（根據實際應用可能為影像、音訊、視訊、文字）
    #param_data：單次推理介面應用傳遞過來的參數，不能為空
    if param_data != None:
        # 此處為人臉註冊程式，省略
        # 應用請求介面：@__app.route('/stream/<action>')
        #image：讀取的攝影機即時視訊影像
        #param_data：即時推理介面，必須為 None
    else:
        # 呼叫介面進行人臉比對
        rects = self.face_id(image, self.face_cascade)
        for x, y, w, h in rects:
        crop = image[y: y + h, x: x + w]
        # 視訊流中人臉特徵編碼
        img_encoding = face_recognition.face_encodings(crop)
        if len(img_encoding) != 0:
            # 獲取人臉特徵編碼
            img_encoding = img_encoding[0]
            # 與註冊的人臉特徵進行對比
            result = face_recognition.compare_faces(self.features, img_encoding,
tolerance=0.4)
            if True in result:
```

```
                result = int(np.argmax(np.array(result, np.uint8)))
                rec_result = self.feature_names[result]
                cv.putText(image, rec_result, (x, y), cv.FONT_HERSHEY_SIMPLEX, 1.2, (0, 
255, 0), thickness=1)
                cv.rectangle(image, (x, y), (x + w, y + h), (0, 255, 0), 1)
                return_result["result_data"]=rec_result
            else:
                cv.putText(image, 'unknown', (x, y), cv.FONT_HERSHEY_SIMPLEX, 1.2, (0, 
255, 0), thickness=1)
                cv.rectangle(image, (x, y), (x + w, y + h), (0, 255, 0), 1)
                return_result["result_data"]='unknown'


        else:
            cv.putText(image, 'unknown', (x, y), cv.FONT_HERSHEY_SIMPLEX, 1.2, (0, 
255, 0), thickness=1)
            cv.rectangle(image, (x, y), (x + w, y + h), (0, 255, 0), 1)
            return_result["result_data"]='unknown'
            return_result["result_image"] = self.image_to_base64(image)
    return return_result
```

3）單元測試

演算法檔案 algorithm\image_face_recognition\image_face_recognition.py 提供了單元測試程式，透過傳參 0 實現人臉註冊，透過傳參 1 實現人臉比對。關鍵程式如下：

```
# 單元測試，如果處理類別中引用了檔案，則在單元測試中要修改檔案路徑
if __name__=='__main__':
    mode = sys.argv[1]
    c_dir = os.path.split(os.path.realpath(__file__))[0]
    # 建立影像處理物件
    img_object = ImageFaceRecognition(c_dir)

    # 讀取測試影像
    image = cv.imread(c_dir+"/test.jpg")
    # 將影像編碼成資料流程
```

```
img = cv.imencode('.jpg', image, [cv.IMWRITE_JPEG_QUALITY, 60])[1]

# 設定參數
addUser_data = {"type":0, "reg_name":"lilianjie"}

# 呼叫介面進行人臉註冊
if mode == '0':
    result = img_object.inference(img, addUser_data)
    print(result)

# 呼叫介面進行人臉辨識
elif mode == '1':
    result = img_object.inference(image, None)
    frame = img_object.base64_to_image(result["result_image"])
    print(result)

    # 影像顯示
    cv.imshow('frame',frame)
    while True:
        key=cv.waitKey(1)
        if key==ord('q'):
            break
    cv.destroyAllWindows()
else:
    print(" 參數錯誤！ ")
```

2.2.2 開發步驟與驗證

2.2.2.1 專案部署

1）硬體部署

（1）為邊緣計算閘道 GW3588 連接 Wi-Fi 天線、攝影機。

（2）啟動 AiCam 平臺，連接區域網內的 Wi-Fi 網路，記錄 AiCam 平臺的 IP 位址，如 192.168.100.200。

2）專案部署

（1）執行 MobaXterm 工具，透過 SSH 登入到邊緣計算閘道。

（2）在 SSH 終端執行以下命令，建立開發專案目錄。

```
$ mkdir -p ~/aiedge-exp
```

（3）透過 SSH 將開發專案程式上傳到 ~/aicam-exp 目錄下，並採用 unzip 命令進行解壓縮，程式如下：

```
$ cd ~/aiedge-exp
$ unzip image_face_recognition.zip
```

2.2.2.2 演算法測試

1）人臉註冊

（1）在 SSH 終端輸入以下命令：

```
$ cd ~/aiedge-exp/image_face_recognition
$ conda activate py36_tf114_torch15_cpu_cv345    //Ubuntu 20.04 作業系統下需要切換環境
$ python3 image_face_recognition.py 0
```

進行人臉註冊的單元測試，本測試將讀取影像，並提交給演算法介面進行人臉註冊，傳回的註冊結果資訊為：

```
// 已儲存人臉
{'result_data': None, 'msg': '註冊成功！', 'code': 200, 'result_image': None, 'origin_
image': None}
```

（2）在演算法資料夾下可以看到生成的人臉特徵檔案 lilianjie.npy。

2）人臉比對

在 SSH 終端輸入以下命令：

```
$ cd ~/aiedge-exp/image_face_recognition
$ conda activate py36_tf114_torch15_cpu_cv345    //Ubuntu 20.04 作業系統下需要切換環境
$ python3 image_face_recognition.py 1
```

進行人臉比對的單元測試，本測試將讀取影像，並提交給演算法介面進行人臉比對，比對完成後將結果影像在視窗中顯示，傳回的比對結果資訊為：


```
{'msg': None, 'result_image': '/9j/4AAQSkZJRgABAQAAAQABAAD//9k=', 'result_data':
'lilianjie', 'origin_image': None, 'code': 200}
```


人臉對比的單元測試結果如圖 2.24 所示。

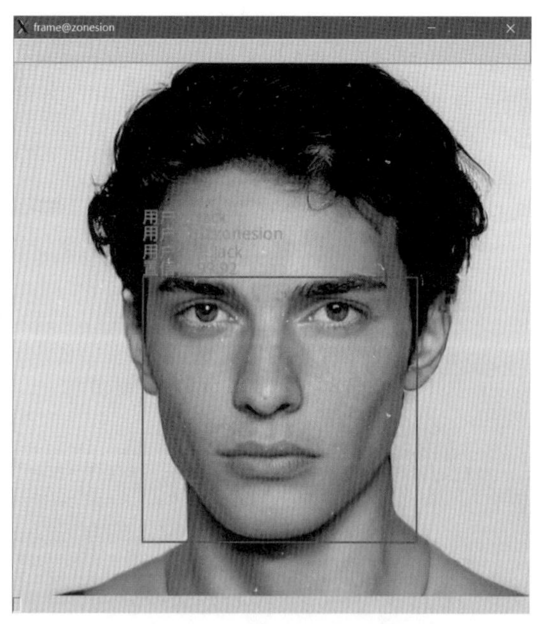

▲ 圖 2.24 人臉比對的單元測試結果

2.2.3 本節小結

本節首先介紹了機器視覺應用導向的邊緣計算（邊緣視覺）框架，然後介紹了邊緣計算的演算法介面和演算法設計，最後介紹了邊緣計算的應用的開發步驟和驗證。

2.2.4 思考與擴充

（1）AiCam 平臺即時推理的基本步驟有哪些？

（2）AiCam 平臺單次推理的基本步驟有哪些？

2.3 邊緣計算的硬體設計

邊緣計算的硬體設計需要綜合考慮性能、功耗、即時性和安全性等多方面的要求，以適應邊緣應用的多樣性和特殊性。以下是邊緣計算硬體設計的特點：

（1）低功耗設計：邊緣裝置通常是行動裝置或嵌入式裝置，因此需要採用低功耗的裝置，以延長電池壽命。低功耗裝置包括功耗最佳化的處理器、能效高的元件和最佳化的電源管理。

（2）小尺寸和輕量化：邊緣裝置可能需要攜帶或嵌入在各種環境中，因此需要考慮硬體的小尺寸和輕量化，以適應各種應用場景。

（3）即時性能：邊緣計算通常需要在即時或近即時條件下進行資料處理和決策。因此，邊緣硬體需要具備足夠的即時性能，能夠快速回應和處理感測器資料。

（4）本機存放區和快取：由於邊緣計算強調在本地裝置上進行一部分資料處理，因此需要足夠的本機存放區和快取，這可以減少對雲端的依賴，提高性能和降低延遲。

（5）多模組設計：邊緣裝置通常會整合多種感測器、通訊模組和其他元件，以支援多樣化的應用，因此在設計硬體時需要考慮模組化，以方便擴充和訂製。

（6）安全硬體模組：邊緣計算需要處理分佈在多個裝置上的資料，因此安全性是一個至關重要的因素。在設計硬體時，需要考慮安全硬體模組，用於加密、身份驗證和其他安全功能。

（7）通訊介面：邊緣裝置通常需要與其他裝置或雲端進行通訊，因此在設計硬體時需要考慮多種通訊介面（如 Wi-Fi、藍牙、LoRa 等），以滿足不同的通訊需求。

本節的基礎知識如下：

- 掌握邊緣計算導向的智慧物聯網平臺框架。
- 結合智慧物聯網平臺，掌握應用程式開發框架的應用介面、通訊協定和開發工具。
- 結合智慧產業套件範例，掌握邊緣計算的硬體設計方法。

2.3.1 原理分析與開發設計

2.3.1.1 邊緣計算導向的智慧物聯網平臺框架

物聯網利用有線或無線通訊方式，實現了人與物、物與物之間的數位化連接。物聯網的智慧化能夠釋放物聯網的底層能量，開拓創新應用空間。傳統的物聯網包括感知層、網路層、平臺層、應用層四個部分；智慧物聯網融入了人工智慧演算法，擴大了應用邊界，實現了從連接萬物到喚醒萬物、從中心化到端邊雲協作、從技術革新到產業革命、從物聯網思維到智聯網思維的變化。

傳統物聯網與智慧物聯網比較如圖 2.25 所示，智慧物聯網可支撐更細粒度的應用場景落地、挖掘巨量異質資料價值。特別是在資料流程的傳輸過程中，新的應用場景對資料的需求不僅停留在資料分析層面，還對資料形態和中間過程提出了要求，需要基於多模資料進行互動。

AiCam 平臺能夠連接巨量的物聯網硬體，可以透過智慧物聯網平臺實現與物聯網硬體的互動。智慧物聯網平臺中的平臺層是資料中樞，同時也為感知層、網路層和應用層提供軟 / 硬體平臺和範例支撐。本書中的智慧物聯網平臺採用的是中智訊（武漢）科技有限公司的智雲端平台（ZCloud），其框架如圖 2.26 所示。

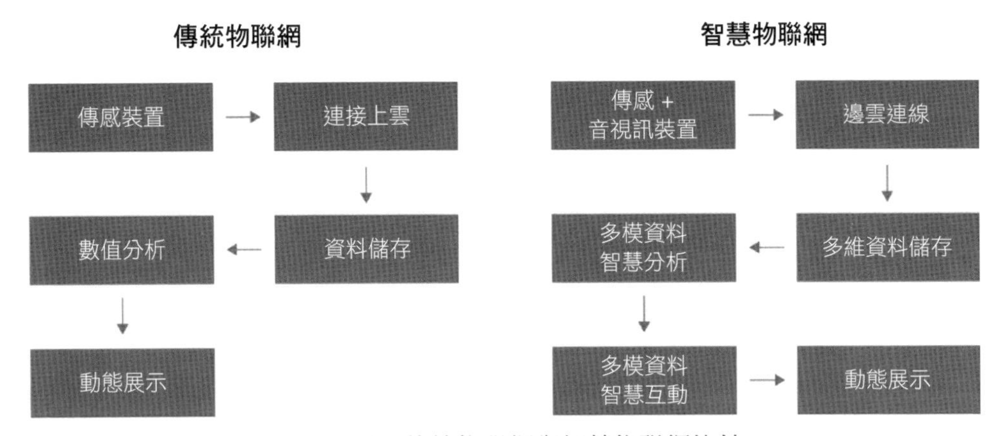

▲ 圖 2.25 傳統物聯網與智慧物聯網比較

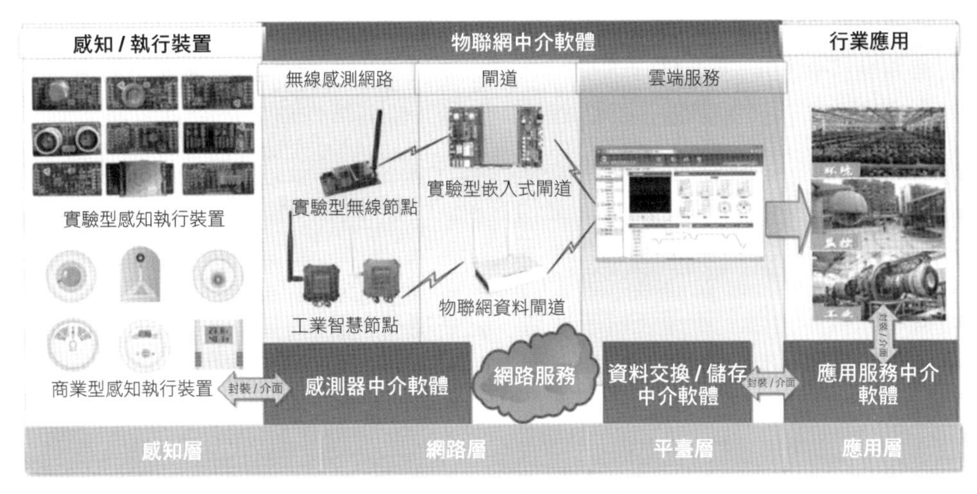

▲ 圖 2.26 智雲端平台的框架

（1）感知層：全面感知。感知層以微控制器、嵌入式技術為核心，賦予物品智慧化、數位化，透過感測器擷取物理世界中的事件和資料，包括各類物理量、標識、音訊、視訊等。物聯網的資料獲取涉及感測器、RFID、多媒體資訊擷取、二維碼和即時定位等技術。智雲端平台提供了各種教學模型、控制節點和豐富的感測器，可滿足感知層的不同應用需求。

（2）網路層：網路傳輸。網路層包括感測網路和網際網路兩部分。感測網路透過無線技術實現物品之間的網路拓樸、通訊，以及資料的傳輸。不同的應用場合應選擇不同類型的協定，如 ZigBee、BLE、Wi-Fi、NB-IoT、LoRa、LTE 等。透過智慧閘道 / 路由裝置，可以實現感測網路和網際網路之間的資料互動。智雲端平台提供了各種教學模型、有線節點、無線節點和閘道，支援異質網路的融合，採用的是 ZXBee 輕量級通訊協定（基於易懂易學的 JSON 資料通信格式），可滿足教學和不同應用的需求。

（3）平臺層：資料中樞。平臺層是智慧物聯網平臺的資料匯聚中心。透過雲端運算、巨量資料等技術，平臺層可以對感知到的資料進行無障礙、高可靠性、高安全性的處理和傳輸，並為上層應用提供資料服務。智雲端平台支持巨量物聯網資料的連線、分類儲存、資料決策、資料分析及資料探勘；採用了分散式巨量資料技術，具備資料的即時訊息推送處理、資料倉儲儲存與資料探勘等功能。智雲端業務管理平臺是基於 B/S 架構的背景分析管理系統，可透過 Web 應用對資料中心進行管理並監控智雲端平台的營運，主要的功能模組有訊息推送、資料儲存、資料分析、觸發邏輯、應用資料、位置服務、簡訊通知、視訊傳輸等。

（4）應用層：應用服務。應用層可透過移動 App、Web 應用、小程式、多種終端設備（手機、電視、電子螢幕、智慧喇叭）等操作互動。根據物聯網應用場景和業務的不同，應用層提供了智慧交通、智慧醫療、智慧家居、智慧物流、智慧電力等行業應用服務。

2.3.1.2 專案模型

在基於智雲端平台開發的智慧物聯網專案中：

（1）各種智慧裝置透過 ZigBee、BLE、Wi-Fi、NB-IoT、LoRa、LTE 等無線網路聯繫在一起，協調器 / 匯聚器是整個網路的資料匯聚中心。

（2）協調器 / 匯聚器與智雲閘道進行互動，透過智雲閘道上執行的服務程式實現了感測網路與網際網路之間的資料交換，既可將資料推送到智雲端平台中心，也可將資料推送到本地區域網。

（3）智雲端資料中心提供了資料儲存、資料推送、資料決策、攝像監控等服務應用介面，本機服務僅支援資料推送服務。

（4）透過智雲 API 可開發智慧物聯網專案的具體應用，能夠對感測網路內的資料進行擷取、控制、決策等。

基於智雲端平台的典型智慧物聯網專案如圖 2.27 所示。

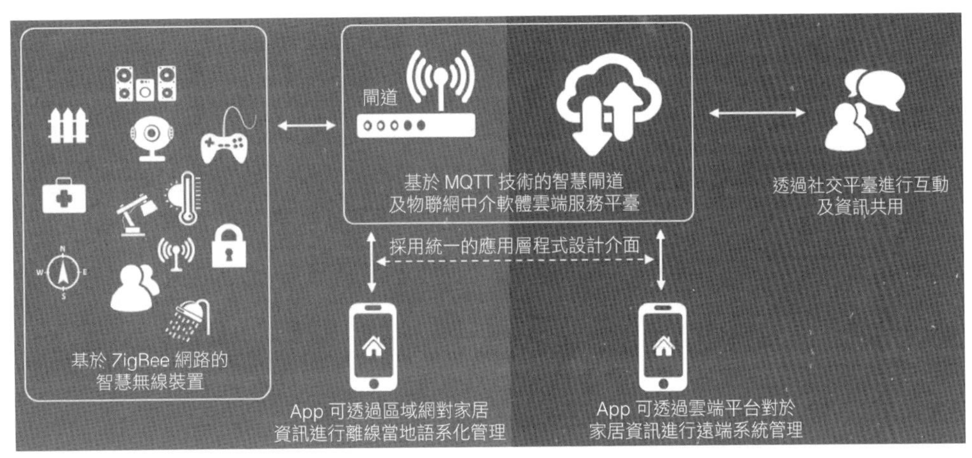

▲ 圖 2.27 基於智雲端平台的典型智慧物聯網專案

2.3.1.3 應用介面

1）應用介面的框架

智雲端平台提供五大類應用介面供開發者使用，包括即時連接（WSNRT Connect）、歷史資料（WSNHistory）、攝像監控（WSNCamera）、自動控制（WSNAutoctrl）、使用者資料（WSNProperty）。智雲端平台應用介面的框架如圖 2.28 所示。

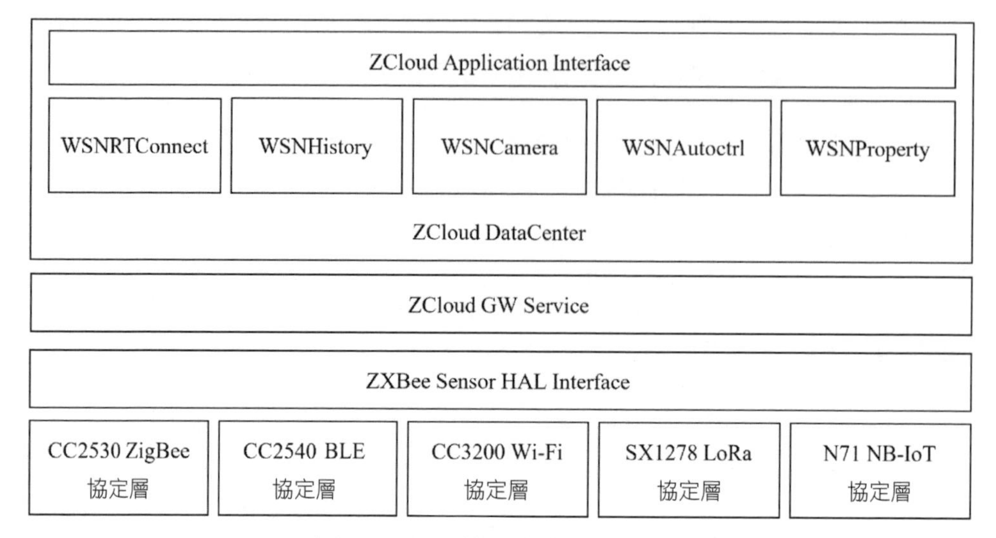

▲ 圖 2.28 智雲端平台應用介面的框架

2）應用介面的說明

針對 Web 應用程式開發，智雲端平台提供了 JavaScript 介面函式庫，使用者呼叫相應的介面即可完成簡單的 Web 應用程式開發。這裡重點介紹即時連接應用介面和歷史資料應用介面。

（1）即時連接應用介面。即時連接應用介面如表 2.5 所示。

▼ 表 2.5 即時連接應用介面

函式	參數說明	功能
new WSNRTConnect(ID, Key);	ID：智雲帳號。Key：智雲金鑰	建立即時資料實例，初始化 ID 及 Key
connect()	無	建立即時資料服務連接
disconnect()	無	斷開即時資料服務連接
onConnect()	無	監聽連接智雲端服務成功
onMessageArrive(mac, dat)	mac：感測器的 MAC 位址。dat：發送的訊息	監聽收到的資料

（續表）

函式	參數說明	功能
sendMessage(mac, dat)	mac：感測器的 MAC 位址。dat：發送的訊息	發送訊息
setServerAddr(sa)	sa：資料中心伺服器位址及通訊埠	設定 / 改變資料中心伺服器位址及通訊埠

（2）歷史資料應用介面。歷史資料應用介面如表 2.6 所示。

▼ 表 2.6 歷史資料應用介面

函式	參數說明	功能
new WSNRTConnect (ID, Key);	ID：智雲帳號。Key：智雲金鑰	初始化歷史資料物件、ID 及 Key
queryLast1H(ch, cal);	ch：感測器資料通道。cal：回呼函式（處理歷史資料）	查詢最近 1 小時的歷史資料
queryLast6H(ch, cal);	ch：感測器資料通道。cal：回呼函式（處理歷史資料）	查詢最近 6 小時的歷史資料
queryLast12H(ch, cal);	ch：感測器資料通道。cal：回呼函式（處理歷史資料）	查詢最近 12 小時的歷史資料
queryLast1D(ch, cal);	ch：感測器資料通道。cal：回呼函式（處理歷史資料）	查詢最近 1 天的歷史資料
queryLast5D(ch, cal);	ch：感測器資料通道。cal：回呼函式（處理歷史資料）	查詢最近 5 天的歷史資料
queryLast14D(ch, cal);	ch：感測器資料通道。cal：回呼函式（處理歷史資料）	查詢最近 14 天的歷史資料
queryLast1M(ch, cal);	ch：感測器資料通道。cal：回呼函式（處理歷史資料）	查詢最近 1 個月（30 天）的歷史資料
queryLast3M(ch, cal);	ch：感測器資料通道。cal：回呼函式（處理歷史資料）	查詢最近 3 個月（90 天）的歷史資料
queryLast6M(ch, cal);	ch：感測器資料通道。cal：回呼函式（處理歷史資料）	查詢最近 6 個月（180 天）的歷史資料

（續表）

函式	參數說明	功能
queryLast1Y(ch, cal);	ch：感測器資料通道。cal：回呼函式（處理歷史資料）	查詢最近1年(365天)的歷史資料
query(cal);	cal：回呼函式（處理歷史資料）	獲取所有通道最後一次資料
query(ch, cal);	ch：感測器資料通道。cal：回呼函式（處理歷史資料）	獲取指定通道下最後一次資料
query(ch, start, end, cal);	ch：感測器資料通道。cal：回呼函式（處理歷史資料）。start：起始時間。end：結束時間。時間為 ISO 8601 格式的日期，如 2010-05-20T11:00:00Z	透過起止時間查詢指定時間段的歷史資料（根據時間範圍選擇預設採樣間隔）
query(ch, start, end, interval, cal);	ch：感測器資料通道。cal：回呼函式（處理歷史資料）。start：起始時間。end：結束時間。interval：採樣點的時間間隔。時間為 ISO 8601 格式的日期，如 2010-05-20T11:00:00Z	透過起止時間查詢指定時間段指定時間間隔的歷史資料
setServerAddr(sa)	sa：資料中心伺服器位址及通訊埠	設定 / 改變資料中心伺服器位址及通訊埠

2.3.1.4 通訊協定

1）協定說明

智雲端平台支援感測網路資料的無線連線，並定義了物聯網資料通訊的規範。智雲端平台採用輕量級的 ZXBee 通訊協定，該協定採用 JSON 資料格式，更加清晰易懂。

ZXBee 通訊協定對物聯網底層到上層的資料做了定義，具有以下特點：

- 資料格式的語法簡單、語義清晰，參數少而精；

- 參數名稱合乎邏輯、見名知義，變數和命令的分工明確；
- 參數的讀寫許可權分配合理，可以有效防止不合理的操作，可最大限度地確保資料安全；
- 變數能對值進行查詢，方便應用程式偵錯；
- 命令是對位操作的，能夠避免記憶體資源的浪費。

2）協定簡介

（1）通訊協定資料格式。通訊協定資料格式為「{[參數]=[值],{[參數]=[值],……}」。

① 每筆資料以「{」作為起始字元，以「}」作為結束字元。

②「{}」內的多個參數以「,」分隔。

注意：資料格式中的字元均為英文半形符號。

範例如下：

```
{CD0=1,D0=?}
```

（2）通訊協定參數說明如下：

① 參數名稱定義為：

- 變數：A0 ～ A7、D0、D1、V0 ～ V3。
- 命令：CD0、OD0、CD1、OD1。
- 特殊參數：ECHO、TYPE、PN、PANID、CHANNEL。

② 變數可以對值進行查詢。範例如下：

```
{A0=?}
```

③ 變數 A0 ～ A7 在物聯網雲端資料中心可以儲存為歷史資料。

④ 命令是對位操作的。

3）參數說明

具體的參數說明如下：

（1）A0 ～ A7：用於傳輸感測器數值或攜帶的資訊量，只能透過賦值「?」來查詢變數的當前值，可上傳到物聯網雲端資料中心儲存。

① 溫濕度感測器用 A0 表示溫度值、用 A1 表示濕度值，數值型態為浮點型，精度為 0.1。

② 火焰警告感測器用 A3 表示警示狀態，數值型態為整數，0 表示未檢測到火焰，1 表示檢測到火焰。

③ 高頻 RFID 模組用 A0 表示卡片的 ID，數值型態為字串。

（2）D0：D0 的 Bit0 ～ Bit7 分別對應 A0 ～ A7 的狀態是否主動上傳，只能透過賦值「?」來查詢變數的當前值，0 表示禁止上傳，1 表示允許主動上傳。

① 溫濕度感測器用 A0 表示溫度值，A1 表示濕度值，D0=0 表示不上傳溫度值和濕度值，D0=1 表示主動上傳溫度值，D0=2 表示主動上傳濕度值，D0=3 表示主動上傳溫度和濕度值。

② 火焰警告感測器用 A3 表示警示狀態，D0=0 表示不檢測火焰，D0=1 表示檢測火焰。

③ 高頻 RFID 模組用 A0 表示卡片 ID，D0=0 表示刷卡時不上報 ID，D0=1 表示刷卡時上報 ID。

（3）CD0/OD0：對 D0 進行位元操作，CD0 表示清零操作，OD0 表示置 1 操作。

① 溫濕度感測器用 A0 表示溫度值，A1 表示濕度值，CD0=1 表示關閉 A0 溫度值的主動上報。

② 火焰警告感測器用 A3 表示警示狀態，OD0=1 表示開啟火焰檢測，當檢測到火焰時，會主動上報 A0 的數值。

（4）D1：D1 表示控制編碼，只能透過賦值「?」來查詢變數的當前值，使用者可根據感測器的屬性來自訂 D1 的功能。

① 溫濕度感測器用 D1 的 Bit0 表示電源開關狀態，如 D1=0 表示電源處於關閉狀態，D1=1 表示電源處於開啟狀態。

② 繼電器用 D1 的 Bit0 ～ Bit1 表示各路繼電器狀態，如 D1=0 表示關閉繼電器 S1 和 S2，D1=1 表示開啟繼電器 S1，D1=2 表示開啟繼電器 S2，D1=3 表示開啟繼電器 S1 和 S2。

③ 風扇用 D1 的 Bit0 表示電源開關狀態，用 D1 的 Bit1 表示正轉或反轉，如 D1=0 或 D1=2 表示風扇停止轉動（電源斷開），D1=1 表示風扇處於正轉狀態，D1=3 表示風扇處於反轉狀態。

④ 紅外電器遙控用 D1 的 Bit0 表示電源開關狀態，用 D1 的 Bit1 表示工作模式 / 學習模式，如 D1=0 或 D1=2 表示電源處於關閉狀態，D1=1 表示電源處於開啟狀態且為工作模式，D1=3 表示電源處於開啟狀態且為學習模式。

（5）CD1/OD1：對 D1 進行位元操作，CD1 表示清零操作，OD1 表示置 1 操作。

（6）V0 ～ V3：用於表示感測器的參數，使用者可根據感測器的屬性自訂參數功能，這些參數的許可權為讀寫。

① 溫濕度感測器用 V0 ～ V3 表示自動上傳資料的時間間隔。

② 風扇用 V0 ～ V3 表示風扇轉速。

③ 紅外電器遙控用 V0 ～ V3 表示紅外學習的鍵值。

④ 語音合成用 V0 ～ V3 表示需要合成的語音字元。

（7）特殊參數：ECHO、TYPE、PN、PANID、CHANNEL。

① ECHO：用於檢測節點是否線上，將發送的值進行回顯，如發送「{ECHO=test}」，若節點線上則回覆資料「{ECHO=test}」。

② TYPE：表示節點類型，該參數包含了節點類別、節點類型、節點名稱，只能透過賦值「?」來查詢該參數的當前值。TYPE 的值由 5 個 ASCII 代碼位元組表示，第 1 位元組表示節點類別，1 表示 ZigBee、2 表示 RF433、3 表示 Wi-Fi、4 表示 BLE、5 表示 IPv6、9 表示其他；第 2 位元組表示節點類型，0 表示匯聚節點、1 表示路由器 / 中繼節點、2 表示終端節點；第 3 ～ 5 個位元組表示節點名稱，編碼由使用者自訂。

③ PN：表示某節點的上行節點位址資訊和鄰居節點位址資訊，只能透過賦值「?」來查詢該參數的當前值。PN 的值是上行節點位址資訊和鄰居節點位址資訊的組合，其中每 4 個位元組表示一個節點的位址後 4 位元，第一個 4 位元組表示該節點上行節點位址的後 4 位元，第 2 ～ n 個 4 位元組表示當前節點鄰居節點位址的後 4 位元。

4）參數定義

ZXBee 通訊協定的參數定義如表 2.7 所示。

▼ 表 2.7 ZXBee 通訊協定的參數定義

節點名稱	TYPE	參數	屬性	許可權	說明
Sensor-A 擷取類感測器	601	A0	溫度	R	溫度值，浮點型，精度為 0.1，範圍為 -40.0 ～ 105.0，單位為℃
		A1	濕度	R	濕度值，浮點型，精度為 0.1，範圍為 0 ～ 100.0，單位為 %RH
		A2	光照度	R	光照度值，浮點型，精度為 0.1，範圍為 0 ～ 65535.0，單位為 lx

節點名稱	TYPE	參數	屬性	許可權	說明
Sensor-A 擷取類感測器	601	A3	空氣品質	R	空氣品質值，表徵空氣污染程度，整數，範圍為 0 ～ 20000，單位為 ppm
		A4	大氣壓力	R	大氣壓力值，浮點型，精度為 0.1，範圍為 800.0 ～ 1200.0，單位為 hPa
		A5	跌倒狀態	R	透過三軸感測器計算出跌倒狀態，0 表示未跌倒，1 表示跌倒
		A6	距離	R	距離值，浮點型，精度為 0.1，範圍為 10.0 ～ 80.0，單位為 cm
		D0(OD0/CD0)	上報狀態	R/W	D0 的 Bit0 ～ Bit7 分別代表 A0 ～ A7 的上報狀態，1 表示主動上報，0 表示不上報
		D1(OD1/CD1)	繼電器	R/W	D1 的 Bit6 ～ Bit7 分別代表繼電器 K1、K2 的開關狀態，0 表示斷開，1 表示吸合
		V0	上報時間間隔	R/W	A0 ～ A7 和 D1 的主動上報時間間隔，預設為 30，單位為 s
Sensor-B 控制類感測器	602	D1(OD1/CD1)	RGB	R/W	D1 的 Bit0 ～ Bit1 代表 RGB 三色燈的顏色狀態，00 表示關、01 表示紅色、10 表示綠色、11 表示藍色
		D1(OD1/CD1)	步進馬達	R/W	D1 的 Bit2 表示步進馬達的正 / 反轉狀態，0 表示正轉，1 表示反轉

（續表）

節點名稱	TYPE	參數	屬性	許可權	說明
Sensor-B 控制類 感測器	602	D1(OD1/CD1)	風扇 / 蜂鳴器	R/W	D1 的 Bit3 表示風扇 / 蜂鳴器的開關狀態，0 表示關閉，1 表示開啟
		D1(OD1/CD1)	LED	R/W	D1 的 Bit4 ～ Bit5 表示 LED1、LED2 的開關狀態，0 表示關閉，1 表示開啟
		D1(OD1/CD1)	繼電器	R/W	D1 的 Bit6 ～ Bit7 表示繼電器 K1、K2 的開關狀態，0 表示斷開，1 表示吸合
		V0	上報間隔	R/W	A0 ～ A7 和 D1 的循環上報時間間隔
Sensor-C 保全類 感測器	603	A0	人體紅外 / 觸控	R	人體紅外 / 觸控感測器狀態，設定值為 0 或 1，1 表示有人體活動 / 觸控動作，0 表示無人體活動 / 觸控動作
		A1	震動	R	震動狀態，設定值為 0 或 1，1 表示檢測到震動，0 表示未檢測到震動
		A2	霍爾	R	霍爾狀態，設定值為 0 或 1，1 表示檢測到磁場，0 表示未檢測到磁場
		A3	火焰	R	火焰狀態，設定值為 0 或 1，1 表示檢測到火焰，0 表示未檢測到火焰

（續表）

節點名稱	TYPE	參數	屬性	許可權	說明
Sensor-C 保全類感測器	603	A4	瓦斯	R	瓦斯洩漏狀態，設定值為 0 或 1，1 表示檢測到瓦斯洩漏，0 表示未檢測到瓦斯洩漏
		A5	光柵	R	光柵（紅外對射）狀態值，設定值為 0 或 1，1 表示檢測到阻擋，0 表示未檢測到阻擋
		D0(OD0/CD0)	上報狀態	R/W	D0 的 Bit0 ～ Bit7 分別表示 A0 ～ A7 的上報狀態，1 表示主動上報，0 表示不上報
		D1(OD1/CD1)	繼電器	R/W	D1 的 Bit6 ～ Bit7 分別表示繼電器 K1、K2 的開關狀態，0 表示斷開，1 表示吸合
		V0	上報間隔	R/W	A0 ～ A7 和 D1 的循環上報時間間隔
Sensor-D 顯示類感測器	604	五向開關狀態	A0	R	觸發上報，1 表示上（UP）、2 表示左（LEFT）、3 表示下（DOWN）、4 表示右（RIGHT）、5 表示中心（CENTER）
		OLED 背光開關	D1(OD1/CD1)	R/W	D1 的 Bit0 代表 LCD 的背光開關狀態，1 表示開啟背光開關，0 表示關閉背光開關
		數位管背光開關	D1(OD1/CD1)	R/W	D1 的 Bit1 代表數位管的背光開關狀態，1 表示開啟背光開關，0 表示關閉背光開關

（續表）

節點名稱	TYPE	參數	屬性	許可權	說明
Sensor-D 顯示類 感測器	604	上報間隔	V0	R/W	A0 值的循環上報時間間隔
		車牌 / 儀表	V1	R/W	車牌號碼 / 儀表值
		車位數	V2	R/W	停車場空閒車位數
		模式設定	V3	R/W	1 表示停車模式，2 表示抄表模式
Sensor-EL 低頻 辨識類	605	卡號	A0	—	字串（主動上報，不可查詢）
		卡類型	A1	R	整數，0 表示 125 kHz 卡片，1 表示 13.56 MHz 的卡片
		卡餘額	A2	R	整數，範圍 0 ～ 8000，手動查詢
		裝置餘額	A3	R	浮點型，裝置餘額
		裝置單次消費金額	A4	R	浮點型，裝置本次消費扣款金額
		裝置累計消費金額	A5	R	浮點型，裝置累計扣款金額
		門鎖狀態	D1(OD1/CD1)	R/W	D1 的 Bit0 表示門鎖的開關狀態，0 表示關閉，1 表示開啟
		充值金額	V1	R/W	傳回充值狀態，設定值為 0 或 1，1 表示操作成功，0 表示操作不成功
		扣款金額	V2	R/W	傳回扣款狀態，設定值為 0 或 1，1 表示操作成功，0 表示操作不成功

（續表）

節點名稱	TYPE	參數	屬性	許可權	說明
Sensor-EH 高頻辨識類	606	卡號	A0	R	字串（主動上報，不可查詢）
		卡餘額	A2	R	整數，範圍 0 ～ 8000，手動查詢
		ETC 桿開關	D1(OD1/CD1)	R/W	D1 的 Bit0 表示 ETC 欄桿的狀態，0 表示落下，1 表示抬起，3 s 後自動落下並將 Bit0 清零
		充值金額	V1	R/W	傳回充值狀態，設定值為 0 或 1，1 表示操作成功，0 表示操作不成功
		扣款金額	V2	R/W	傳回扣款狀態，設定值為 0 或 1，1 表示操作成功，0 表示操作不成功

2.3.1.5 開發工具

1）模擬工具

智慧物聯網平臺的虛擬模擬系統如圖 2.29 所示，可滿足物聯網感測網路與網際網路的互聯、互通和互動。虛擬模擬系統主要包括 VR 虛擬體驗系統、三維場景模擬系統、圖形設定應用系統、智雲中介軟體平臺、智雲硬體物元模擬平臺、物聯網硬體平臺，不僅能夠與物聯網硬體平臺實現虛實結合，還可以進行物聯網感測協定模擬和物聯網資訊安全模擬。

智雲硬體物元模擬平臺主要完成物聯網的硬體資料模擬，基於智慧物聯網平臺框架的硬體系統可以輕鬆連線並模擬物聯網物元的屬性，根據指令稿及配置產生物元資料、執行狀態等，為物聯網應用提供硬體資料來源支撐。智雲硬體物元模擬平臺不僅可以模擬各種感測器、執行器等常用硬體，也可以對貴重裝置進行資料或狀態的模擬，大大降低開發成本，還可以對複雜場合運行維護的裝置（如對人體危害領域）進行有效的模擬，降低開發風險。

2）偵錯工具

ZCloudWebTools 是一款硬體偵錯工具，不僅可以查看網路拓撲結構，還可以分析節點的即時資料和歷史資料並以曲線的形式進行展示。透過 ZCloudWebTools 可以進行 ZXBee 協定的偵錯。

▲ 圖 2.29 智慧物聯網平臺的虛擬模擬系統

ZCloudWebTools 提供了即時資料功能，不僅能夠透過雲端平台向感測器節點發送命令，也能夠接收感測器節點主動上報的資料，用於資料分析和偵錯。在實際使用 ZCloudWebTools 時，首先將其切換到即時資料選項，然後將閘道的使用者帳號和金鑰填寫到相應的帳號和金鑰位置，按一下「連接」按鈕即可。成功連接 ZCloudWebTools 後，在位址處填寫感測器節點的 MAC 位址，在資料處填寫 ZXBee 協定的命令，按一下「發送」按鈕後，下方會顯示感測器節點傳回資料。ZCloudWebTools 的介面如圖 2.30 所示。

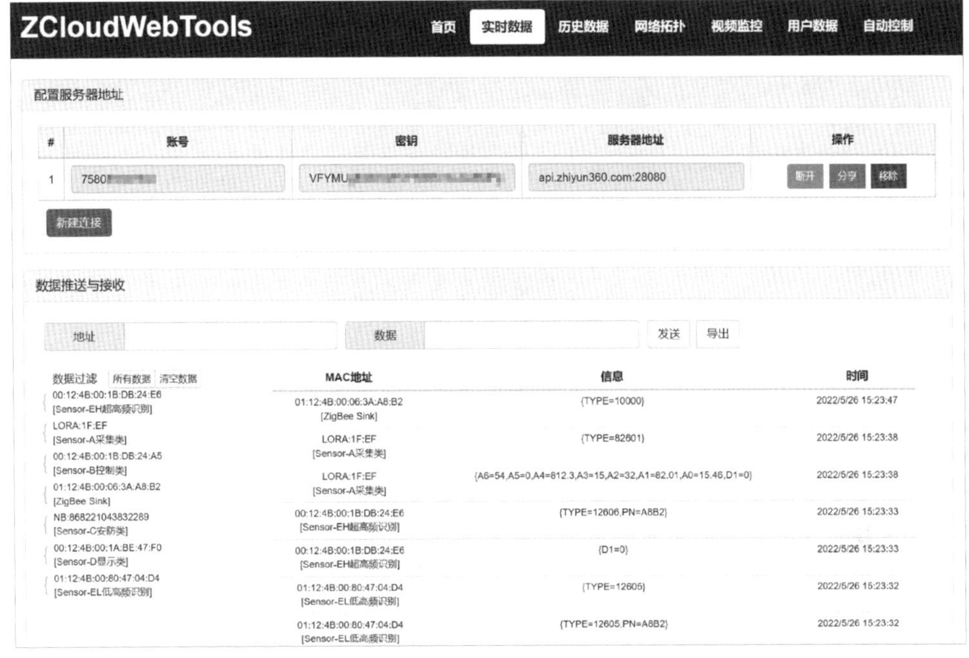

▲ 圖 2.30 ZCloudWebTools 的介面

2.3.2 開發步驟與驗證

2.3.2.1 原型部署

1）登入虛擬模擬平臺

透過實際的硬體或智雲硬體物元模擬平臺可架設一個智慧物聯網專案原型，從而幫助讀者學習邊緣硬體的開發。本節採用智雲硬體物元模擬平臺來架設智慧物聯網專案原型。

使用 Chrome 瀏覽器登入智雲硬體物元模擬平臺，並註冊使用者，如圖 2.31 所示。

▲ 圖 2.31 登入智雲硬體物元模擬平臺並進行註冊

2）建立「智慧產業專案」

按一下左側導覽列的專案管理，在專案清單中按一下新建專案：

（1）名稱：填寫專案名稱，如「智慧產業專案」。

（2）使用者 ID、使用者金鑰使用智雲授權 ID 和金鑰，伺服器位址使用預設的位址即可。

（3）感測器：增加專案中需要使用的感測器，勾選左邊欄中 Sensor-A、Sensor-B、Sensor-C、Sensor-D、Sensor-EL、Sensor-EH，並按一下「增加」按鈕，將選中的感測器增加到右側。

（4）專案配置完成之後，按一下「立即建立」按鈕，即可建立「智慧產業專案」。

建立「智慧產業專案」的步驟如圖 2.32 和圖 2.33 所示。

3）執行「智慧產業專案」

（1）在「專案管理」選單中找到新建的「智慧產業專案」，進入該專案後等待載入完成；按一下右上角的「啟動」按鈕啟動專案，依次按一下每個感測器的「開啟」按鈕啟動裝置，大約 30 s 後感測器資料就會開始上傳並更新。執行「智慧產業專案」如圖 2.34 所示。

▲ 圖 2.32 建立「智慧產業專案」的步驟一

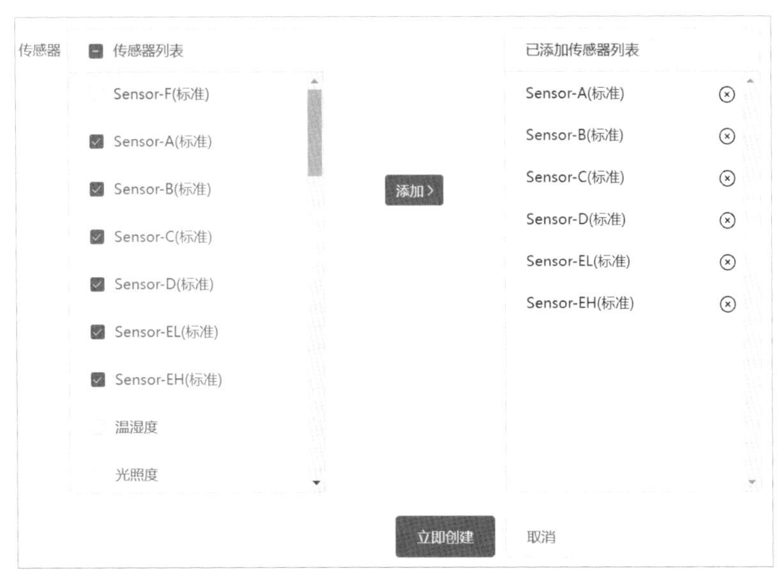

▲ 圖 2.33 建立「智慧產業專案」的步驟二

▲ 圖 2.34 執行「智慧產業專案」

（2）透過應用軟體與「智慧產業專案」虛擬硬體進行資料獲取和控制。

2.3.2.2 硬體偵錯

開啟 ZcloudWebTools 軟體，按一下「即時資料」，填寫「智慧產業專案」的帳號和金鑰資訊，按一下「連接」按鈕連接專案，可以看到當前專案中裝置的即時資料。

下面以「智慧產業專案」中的 Sensor-A、Sensor-B、Sensor-C 為例介紹 ZXBee 協定。

1）Sensor-A

Sensor-A 節點是採用 ZigBee 網路連接的節點，能夠準確測量室內溫 / 濕度、光照度、空氣品質、大氣壓力等資料。根據智雲端平台採用的 ZXBee 協定可知，Sensor-A 節點支援的操作如表 2.8 所示。

▼ 表 2.8 Sensor-A 節點支援的操作

發送命令	接收結果	含義
{A0=?}	{A0=XX}	溫度值，浮點型，精度為 0.1，單位℃
{A1=?}	{A1=XX}	濕度值，浮點型，精度為 0.1，單位為 %RH
{A2=?}	{A2=XX}	光照度值，浮點型，精度為 0.1，單位為 lx
{A3=?}	{A3=XX}	空氣品質值，整數，單位為 ppm
{A4=?}	{A4=XX}	大氣壓力值，浮點型，精度為 0.1，單位為 hPa
{A5=?}	{A5=XX}	跌倒狀態，0 表示未跌倒，1 表示跌倒
{A6=?}	{A6=XX}	距離值，浮點型，精度為 0.1，單位為 cm
{D0=?}	{D0=XX}	D0 的 Bit0 ～ Bit7 對應 A0 ～ A7 的主動上報功能是否啟用，0 表示允許主動上報，1 表示不允許主動上報
{D1=?}	{D1=XX}	D1 的 Bit6 ～ Bit7 表示繼電器 K1、K2 的開關狀態，0 表示斷開，1 表示吸合
{A0=?,A1=?,A2=?,A3=?,A4=?,A5=?,A6=?,D0=?,D1=?}	回覆 A0 ～ A6、D0 ～ D1 所有資料	查詢所有資料
{CD1=XXX,D1=?} {OD1=XXX,D1=?}	{D1=XX}	CD1 表示位元清零，OD1 表示位置 1。D1 的位元用來控制裝置：Bit6 用於控制繼電器 K1 開關，0 表示關，1 表示開；Bit7 用於控制繼電器 K2 開關，0 表示關，1 表示開
{V0=?} {V0=XXX,V0=?}	{V0=XXX}	查詢 / 設定 A0 ～ A7、D1 的主動上報時間間隔，預設為 30，單位為 s

以查詢即時溫度資料為例，在「位址」欄輸入 Sensor-A 節點的 MAC 位址，在「資料」欄輸入命令 {A0=?}，然後按一下「發送」按鈕，在資訊傳回欄可以看到傳回的即時溫度資料，如圖 2.35 所示。

2）Sensor-B

根據智雲端平台採用的 ZXBee 協定可知，Sensor-B 節點支援的操作如表 2.9 所示。

▲ 圖 2.35 查詢即時溫度範例

▼ 表 2.9 Sensor-B 節點支援的操作

發送命令	接收結果	含義
{D1=?}	{D1=XX}	查詢所有資料
{CD1=XXX,D1=?} {OD1=XXX,D1=?}	{D1=XX}	CD1 表示位元清零，OD1 表示位置 1。D1 的位元用於控制裝置，Bit0 ～ Bit1 用於控制 RGB 燈的顏色，00 表示關閉，01 表示紅色，10 表示綠色，11 表示藍色

（續表）

發送命令	接收結果	含義
{CD1=XXX,D1=?} {OD1=XXX,D1=?}	{D1=XX}	CD1 表示位元清零，OD1 表示位置 1。D1 的位元用於控制裝置，Bit2 用於控制步進馬達的正反轉，0 表示正轉，1 表示反轉
{CD1=XXX,D1=?} {OD1=XXX,D1=?}	{D1=XX}	CD1 表示位元清零，OD1 表示位置 1。D1 的位元用於控制裝置，Bit3 用於控制風扇 / 蜂鳴器的開關，0 表示關，1 表示開
{CD1=XXX,D1=?} {OD1=XXX,D1=?}	{D1=XX}	CD1 表示位元清零，OD1 表示位置 1。D1 的位元用於控制裝置，Bit4 用於控制 LED1 的開關，0 表示關，1 表示開；Bit5 用於控制 LED2 的開關，0 表示關，1 表示開
{CD1=XXX,D1=?} {OD1=XXX,D1=?}	{D1=XX}	CD1 表示位元清零，OD1 表示位置 1。D1 的位元用於控制裝置，Bit6 用於控制繼電器 K1 的開關，0 表示斷開，1 表示吸合；Bit7 用於控制繼電器 K2 的開關，0 表示斷開，1 表示吸合
{V0=?} {V0=XXX,V0=?}	{V0=XXX}	查詢 / 設定 D1 的主動上報時間間隔，預設為 30，單位為 s

以控制窗簾（步進馬達）為例，在「位址」欄輸入 Sensor-B 節點的 MAC 位址，在「資料」欄輸入命令 {OD1=4,D1=?} 開啟窗簾，然後按一下「發送」按鈕，在資訊傳回欄可以看到窗簾圖示發生對應的變化。

未發送命令之前窗簾處於關閉狀態，如圖 2.36 所示。

▲ 圖 2.36 未發送命令之前窗簾處於關閉狀態

使用 ZCloudWebTools 發送命令，如圖 2.37 所示。

数据推送与接收

地址 00:12:4B:00:1B:DB:24:A5　数据 {OD1=4,D1=?}　发送　导出

数据过滤　所有数据　清空数据

01:12:4B:00:06:3A:A8:B2 [ZigBee Sink]
LORA:1F:EF [Sensor-A采集类]
00:12:4B:00:1B:DB:24:A5 [Sensor-B控制类]
NB:868221043832289 [Sensor-C安防类]
00:12:4B:00:1A:BE:47:F0 [Sensor-D显示类]
01:12:4B:00:80:47:04:D4 [Sensor-EL低高频识别]
00:12:4B:00:1B:DB:24:E6 [Sensor-EH超高频识别]

MAC地址	信息	时间
00:12:4B:00:1B:DB:24:A5 [Sensor-B控制类]	{D1=4}	2022/5/26 15:36:25
00:12:4B:00:1B:DB:24:A5 [Sensor-B控制类]	{TYPE=12602,PN=A8B2}	2022/5/26 15:36:23
00:12:4B:00:1B:DB:24:A5 [Sensor-B控制类]	{D1=6}	2022/5/26 15:36:23
00:12:4B:00:1B:DB:24:A5 [Sensor-B控制类]	{D1=6}	2022/5/26 15:36:18
00:12:4B:00:1B:DB:24:A5 [Sensor-B控制类]	{D1=4}	2022/5/26 15:36:12
00:12:4B:00:1B:DB:24:A5 [Sensor-B控制类]	{D1=6}	2022/5/26 15:35:58
00:12:4B:00:1B:DB:24:A5 [Sensor-B控制类]	{D1=2}	2022/5/26 15:35:53

▲ 圖 2.37 使用 ZCloudWebTools 發送命令

發送命令後可以看到窗簾開啟，如圖 2.38 所示。

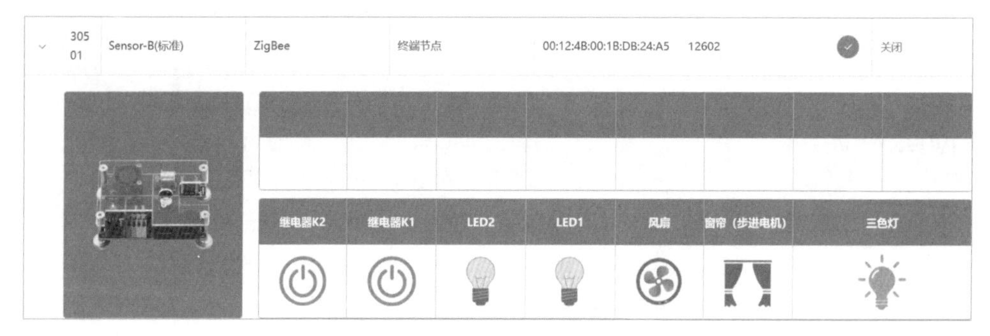

▲ 圖 2.38 發送命令之後窗簾處於開啟狀態

3）Sensor-C

根據智雲端平台採用的 ZXBee 協定可知，Sensor-C 節點支援的操作如表 2.10 所示。

▼ 表 2.10 Sensor-C 節點支援的操作

發送命令	接收結果	含義
{A0=?}	{A0=XX}	人體紅外 / 觸控感測器狀態，1 表示檢測到人體活動 / 觸控動作，0 表示未檢測到人體活動 / 觸控動作
{A1=?}	{A1=XX}	震動狀態，1 表示檢測到震動，0 表示未檢測到震動
{A2=?}	{A2=XX}	霍爾狀態，1 表示檢測到磁場，0 表示未檢測到磁場
{A3=?}	{A3=XX}	火焰狀態，1 表示檢測到火焰，0 表示未檢測到火焰
{A4=?}	{A4=XX}	瓦斯洩漏狀態，1 表示檢測到瓦斯洩漏，0 表示未檢測到瓦斯洩漏
{A5=?}	{A5=XX}	光柵（紅外對射）狀態值，1 表示檢測到阻擋，0 表示未檢測到阻擋
{D0=?}	{D0=XX}	D0 的 Bit0 ～ Bit7 表示 A0 ～ A7 的主動上報功能是否啟用，0 表示允許主動上報，1 表示不允許主動上報
{D1=?}	{D1=XX}	D1 的 Bit6 ～ Bit7 用於控制繼電器 K1、K2 的開關狀態，0 表示斷開，1 表示吸合
{A0=?,A1=?,A2=?,A3=?,A4=?,A5=?, D0=?,D1=?}	A0 ～ A5、D0 ～ D1 所有資料	查詢所有資料
{CD1=XXX,D1=?} {OD1=XXX,D1=?}	{D1=XX}	CD1 表示位元清零，OD1 表示位置 1。D1 的位元用於控制裝置，Bit6 用於控制繼電器 K1 的開關，0 表示關，1 表示開；Bit7 用於控制繼電器 K2 的開關，0 表示關，1 表示開
{V0=?} {V0=XXX,V0=?}	{V0=XXX}	查詢 / 設定 A0 ～ A7、D1 的主動上報時間間隔，預設為 30，單位為 s

以光柵感測器為例，在虛擬模擬平臺手動給光柵輸出「1」進行警告，如圖 2.39 所示。

▲ 圖 2.39 在虛擬模擬平臺手動給光柵輸出「1」進行警告

在 ZCloudTools 的資訊傳回欄可以看到光柵上報的「{A5=1}」的資訊，如圖 2.40 所示。

数据推送与接收

地址 NB:868221043832289　数据 {OD1=4,D1=?}　发送　导出

数据过滤　所有数据　清空数据
- 01:12:4B:00:06:3A:A8:B2 [ZigBee Sink]
- LORA:1F:EF [Sensor-A采集类]
- 00:12:4B:00:1B:DB:24:A5 [Sensor-B控制类]
- NB:868221043832289 [Sensor-C安防类]
- 00:12:4B:00:1A:BE:47:F0 [Sensor-D显示类]
- 01:12:4B:00:80:47:04:D4 [Sensor-EL低高频识别]
- 00:12:4B:00:1B:DB:24:E6 [Sensor-EH超高频识别]

MAC地址	信息	时间
NB:868221043832289 [Sensor-C安防类]	{A5=1}	2022/5/26 15:51:4
NB:868221043832289 [Sensor-C安防类]	{A5=0}	2022/5/26 15:51:1
NB:868221043832289 [Sensor-C安防类]	{A5=1,A4=0,A3=0,A2=0,A1=0,A0=0,D1=0}	2022/5/26 15:50:54
NB:868221043832289 [Sensor-C安防类]	{TYPE=72603}	2022/5/26 15:50:54
NB:868221043832289 [Sensor-C安防类]	{TYPE=72603}	2022/5/26 15:50:23
NB:868221043832289 [Sensor-C安防类]	{A5=1,A4=0,A3=0,A2=0,A1=0,A0=0,D1=0}	2022/5/26 15:50:23
NB:868221043832289 [Sensor-C安防类]	{A5=1}	2022/5/26 15:50:18

▲ 圖 2.40 在 ZCloudTools 的資訊傳回欄查看上報資訊

2.3.2.3 應用程式開發

1）即時連接範例

（1）解壓縮 RTConnectDemo-Web.zip，用記事本開啟即時連接範例程式 RTConnectDemo-Web\js\script.js，填寫「智慧產業專案」的智雲帳號、金鑰，以及 Sensor-A、Sensor-B 節點的 MAC 位址，程式如下：

```
/*********************************************************************************
* 初始化變數
*********************************************************************************/
var myZCloudID = "12345678";                              // 智雲帳號
var myZCloudKey = "12345678";                             // 智雲金鑰
var SensorA = "01:12:4B:00:E3:7D:D6:64";                  //Sensor-A 節點 MAC 位址
var SensorB = "01:12:4B:00:27:22:AC:4E";                  //Sensor-B 節點 MAC 位址
var rtc = new WSNRTConnect(myZCloudID, myZCloudKey); // 建立資料連接服務物件

/*********************************************************************************
* 與智雲端服務連接，監聽和解析即時資料並顯示
*********************************************************************************/
$(function () {
    rtc.setServerAddr("api.zhiyun360.com:28080");     // 設定伺服器位址
    rtc.connect();                                    // 資料推送服務連接
    rtc.onConnect = function () {                     // 連接成功回呼函式
        rtc.sendMessage(SensorA, "{A0=?,A1=?}");      // 查詢溫濕度的初始值
        rtc.sendMessage(SensorB, "{D1=?}");           // 查詢燈光初始值
        $("#ConnectState").text(" 資料服務連接成功！ ");
    };
    rtc.onConnectLost = function () {                 // 資料服務掉線回呼函式
        $("#ConnectState").text(" 資料服務掉線！ ");
    };
    rtc.onmessageArrive = function (mac, dat) {       // 訊息處理回呼函式
        console.log(mac+" >>> "+dat);

        if (mac == SensorA) {                         // 判斷是否是 Sensor-A 節點的資料
            if (dat[0] == '{' && dat[dat.length - 1] == '}') {  // 判斷字串首尾是否為 {}
                dat = dat.substr(1, dat.length - 2);          // 截取 {} 內的字串
                var its = dat.split(',');                     // 以 ',' 來分割字串
                for (var x in its) {                          // 迴圈遍歷
                    var t = its[x].split('=');                // 以 '=' 來分割字串
                    if (t.length != 2) continue;              // 滿足條件時結束當前迴圈
                    if (t[0] == "A0") {                       // 判斷參數 A0
                        var tem = parseFloat(t[1]);           // 讀取溫度資料
                        $("#currentTem").text(tem + "℃ ");   // 在頁面顯示溫度資料
                    }
                    if (t[0] == "A1") {                       // 判斷參數 A1
                        var hum = parseFloat(t[1]);           // 讀取濕度資料
```

```
                        $("#currentHum").text(hum + "%");      // 在頁面顯示濕度資料
                    }
                }
            }
        } else if (mac == SensorB) {                       // 判斷是否是 Sensor-B 節點的資料
            if (dat[0] == '{' && dat[dat.length - 1] == '}') {// 判斷字串首尾是否為 {}
                dat = dat.substr(1, dat.length - 2);           // 截取 {} 內的字串
                var its = dat.split(',');                      // 以 ',' 來分割字串
                for (var x in its) {                           // 迴圈遍歷
                    var t = its[x].split('=');                 // 以 '=' 來分割字串
                    if (t.length != 2) continue;               // 滿足條件時結束當前迴圈
            if (t[0] == "D1") {                                // 判斷參數 D1
                        var LightStatus = parseInt(t[1]);      // 根據 D1 的值來進行開關的切換
                        if ((LightStatus & 0x30) == 0x30) {
                            $('#btn_img').attr('src','images/an-on.png')
                        } else if ((LightStatus & 0x30) == 0) {
                            $('#btn_img').attr('src','images/an-off.png')
                        }
                    }
                }
            }
        }

    };
});

/*****************************************************************************************
* 處理按鍵事件
*****************************************************************************************/
$('#btn_img').click(function(){
    if($('#btn_img').attr('src') == 'images/an-on.png'){
        rtc.sendMessage(SensorB, "{CD1=48,D1=?}");             // 發送關閉燈光命令
    }else{
        rtc.sendMessage(SensorB, "{OD1=48,D1=?}");             // 發送開啟燈光命令
    }
})
```

（2）透過 Chrome 瀏覽器開啟即時連接範例程式 RTConnectDemo-Web\index.html，查看即時資料資訊及燈的開關操作。即時連接範例如圖 2.41 所示。

▲ 圖 2.41 即時連接範例

即時連接範例程式 RTConnectDemo-Web\index.html 的程式如下：

```
#HTML 程式如下
<!DOCTYPE html >
<html lang="zh-cmn-Hans">
<head>
<meta charset="UTF-8">
<link rel="stylesheet" href="css/bootstrap.min.css">
<link rel="stylesheet" href="css/index.css">
<title> 即時連接範例 </title>
</head>
<body>
    <!-- 頁頭 -->
    <div class="header">
        <div>
            <div>
                <h1> 即時連接範例 </h1>
            </div>
            <div id="ConnectState"> 資料服務連接中 ...</div>
        </div>
    </div>
    <!-- 頁中 -->
    <div class="main">
        <div>
            <div> 當前溫度：</div>
            <div id="currentTem">--℃ </div>
```

```
        </div>
        <div>
            <div>當前濕度：</div>
            <div id="currentHum">--%</div>
        </div>
        <div>
            <div>燈光開關：</div>
            <img id="btn_img" src="images/an-off.png" alt=""/>
        </div>
    </div>
    <!-- 頁尾 -->
    <div class="footer">開發</div>
    <!--引入js-->
    <script    src="js/jquery-1.11.0.min.js"></script>
    <script    src="js/highcharts.js"></script>
    <script    src="js/WSN/WSNRTConnect.js"></script>
    <script    src="js/script.js"></script>
</body>
</html>
```

（3）在虛擬模擬平臺可以看到「智慧產業專案」感測器相關的變化，如圖 2.42 所示。

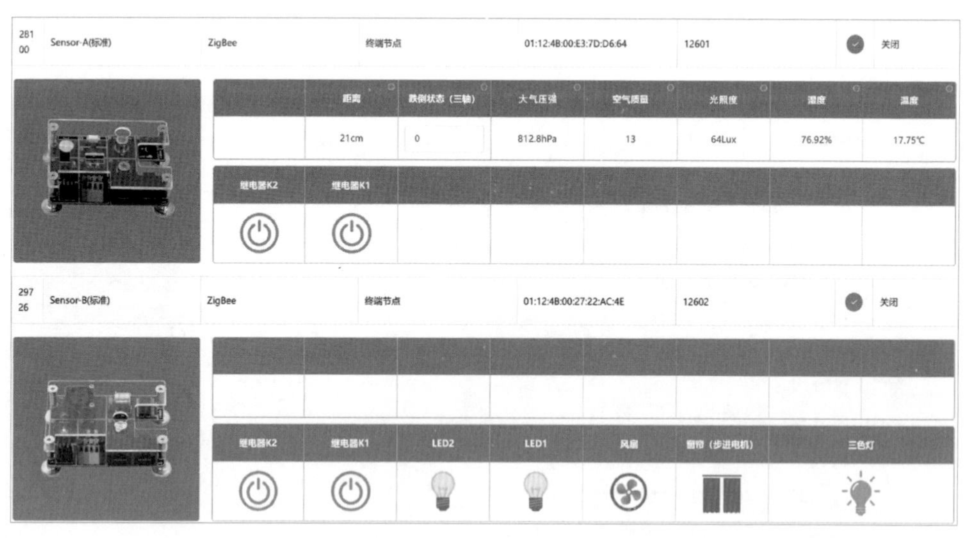

▲ 圖 2.42 在虛擬模擬平臺中查看「智慧產業專案」感測器的變化

2）歷史資料範例

（1）解壓縮 HistoryDemo-Web.zip，用記事本開啟歷史資料範例程式 HistoryDemo-Web\js\script.js，填寫「智慧產業專案」的智雲帳號、金鑰，以及 Sensor-A 節點的 MAC 位址，程式如下：

```
/*********************************************************************************
* 初始化變數
*********************************************************************************/
var myZCloudID = "12345678";                                    // 智雲帳號
var myZCloudKey = "12345678";                                   // 智雲金鑰
var SensorA = "01:12:4B:00:E3:7D:D6:64";                        //Sensor-A 節點 MAC 位址
var channel = `${SensorA}_A2`;                                  // 感測器資料通道

var rtc = new WSNRTConnect(myZCloudID, myZCloudKey);            // 建立資料連接服務物件
var myHisData = new WSNHistory(myZCloudID, myZCloudKey);        // 建立歷史資料服務物件
var LightIntensity;

/*********************************************************************************
* 與智雲端服務連接，監聽和解析即時資料並顯示
*********************************************************************************/
$(function(){
    rtc.setServerAddr("api.zhiyun360.com:28080");               // 設定伺服器位址
    rtc.connect();
    rtc.onConnect = function() {                                // 連接成功回呼函式
        rtc.sendMessage(SensorA, "{A2=?}");                     // 查詢光照度的初始值
        $("#ConnectState").text(" 資料服務連接成功！ ");
    };

    rtc.onConnectLost = function() {                            // 資料服務掉線回呼函式
        $("#ConnectState").text(" 資料服務掉線！ ");
    };

    rtc.onmessageArrive = function(mac, dat) {                  // 訊息處理回呼函式
    console.log(mac+" >>> "+dat);

        if (mac == SensorA) {                                   // 判斷感測器的 MAC 位址
            if (dat[0] == '{' && dat[dat.length - 1] == '}') {    // 判斷字串首尾是否為 {}
                dat = dat.substr(1, dat.length - 2);            // 截取 {} 內的字串
```

```
                var its = dat.split(',');                          //以 ',' 來分割字串
                for (var x in its) {
                    var t = its[x].split('=');                     //以 '=' 來分割字串
                    if (t.length != 2) continue;
                    if (t[0] == "A2") {                            //判斷參數A2
                        LightIntensity = parseInt(t[1]);
                        $("#currentTem").text(LightIntensity + "Lux");  //在頁面顯示
光照度資料
                    }
                }
            }
        }
    };
})

/*********************************************************************************
* 預設呼叫歷史資料圖表，參數為下拉選項初始值
*********************************************************************************/
checkHistory('MessSet', '#line_charts');
/*********************************************************************************
* 透過下拉選項切換歷史資料時間範圍
*********************************************************************************/
$('#MessSet').change(function () {
    checkHistory('MessSet', '#line_charts');
})

/*********************************************************************************
* 名稱：checkHistory(set, tagIndex, hisDiv)
* 功能：連接呼叫歷史資料
* 參數：set：獲取選中的歷史資料時間範圍
*       tagindex：賦值給對應的歷史查詢物件
*       hisdiv：顯示圖表的節點
*********************************************************************************/
function checkHistory(set, hisDiv) {
    var time = $('#' + set).val();                                 //設定時間
    myHisData.setServerAddr("api.zhiyun360.com:8080");             //設定伺服器位址

    console.log('查詢時間為：' + time + '，查詢通道為：' + channel);
    myHisData[time](channel, function (dat) {
```

```
        console.log(dat)                                //輸出查詢到的歷史資料
        if (dat.datapoints.length > 0) {
            var data = DataAnalysis(dat);               //將 JSON 格式的資料轉化為圖表資料
            showChart(hisDiv, 'spline', '', false, eval(data));        //顯示圖表資料曲線
        }
    });
}

/************************************************************************************
* 將 JSON 格式的資料轉換成 [x1,y1],[x2,y2],[x3,y3]... 格式的陣列（與歷史資料圖表相關）
************************************************************************************/
function DataAnalysis(data, timezone) {
    var str = '';
    var value;
    var len = data.datapoints.length;
    if (timezone == null) {
        timezone = "+8";
    }
    var zoneOp = timezone.substring(0, 1);
    var zoneVal = timezone.substring(1);
    var tzSecond = 0;
    $.each(data.datapoints, function (i, ele) {
        if (zoneOp == '+') {
            value = Date.parse(ele.at) + tzSecond;
        }
        if (zoneOp == '-') {
            value = Date.parse(ele.at) - tzSecond;
        }
        if (ele.value.indexOf("http") != -1) {
            str = str + '[' + value + ',"' + ele.value + '"]';
        } else {
            str = str + '[' + value + ',' + ele.value + ']';
        }
        if (i != len - 1)
            str = str + ',';
    });
    return "[" + str + "]";
}
```

```
/**********************************************************************************
* 畫曲線圖的方法（與歷史資料圖表相關）
**********************************************************************************/
function showChart(sid, ctype, unit, step, data) {
    $(sid).highcharts({
        chart: {
            backgroundColor: 'transparent',
            type: ctype,
            animation: false,
            zoomType: 'x'
        },
        legend: {
            enabled: false
        },
        title: {
            text: ''
        },
        xAxis: {
            type: 'datetime',
            labels: {
                style: {
                    color: 'rgb(0, 0, 0)',
                }
            }
        },
        yAxis: {
            title: {
                text: ''
            },
            minorGridLineWidth: 0,
            gridLineWidth: 1,
            alternateGridColor: null,
            labels: {
                style: {
                    color: 'rgb(0, 0, 0)',
                }
            }
        },
        tooltip: {
            formatter: function () {
```

```
            return '' +
            Highcharts.dateFormat('%Y-%m-%d %H:%M:%S', this.x) + '<br><b>' +
                             this.y + unit + '</b>';
        }
    },
    plotOptions: {
        spline: {
            lineWidth: 2,
            states: {
                hover: {
                    lineWidth: 3
                }
            },
            marker: {
                enabled: false,
                states: {
                    hover: {
                        enabled: true,
                        symbol: 'circle',
                        radius: 3,
                        lineWidth: 1
                    }
                }
            }
        },
        line: {
            lineWidth: 1,
            states: {
                hover: {
                    lineWidth: 1
                }
            },
            marker: {
                enabled: false,
                states: {
                    hover: {
                        enabled: true,
                        symbol: 'circle',
                        radius: 3,
                        lineWidth: 1
```

```
                        }
                    }
                }
            }
        },
        series: [{
            marker: {
                symbol: 'square'
            },
            data: data,
            step: step,
        }],
        navigation: {
            menuItemStyle: {
                fontSize: '10px'
            }
        }
    });
}
```

（2）透過 Chrome 瀏覽器開啟歷史資料範例程式 HistoryDemo-Web\index.html，查看光照度即時資料資訊及歷史資料曲線資訊，透過選擇曲線的下拉選項可以查看其他時間範圍的歷史資料曲線。歷史資料範例如圖 2.43 所示。

注意：裝置需要執行一段時間後才可以查看一段時間內的歷史資料。

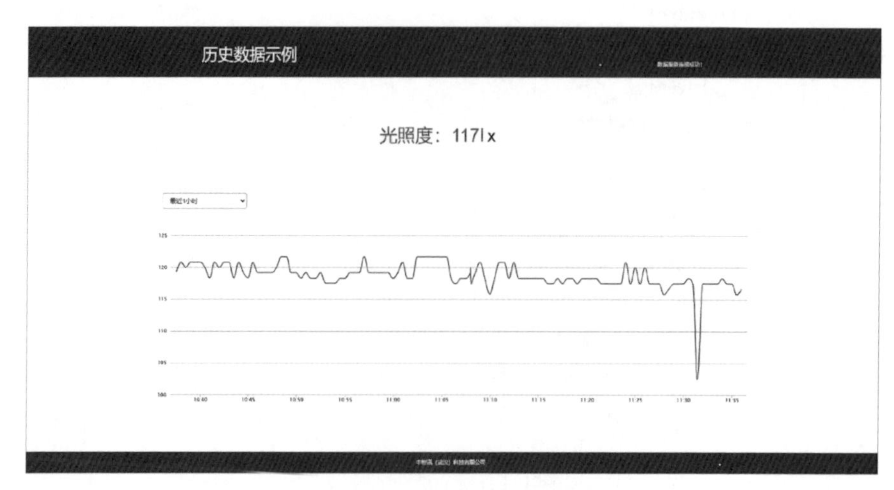

▲ 圖 2.43 歷史資料範例

2.3.3 本節小結

本節主要介紹邊緣計算的硬體設計，首先介紹了智慧物聯網平臺的框架、專案模型、應用介面、通訊協定和開發工具；然後介紹了邊緣計算開發的原型部署、硬體偵錯和應用程式開發。透過本節的學習，讀者可以架設智慧物聯網專案的原型，熟悉邊緣計算的硬體開發。

2.3.4 思考與擴充

（1）如何透過邊緣計算的閘道服務程式獲取智雲帳號金鑰？

（2）結合 AiCam 平臺和智慧物聯網平臺的框架，簡述智慧物聯網與傳統物聯網的異同。

2.4 邊緣計算的應用程式開發

邊緣計算的應用程式開發需要在兼顧性能、即時性、資源限制和安全性的前提下，為特定的邊緣場景設計和最佳化應用程式，需要綜合考慮硬體、軟體和網路環境的特點，以滿足邊緣計算的應用需求。以下是邊緣應用程式開發的主要特點：

（1）本地決策：邊緣計算的應用通常是在邊緣裝置上進行的，往往不依賴於雲端，因此邊緣計算的應用程式開發需要考慮如何在邊緣裝置上實現本地感知和決策，以減小對網路的依賴和降低延遲。

（2）即時性要求：邊緣應用往往是即時或近即時的，如智慧交通管理、工業自動化和醫療監測。邊緣計算的應用程式開發需要關注即時性能，確保邊緣裝置能迅速處理資料。

（3）資源最佳化：邊緣裝置的計算、儲存和電源等通常是有限的，因此邊緣計算的應用程式開發需要考慮如何最佳化演算法和程式，以適應資源限制的情況，避免消耗過多的資源。

（4）邊緣快取：為了提高性能和減小延遲，邊緣應用可能需要在邊緣裝置上快取資料或模型，這就需要合理管理快取內容，以滿足不同應用場景的需求。

（5）安全性和隱私保護：邊緣計算的應用程式開發需要重視安全性和隱私保護，資料可能分佈在多個邊緣裝置上，應用程式必須強化安全機制，如加密、認證和許可權管理，以確保資料的機密性和完整性。

（6）多模組整合：邊緣應用通常需要整合多種感測器、通訊模組和硬體元件，以滿足應用的需求，因此邊緣計算的應用程式開發需要考慮如何有效整合和協調這些模組，以實現應用的功能。

（7）分散式運算：邊緣計算通常涉及分散式運算，資料處理通常是在多個裝置上進行的，因此邊緣計算的應用程式開發需要考慮資料傳輸、協作處理和分散式演算法。

本節的基礎知識如下：

- 掌握機器視覺導向的邊緣計算的應用程式開發框架。
- 基於 AiCam 平臺，掌握人臉辨識即時推理、單次推理的邊緣計算視覺應用程式開發。
- 結合邊緣計算的硬體平臺，掌握雲 - 邊 - 端協作的人工智慧邊緣應用程式開發過程。

2.4.1 原理分析與開發設計

2.4.1.1 邊緣視覺的應用程式開發邏輯

邊緣計算比雲端運算更靠近使用者，因此將邊緣計算和人工智慧技術結合在一起，不僅有利於實現邊緣智慧，還可以提高邊緣伺服器的服務輸送量和資源使用率。面對巨量的複雜資料，機器學習及深度學習可以依靠其強大的學習能力和推理能力從資料中提取有價值的資訊，協助邊緣伺服器進行決策和管理。邊緣計算和人工智慧的結合可以分為以下五類：

- 在邊緣伺服器部署人工智慧，對外提供服務。
- 在邊緣伺服器中進行部分或全部推理，滿足不同服務對準確性和延遲的要求。
- 基於人工智慧的邊緣計算平臺，在網路框架、軟 / 硬體等方面滿足深度學習計算要求。
- 在邊緣伺服器上進行深度學習訓練，加速模型訓練，提高資料的隱私保護能力。
- 最佳化邊緣計算的決策，維護和管理邊緣伺服器的功能。

邊緣視覺的應用程式開發邏輯如圖 2.44 所示，在邏輯上大致分為三層：終端層、邊緣層以及中心雲端層。

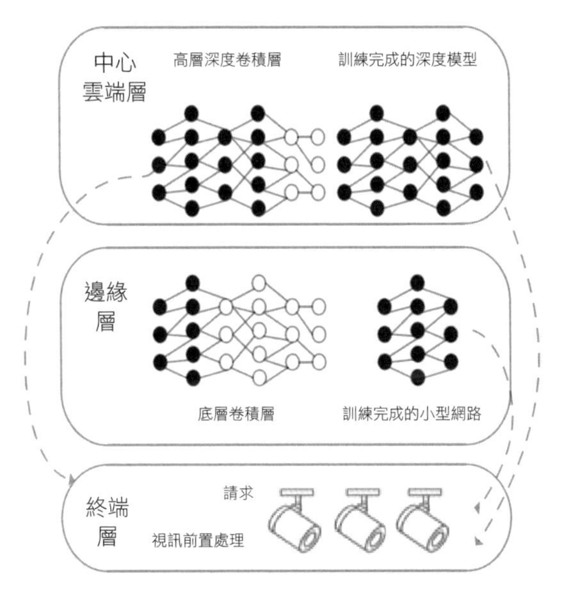

▲ 圖 2.44 邊緣視覺的應用程式開發邏輯

終端層：主要由視訊擷取裝置（如攝影機）組成，其功能是進行媒體資料壓縮、影像前置處理和影像分割等。終端層的算力一般較小，可以訓練一個域感知自我調整的淺層輕量級網路，對提升物件偵測精度有很大的作用。雖然終端層可以承受部分計算任務，但其主要任務還是對來源資料進行前置處理，然後發送給邊緣層。

邊緣層：該層的算力比終端層大，節點之間相互協作可以完成大量的計算，提供給使用者良好的體驗。如果在邊緣節點上壓縮網路模型或部署多工網路模型，則可以在很大程度上減少中心雲端層的計算壓力。

中心雲端層：中心雲端層通常會整合不同的深度模型，以獲得全域知識。當邊緣伺服器無法得到計算結果時，中心雲端層可以使用其算力和全域知識進行最後的處理，協助邊緣節點進行結果計算和參數更新。

2.4.1.2 基於 AiCam 平臺的邊緣視覺應用程式開發框架

AiCam 平臺是一個整合了演算法、模型、硬體的輕量級應用程式開發框架，能夠為每個應用提供邊緣視覺演算法，透過 RESTful 介面供前端應用呼叫。

AiCam 平臺內建了豐富的模型庫、演算法庫、硬體庫和應用案例，可幫助讀者快速掌握邊緣視覺應用程式開發技能，主要包括影像基礎、影像應用、深度學習、雲邊應用、邊緣智慧等內容。基於 AiCam 平臺的邊緣視覺應用程式開發框架如圖 2.45 所示。

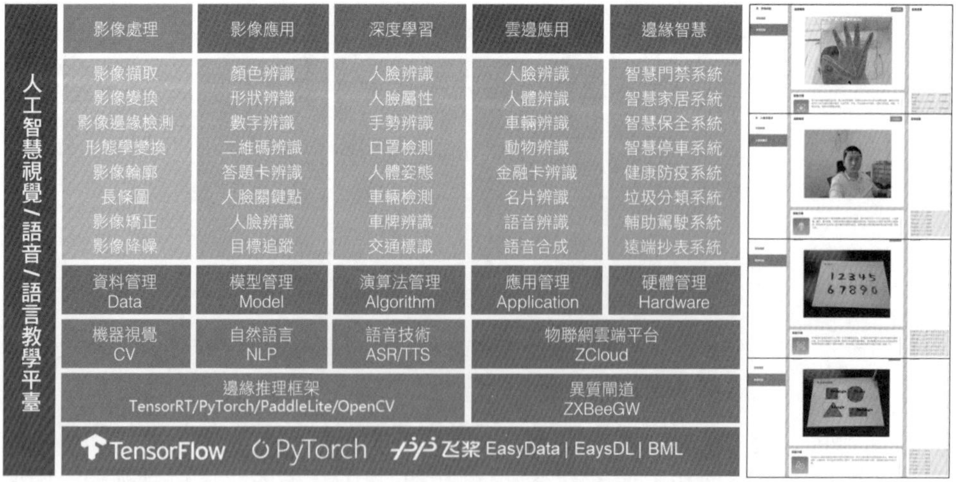

人工智慧視覺 / 語音 / 語言教學平臺

影像處理	影像應用	深度學習	雲邊應用	邊緣智慧
影像擷取 影像變換 影像邊緣檢測 形態學變換 影像輪廓 長條圖 影像矯正 影像降噪	顏色辨識 形狀辨識 數字辨識 二維碼辨識 答題卡辨識 人臉關鍵點 人臉辨識 目標追蹤	人臉辨識 人臉屬性 手勢辨識 口罩檢測 人體姿態 車輛檢測 車牌辨識 交通標識	人臉辨識 人體辨識 車輛辨識 動物辨識 金融卡辨識 名片辨識 語音辨識 語音合成	智慧門禁系統 智慧家居系統 智慧保全系統 智慧停車系統 健康防疫系統 垃圾分類系統 輔助駕駛系統 遠端抄表系統
資料管理 Data	模型管理 Model	演算法管理 Algorithm	應用管理 Application	硬體管理 Hardware
機器視覺 CV	自然語言 NLP	語音技術 ASR/TTS	物聯網雲端平台 ZCloud	
邊緣推理框架 TensorRT/PyTorch/PaddleLite/OpenCV			異質閘道 ZXBeeGW	
TensorFlow	PyTorch	飞桨 EasyData | EaysDL | BML		

12345
67890

▲ 圖 2.45 基於 AiCam 平臺的邊緣視覺應用程式開發框架

2.4.1.3 基於 AiCam 平臺的邊緣視覺應用程式開發介面

1）介面描述

AiCam 平臺透過 RESTful 介面和 Flask 服務為應用層提供了演算法排程。根據實際的 AI 應用邏輯，演算法排程有兩種互動介面，分別用於即時推理和單次推理的應用場景（見圖 2.4）。

2）即時推理

AiCam 平臺提供了攝影機視訊流影像的即時計算推理，並將計算的結果影像和結果資料以資料流程的方式推送給應用層，應用層透過 EventSource 介面獲取即時推送的資料流程。

AiCam 平臺中的 stream-exp 範例呼叫人臉檢測演算法進行即時的人臉檢測，透過存取的 URL 位址（URL 位址為邊緣計算閘道的位址）

```
#http://[gateway_ip:port]/stream/[algorithm_name]?camera_id=0
http://192.168.100.200:4000/stream/image_face_recognition?camera_id=0
```

可獲取 JSON 格式的資料流程，包括結果影像 result_image 和結果資料 result_data。獲取 JSON 格式的資料流程的程式如下：

```
<script>
    // 攝影機視訊的連結
    let linkData = 'http://192.168.100.200:4000/stream/face_detection?camera_id=0'
    let throttle = true                              // 控制辨識結果顯示頻率
    // 請求資料流程資源
    let imgData = new EventSource(linkData)
    // 接收即時資料流
    imgData.onmessage = function (res) {

        // 提取結果影像進行顯示
        let {result_image} = JSON.parse(res.data)
        $('img').attr('src', `data:image/jpeg;base64,${result_image}`)

        // 提取結果資料，每隔 1 s 顯示一次
        let {result_data} = JSON.parse(res.data)
```

```
        if(result_data && throttle){
            throttle = false
            let time = new Date().toLocaleTimeString();        //獲取當前時間
            let html = `<div>${time}————${JSON.stringify(result_data)}</div>`
            $('#info_list').append(html)
            $('#info_list').scrollTop($('#info_list')[0].scrollHeight);
            setTimeout(() => {
                throttle = true
            }, 1000);
        }
    }
</script>
```

3）單次推理

AiCam 平臺的單次推理主要實現了應用層業務需要的單次推理計算請求。應用層將需要計算的影像及配置參數透過 Ajax 介面傳遞給演算法，演算法根據參數進行影像的推理計算，並傳回結果影像（如框出目標位置和辨識內容的影像）和結果資料（如目標座標、目標關鍵點、目標名稱、推理時間、置信度等），供應用層進行展示。

AiCam 平臺中的 file-exp 範例呼叫百度的人臉辨識演算法進行人臉註冊和人臉辨識，前端 Ajax 介面呼叫範例如表 2.11 所示。

▼ 表 2.11 人臉註冊和人臉辨識的前端 Ajax 介面呼叫範例

參數	範例	說明
url	"/file/baidu_face_recognition?camera_id=0"	介面名稱
method	'POST'	呼叫類型
processData	false	資料不做轉換，設定為 false
contentType	false	資料不做轉換，設定為 false
dataType	'json'	要求為 String 類型的參數，預期伺服器傳回的資料型態

（續表）

參數	範例	說明

data	formData 內容：
	// 傳入影像、音訊、視訊、文字
	formData.append('file_name',blob,'image.png');
	// 傳入 JSON 參數（人臉註冊）
	formData.append('param_data', JSON.stringify({
	"APP_ID": user.baidu_id,
	"API_KEY": user.baidu_apikey,
	"SECRET_KEY": user.baidu_secretkey,
	"userId": id,
	"type": 0
	}));
	// 傳入 JSON 參數（人臉比對）
	formData.append('param_data', JSON.stringify({
	"APP_ID": user.baidu_id,
	"API_KEY": user.baidu_apikey,
	"SECRET_KEY": user.baidu_secretkey,
	"type": 1
	}));
		包含 Blob 格式的檔案（影像、音訊、視訊、文字）和 JSON 格式的參數（不能為 None），透過 Flask 服務發送給演算法

success	function(res){} 內容： return_result = {'code': 200, 'msg': None, 'origin_image': None, 'result_image': None, 'result_data': None} 範例： code/msg：200 表示註冊、比對成功。 origin_image/result_image：表示原始影像 / 結果影像	透過 Flask 服務傳回的資料

人臉註冊和人臉辨識的前端 Ajax 介面呼叫範例程式如下：

```
<script>
    // 使用者資訊
    let user = {
        edge_addr:'http://192.168.100.200:4000',
        baidu_id:'12345678',
        baidu_apikey:'12345678',
        baidu_secretkey:'12345678'
    }
    // 攝影機視訊的連結
    let linkData = user.edge_addr + '/stream/index?camera_id=0'
    let throttle = true                                        // 控制辨識結果顯示頻率
    // 請求影像流資源
    let imgData = new EventSource(linkData)
    // 對影像流傳回的資料進行處理
    imgData.onmessage = function (res) {
        let {result_image} = JSON.parse(res.data)
        $('img').attr('src', `data:image/jpeg;base64,${result_image}`)
    }

    // 按一下顯示即時視訊
    $('#video').click(function () {
        $('#register').attr('disabled',false)
        imgData && imgData.close()
        // 請求影像流資源
        imgData = new EventSource(linkData)
        // 對影像流傳回的資料進行處理
        imgData.onmessage = function (res) {
            let {result_image} = JSON.parse(res.data)
            $('img').attr('src', `data:image/jpeg;base64,${result_image}`)
        }
    })

    // 人臉註冊
    $('#register').click(function () {
        let id    = prompt(" 請輸入註冊 ID","");          // 開啟輸入彈窗，獲取 ID
        let img = $('img').attr('src');             // 獲取當前視訊影像
        let blob = dataURItoBlob(img);                 // 轉為 Blob 格式
```

```
    if (id && blob.size > 20) {                    // 若有 ID、影像資料，則進入發起請求環節
        let formData = new FormData();
        formData.append('file_name', blob, 'image.png');
        //type=0 表示人臉註冊
        formData.append('param_data', JSON.stringify({
            "APP_ID": user.baidu_id,
            "API_KEY": user.baidu_apikey,
            "SECRET_KEY": user.baidu_secretkey,
            "userId": id,
            "type": 0
        }));
        $.ajax({
            url: user.edge_addr + "/file/baidu_face_recognition",
            method: 'POST',
            processData: false,                         // 必需的
            contentType: false,                         // 必需的
            dataType: 'json',
            data: formData,
            success: function (res) {
                console.log(res);
                let time = new Date().toLocaleTimeString();       // 獲取當前時間
                // 拼接當前時間與註冊結果
                let html = `<div>${time}————${JSON.stringify(res)}</div>`
                $('#info_list').html(html)                // 插入頁面回應節點
            },error: function (error) {
                console.log(error);
            }
        });
        $('#video').click()
    }
})

// 人臉辨識
$('#result').click(function () {
    imgData && imgData.close()
    let img = $('img').attr('src');                   // 獲取當前視訊影像
    let blob = dataURItoBlob(img);                          // 轉為 Blob 格式
```

```
    var formData = new FormData();
    formData.append('file_name', blob, 'image.png');
    //type=1 表示人臉辨識
    formData.append('param_data', JSON.stringify({
        "APP_ID": user.baidu_id,
        "API_KEY": user.baidu_apikey,
        "SECRET_KEY": user.baidu_secretkey,
        "type": 1
    }));
    $.ajax({
        url:    user.edge_addr + '/file/baidu_face_recognition',
        method: 'POST',
        processData: false,                               // 必需的
        contentType: false,                               // 必需的
        dataType: 'json',
        data: formData,
        success: function (res) {
            console.log(res);
            let time = new Date().toLocaleTimeString();   // 獲取當前時間
            // 拼接當前時間與註冊結果
            let html = `<div>${time}————${JSON.stringify(res)}</div>`
            $('#info_list').html(html)                    // 插入頁面回應節點
            if (res.code == 200) {
                let img = 'data:image/jpeg;base64,' + res.result_image;
                $('#register').attr('disabled',true)
                $('img').attr('src', img)        // 辨識成功後，頁面顯示辨識結果影像
            } else {
                $('#video').click()                                  // 傳回即時視訊
            }
        },
        error: function (error) {
            console.log(error);
            $('#video').click()                                      // 傳回即時視訊
        }
    });
})
// 將 base64 編碼的檔案轉化成 Blob 格式的檔案
function dataURItoBlob(base64Data) {
    var byteString;
```

```
        if (base64Data.split(',')[0].indexOf('base64') >= 0) {
            byteString = atob(base64Data.split(',')[1]);
        } else {
            byteString = unescape(base64Data.split(',')[1]);
        }
        var mimeString = base64Data.split(',')[0].split(':')[1].split(';')[0];
        var ia = new Uint8Array(byteString.length);
        for (var i = 0; i < byteString.length; i++) {
            ia[i] = byteString.charCodeAt(i);
        }
        return new Blob([ia], {
            type: mimeString
        });
    }
</script>
```

2.4.1.4 基於 AiCam 平臺的邊緣視覺應用程式開發工具

AiCamTools 是一款測試 AiCam 平臺演算法的工具，透過該工具可以快速理解演算法的應用互動，實現演算法的呼叫和資料傳回，AiCamTools 的執行介面如圖 2.46 所示。

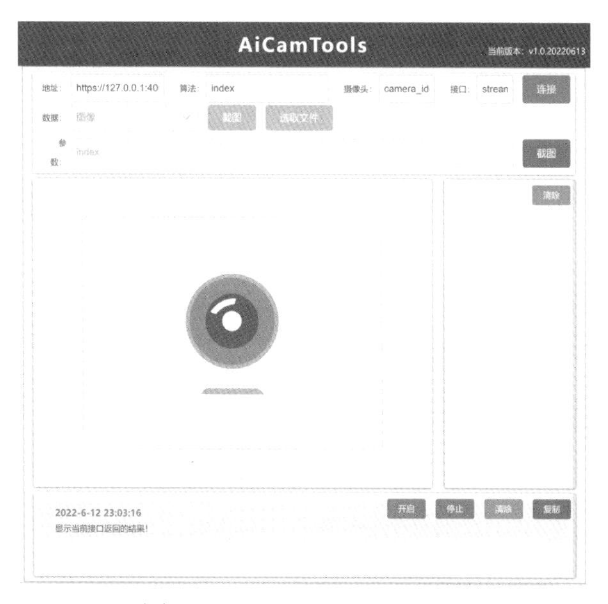

▲ 圖 2.46 AiCamTools 的執行介面

2.4.2 開發步驟與驗證

2.4.2.1 專案部署

1）硬體部署

詳見 2.1.2.1 節。

2）專案部署

（1）執行 MobaXterm 工具，透過 SSH 登入邊緣計算閘道。

（2）在 SSH 終端透過以下命令，建立開發專案目錄。

```
$ mkdir -p ~/aiedge-exp
```

（3）將 aicam 專案壓縮檔上傳到 /home/zonesion/，並透過 unzip 命令進行解壓縮。

```
$ cd ~/aiedge-exp
$ unzip aicam.zip
```

2.4.2.2 專案執行

在 SSH 終端輸入以下命令執行開發專案。

```
$ cd ~/aicam
$ chmod 755 start_aicam.sh
$ conda activate py36_tf114_torch15_cpu_cv345    //Ubuntu 20.04 作業系統下需要切換環境
$ ./start_aicam.sh
// 開始執行指令稿
* Serving Flask app "start_aicam" (lazy loading)
* Environment: production
   WARNING: Do not use the development server in a production environment.
   Use a production WSGI server instead.
* Debug mode: off
* Running on http://0.0.0.0:4000/ (Press CTRL+C to quit)
```

2.4.2.3 介面測試

1）即時推理

人臉檢測演算法（face_detection）基於深度學習實現了視訊流中的人臉即時檢測，透過 AiCam 平臺的即時推理介面可呼叫人臉檢測演算法，相關的呼叫介面如下：

```
// 攝影機視訊的連結
let linkData = 'http://192.168.100.200:4000/stream/face_detection?camera_id=0'
```

開啟 AiCamTools，在「位址」欄中輸入「http://192.168.100.200:4000」，在「演算法」欄中輸入「face_detection」，在「攝影機」欄中輸入「camera_id=0」，在「介面」欄中選擇「stream」，按一下「連接」按鈕即可呼叫人臉檢測演算法的即時推理介面，並在視窗中顯示即時的推理視訊影像，以及傳回的結果資料。透過 AiCamTools 顯示視訊影像如圖 2.47 所示，AiCamTools 可以截圖、清除、停止、複製等操作。

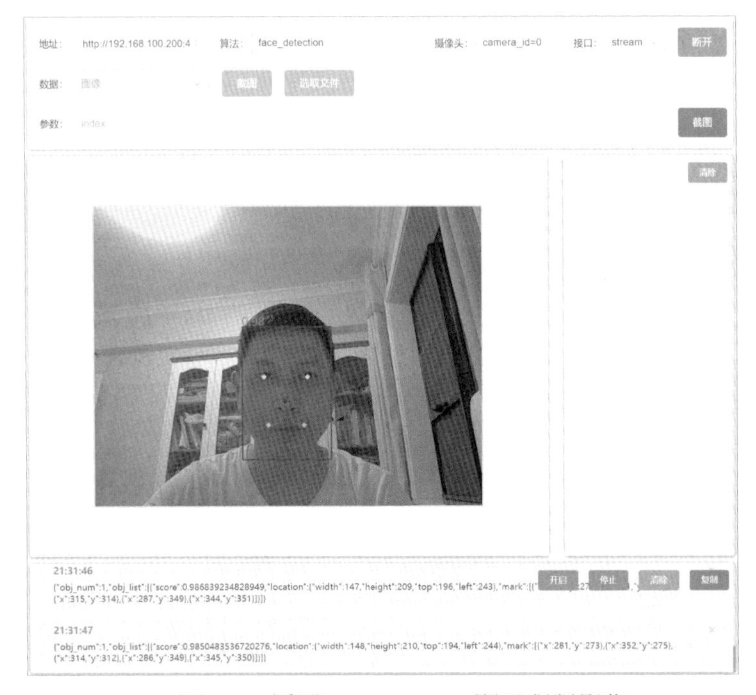

▲ 圖 2.47 透過 AiCamTools 顯示視訊影像

2）單次推理

百度的人臉辨識演算法（baidu_face_recognition）可實現人臉註冊和人臉辨識，透過 AiCam 平臺的單次推理介面可呼叫百度的人臉辨識演算法。

（1）人臉註冊。在前端應用中截取影像，透過 Ajax 介面將影像，以及包含百度帳號資訊的資料傳遞給演算法進行人臉註冊。百度人臉辨識演算法中人臉註冊呼叫的參數如表 2.12 所示。

▼ 表 2.12 百度人臉辨識演算法中人臉註冊呼叫的參數

參數	範例
url	"/file/baidu_face_recognition?camera_id=0"
method	'POST'
processData	false
contentType	false
dataType	'json'

data	let img = $('#face').attr('src')
	let id = $('#userName').val()
	let blob = dataURItoBlob(img);
	let formData = new FormData();
	formData.append('file_name',blob,'image.png');
	//type=0 表示人臉註冊
	formData.append('param_data',JSON.stringify({"APP_ID":config.user.baidu_id, "API_KEY":config.user.baidu_apikey, "SECRET_KEY":config.user.baidu_secretkey,"userId":id,"type":0}));

success	function(res){} 內容： return_result = {'code': 200, 'msg': None, 'origin_image': None, 'result_image': None, 'result_data': None} 範例： code/msg：200 表示人臉註冊成功，404 表示沒有檢測到人臉，408 表示註冊逾時，500 表示人臉介面呼叫失敗

開啟 AiCamTools，在「位址」欄中輸入 http://192.168.100.200:4000，在「演算法」欄中輸入 baidu_face_recognition，在「攝影機」欄中輸入「camera_id=0」，在「介面」欄中選擇「file」，按一下「連接」按鈕即可顯示原始視訊流影像，如圖 2.48 所示。

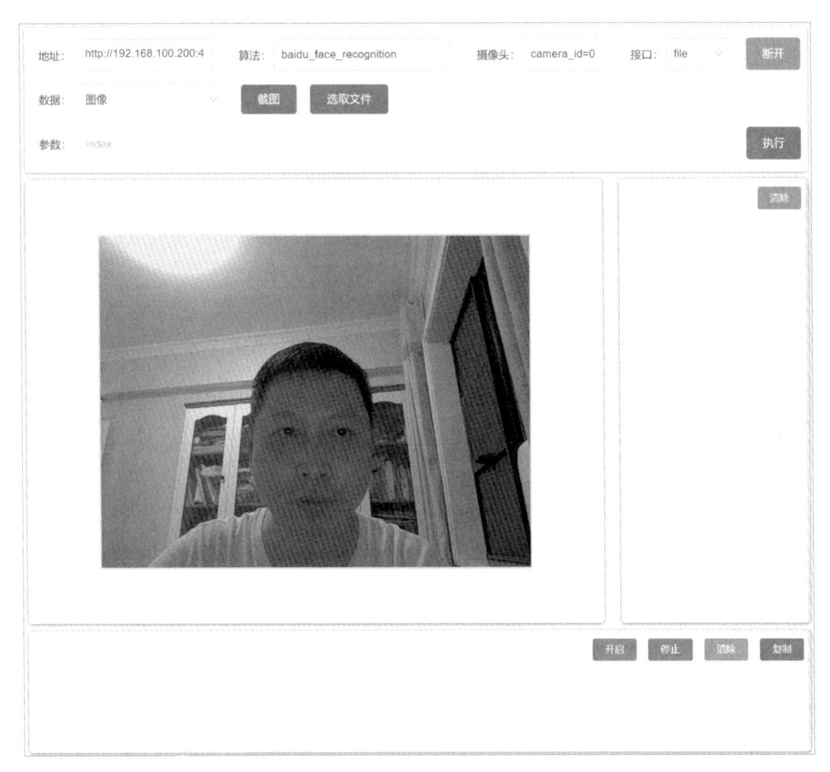

▲ 圖 2.48 原始視訊流影像

在「資料」欄中選擇「影像」後，按一下「截圖」按鈕即可將當前視訊截圖作為需要註冊的人臉並儲存下來，在「參數」欄中輸入「{"APP_ID":"***", "API_KEY":"***", "SECRET_KEY":"***","userId":"***","type":0}」（*** 因不同的使用者或應用而不同），按一下「執行」按鈕即可呼叫百度的人臉辨識演算法和單次推理介面進行人臉註冊，並顯示傳回的結果資料，如圖 2.49 所示。

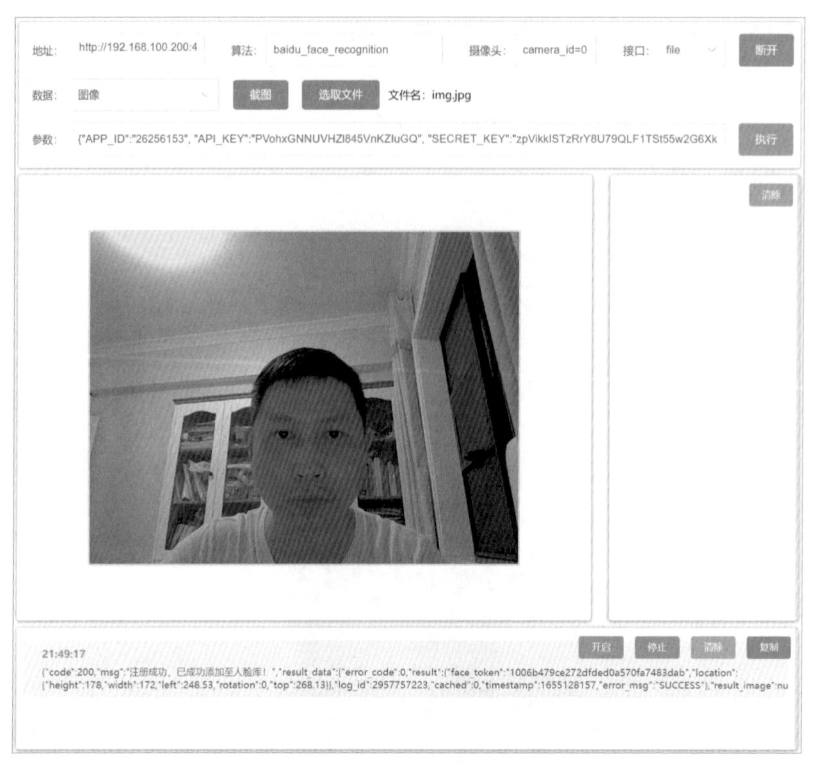

▲ 圖 2.49 人臉註冊時傳回的結果資料

（2）人臉辨識。在前端應用中截取影像，透過 Ajax 介面將影像和包含百度帳號資訊的資料傳遞給百度人臉辨識演算法進行人臉辨識。百度人臉辨識演算法中人臉辨識呼叫的參數如表 2.13 所示。

▼ 表 2.13 百度人臉辨識演算法中人臉辨識呼叫的參數

參數	範例
url	"/file/baidu_face_recognition?camera_id=0"
method	'POST'
processData	false
contentType	false
dataType	'json'

<table>
<tr><th>參數</th><th>範例</th></tr>
<tr><td>data</td><td>let img = $('.camera>img').attr('src')
let blob = dataURItoBlob(img)
var formData = new FormData();
formData.append('file_name',blob,'image.png');
//type=1 表示人臉辨識
formData.append('param_data', JSON.stringify({"APP_ID":config.user.baidu_id, "API_KEY":config.user. baidu_apikey, "SECRET_KEY":config.user.baidu_secretkey,"type":1}));</td></tr>
<tr><td>success</td><td>function(res){} 內容：
return_result = {'code': 200, 'msg': None, 'origin_image': None, 'result_image': None, 'result_data': None}
範例：
code/msg：200 表示人臉註冊成功，404 表示沒有檢測到人臉，500 表示人臉介面呼叫失敗。
origin_image/result_image：原始影像和結果影像。
result_data：傳回的人臉資訊</td></tr>
</table>

開啟 AiCamTools，在「位址」欄中輸入「http://192.168.100.200:4000」，在「演算法」欄中輸入「baidu_face_recognition」，在「攝影機」欄中輸入「camera_id=0」，在「介面」欄中選擇「file」，按一下「連接」按鈕即可顯示原始視訊流影像。

在「資料」欄中選擇「影像」，按一下「截圖」按鈕即可將當前視訊截圖作為需要辨識的物件，在「參數」欄中輸入「{"APP_ID":"***", "API_KEY":"***", "SECRET_KEY":"***", "userId":"***","type":1}」（*** 因不同的使用者或應用而不同），按一下「執行」按鈕即可呼叫百度的人臉辨識演算法和單次推理介面進行人臉辨識，並在右側邊窗中顯示原始影像和結果影像，在下方顯示傳回的結果資料，如圖 2.50 所示。

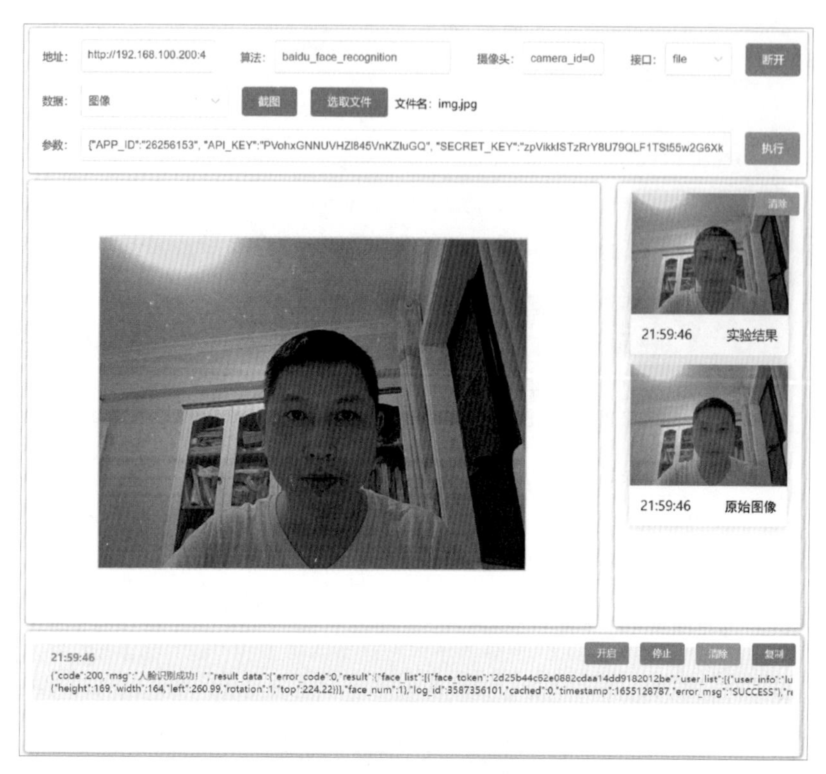

▲ 圖 2.50 人臉辨識結果

2.4.2.4 即時推理（人臉檢測）

AiCam 平臺中的 stream-exp 範例透過呼叫 AiCam 平臺內建的人臉檢測演算法，實現了即時的人臉檢測，並將辨識的影像結果和資料結果顯示在頁面中。

解壓縮該範例的程式 stream-exp.zip，用記事本開啟 index.html 檔案，將 AiCam 平臺的服務連結修改為邊緣計算閘道位址：

```
// 攝影機視訊的連結
let linkData = 'http://192.168.100.200:4000/stream/face_detection?camera_id=0'
```

透過 Chrome 瀏覽器開啟 index.html 檔案，即可呼叫人臉檢測演算法（face_detection）進行即時的人臉檢測，如圖 2.51 所示。

▲ 圖 2.51 透過 Chrome 瀏覽器呼叫人臉檢測演算法進行即時的人臉檢測

2.4.2.5 單次推理（人臉檢測）

1）專案部署

AiCam 平臺中的 file-exp 範例透過呼叫 AiCam 平臺內建的百度人臉辨識演算法，實現了人臉註冊和人臉辨識，並將辨識的影像結果和資料結果顯示在頁面中。

解壓縮 file-exp 範例的程式 file-exp.zip，用記事本開啟 index.html 檔案，修改 AiCam 平臺的服務位址、百度帳號資訊。

```
// 使用者資訊
let user = {
    edge_addr:'http://192.168.100.200:4000',
    baidu_id:'12345678',
    baidu_apikey:'12345678',
    baidu_secretkey:'12345678'
}
```

透過 Chrome 瀏覽器開啟 index.html 檔案，即可顯示原始視訊流影像，如圖 2.52 所示。

2）人臉註冊

（1）按一下「人臉註冊」按鈕，將呼叫百度的人臉辨識演算法（baidu_face_recognition）進行人臉註冊。

（2）在彈出的對話方塊中填寫需要註冊的人臉名稱 ID，如圖 2.53 所示，填寫完成後按一下「確認」按鈕退出對話方塊，當前的視訊截圖會被上傳到演算法層進行人臉註冊。

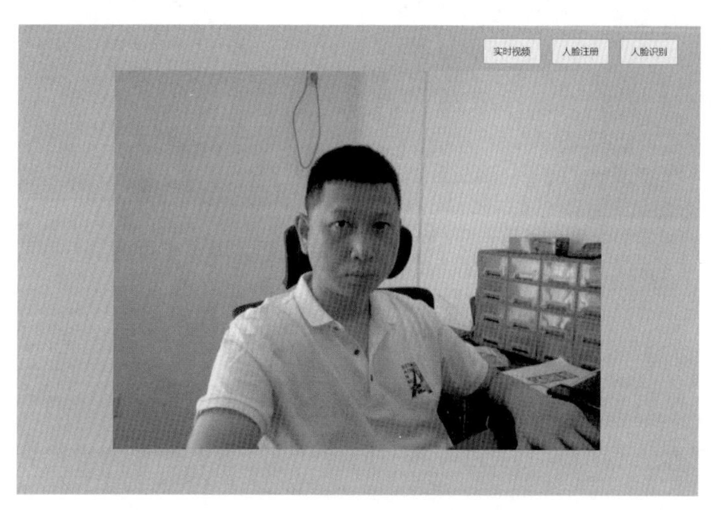

▲ 圖 2.52 透過 Chrome 瀏覽器顯示的原始視訊流影像

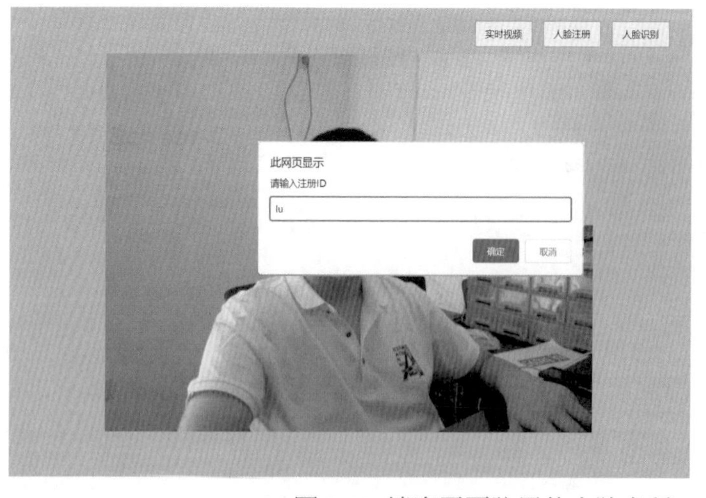

▲ 圖 2.53 填寫需要註冊的人臉名稱 ID

（3）人臉註冊的結果資訊將在頁面的右側視窗中，資訊如下：

```
08:32:13————{"origin_image":null,"result_image":null,"code":200,"result_data
":{"timestamp":1654475532,"cached":0,"error_code":0,"log_id":1932140814,"error_
msg":"SUCCESS","result":{"face_token":"cf4186d79250fdd248bbce074ee14edc","location":{"
width":121,"rotation":0,"top":177.79,"height":115,"left":250.72}}},"msg":"註冊成功，已
成功增加至人臉庫！"}
```

3）人臉辨識

按一下「人臉辨識」按鈕，將呼叫百度的人臉辨識演算法（baidu_face_recognition）進行人臉比對，如果成功辨識人臉，則顯示辨識到的人臉結果影像，並傳回辨識的結果資訊，如圖 2.54 所示。

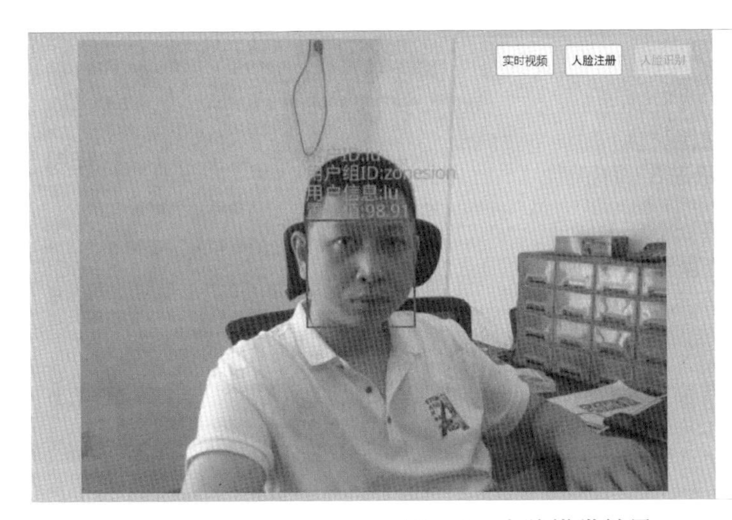

▲ 圖 2.54 人臉辨識結果

2.4.3 本節小結

本節首先介紹了邊緣視覺的應用程式開發邏輯，以及基於 AiCam 平臺的邊緣視覺應用程式開發框架、開發介面和開發工具；然後介紹了具體的開發步驟和驗證步驟，包括專案部署、專案執行、介面測試等內容。透過本節的學習，讀者可以了解邊緣視覺的應用程式開發邏輯，掌握基於 AiCam 平臺的邊緣視覺開發過程，學習 AiCamTools 的使用方法，實現人臉辨識的邊緣視覺應用。

2.4.4 思考與擴充

（1）簡述基於 AiCam 的邊緣視覺應用程式開發邏輯框架。

（2）在 AiCam 平臺上，如何實現演算法的應用互動、演算法排程和資料傳回？

MEMO

MEMO

邊緣計算與人工智慧模型開發

本章結合 AiCam 平臺學習邊緣計算與人工智慧模型開發全流程。本章內容包括：

（1）資料獲取與標註：掌握資料獲取和標註的過程，以及基於 CreateImg 進行影像擷取、基於 labelImg 進行影像標註的過程。

（2）YOLOv3 模型的訓練與驗證：掌握模型訓練和驗證的作用，了解主流深度學習開發框架、深度學習物件辨識演算法原理、YOLO 的原理，結合 YOLOv3 和 Darknet 框架對交通號誌辨識模型進行訓練與驗證。

（3）YOLOv5 模型的訓練與驗證：掌握 YOLOv5 原理，結合 PyTorch 框架和 YOLOv5 對口罩檢測模型進行訓練與驗證。

（4）YOLOv3 模型的推理與驗證：掌握從開發框架到推理框架的轉換過程，了解典型的常用行動端邊緣推理框架，結合 YOLOv3 和 NCNN 框架對交通號誌辨識模型進行推理與驗證。

（5）YOLOv5 模型的推理與驗證：掌握 RKNN 框架的工作原理，結合 RKNN 框架和 YOLOv5 對口罩檢測模型進行推理與驗證。

（6）YOLOv3 模型的介面應用：了解常用的模型介面，掌握 NCNN 框架的模型介面設計，結合 YOLOv3 和 NCNN 框架設計交通號誌辨識模型的介面。

（7）YOLOv5 模型的介面應用：掌握 RKNN 框架的模型介面設計，結合 YOLOv5 和 RKNN 框架設計口罩檢測模型的介面。

（8）YOLOv3 模型的演算法設計：掌握 AiCam 平臺的開發框架，結合 AiCam 平臺和 YOLOv3 設計交通號誌辨識演算法。

（9）YOLOv5 模型的演算法設計：結合 AiCam 平臺和 YOLOv5 設計交通號誌辨識演算法。

3.1 資料獲取與標註

資料獲取與標註是深度學習和人工智慧開發中的關鍵步驟，對建構高性能、堅固性強的模型至關重要。資料獲取與標註的作用如下：

（1）訓練模型：深度學習需要大量的資料來訓練模型，以便從中學到模式和規律。透過擷取並標註有代表性的資料集，可以幫助模型更進一步地理解輸入資料的特徵和關係。

（2）驗證模型：擷取的資料集可以用來驗證模型的性能，標註的資料集可以用來測試模型的準確性、堅固性和泛化能力，從而評估模型在真實世界中的表現。

（3）改進模型：透過分析模型在擷取和標註的資料集上的表現，可以辨識模型的弱點和錯誤，並進一步最佳化和改進模型的性能，有助不斷提升模型的品質。

（4）解決樣本偏差：資料獲取與標註有助解決樣本偏差問題，透過在資料集中包含各種各樣的樣本，可以更進一步地確保模型在各種情境下都能有良好的表現。

（5）應對標籤不平衡：在某些應用中，不同類別的樣本可能具有不平衡的標籤分佈，透過擷取並標註標籤分佈平衡的資料，可以幫助模型更進一步地應對標籤的不平衡。

本節的基礎知識如下：

- 掌握資料獲取與標註的過程。
- 結合邊緣計算平臺掌握透過 CreateImg 擷取影像的步驟。
- 結合邊緣計算平臺掌握透過 labelImg 標註影像的步驟。

3.1.1 原理分析與開發設計

3.1.1.1 資料獲取和標註流程

資料獲取和標註需要確保資料的品質和準確性，從而為邊緣視覺應用提供可靠的資料基礎。資料獲取和標註的一般流程如下：

（1）確定資料需求：在進行資料獲取和標註之前，要明確所需資料的類型和數量，並確保資料和業務需求、演算法模型要求相匹配。

（2）收集原始資料：根據需求收集原始資料，原始資料可以來自各種通路，如網際網路、感測器、攝影機等。

（3）清洗資料：對原始資料進行清洗，去除雜訊、重複和無效資料，確保資料的品質和準確性。

（4）標註資料：將清洗後的資料標註為所需的類別或屬性，以便演算法模型能夠辨識和學習。標註可以手動進行或自動進行，手動標註需要專業的標註人員，自動標註需要訓練好的演算法模型。

（5）驗證資料：資料標註完成後，需要對資料進行驗證和校對，以確保資料的一致性和準確性。

（6）儲存和管理資料：資料驗證完成後，需要將資料儲存在資料庫或雲端，以便進行管理。

（7）更新和迭代資料：資料獲取與標註是一個持續進行的過程，隨著業務需求和演算法模型的變化，需要不斷更新和迭代資料，以保證資料的時效性和準確性。

3.1.1.2 影像擷取

CreateImg 是一款可以自動擷取影像工具，該工具透過 OpenCV 呼叫攝影機自動進行視訊的錄製和影像的擷取，資料獲取完成後被儲存在 AILabelImg/dataset/JPEGImages 資料夾內。

錄製的視訊要儘量在真實場景中進行，將需要檢測的物品放置在攝影機能夠拍攝到的視窗位置，並在不同方位、不同角度、不同光線的條件下錄製物品，儘量保持資料樣本的多樣性。

3.1.1.3 影像標註

labelImg 是一個視覺化的影像標註工具，Faster R-CNN、YOLO、SSD 等物件辨識網路所需的資料集，均可採用 labelImg 標註的影像。labelImg 在完成資料標註後會自動生成描述影像的 XML 檔案，生成的 XML 檔案遵循 PASCAL VOC 格式標準。

3.1.2 開發步驟與驗證

3.1.2.1 影像擷取

將攝影機連線邊緣計算或電腦，透過 CreateImg 擷取影像的步驟如下：

（1）將需要辨識的物品放置在攝影機正前方的合適位置。

（2）開啟終端，輸入以下命令執行 CreateImg：

```
#Linux 環境下執行以下命令
$ conda activate py36_tf114_torch15_cpu_cv345        //Ubuntu 20.04 作業系統下需要切換環境
$ python3 CreateImg.py

#Windows 環境下執行以下命令（需要安裝必要的環境）
$ python CreateImg.py
```

（3）CreateImg 將顯示攝影機的即時影像（見圖 3.1），並將即時影像儲存到資料集目錄 AILabelImg\dataset\JPEGImages。

▲ 圖 3.1 攝影機的即時影像

（4）在不同方位、不同角度、不同光線下拍攝物品，儘量使資料樣本保持多樣性。

（5）錄製一段時間後，按下「Q」退出 CreateImg，在資料集目錄中可以看到擷取的影像。

3.1.2.2 影像標註

1）匯入影像

（1）進入 labelImg 所在的目錄，執行 labelImg.exe。labelImg 的執行介面如圖 3.2 所示。

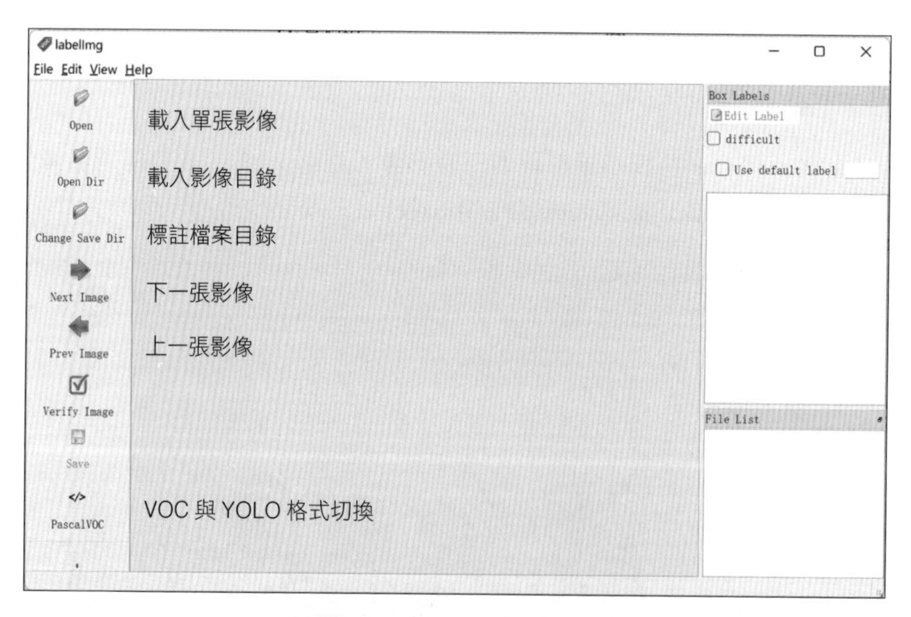

▲ 圖 3.2 labelImg 的執行介面

（2）按一下「Open Dir」按鈕，選擇擷取的影像目錄 AILabelImg\dataset\JPEGImages（可以使用前面擷取的資料集影像，也可以使用本專案程式內提供的交通號誌資料集影像），如圖 3.3 所示。

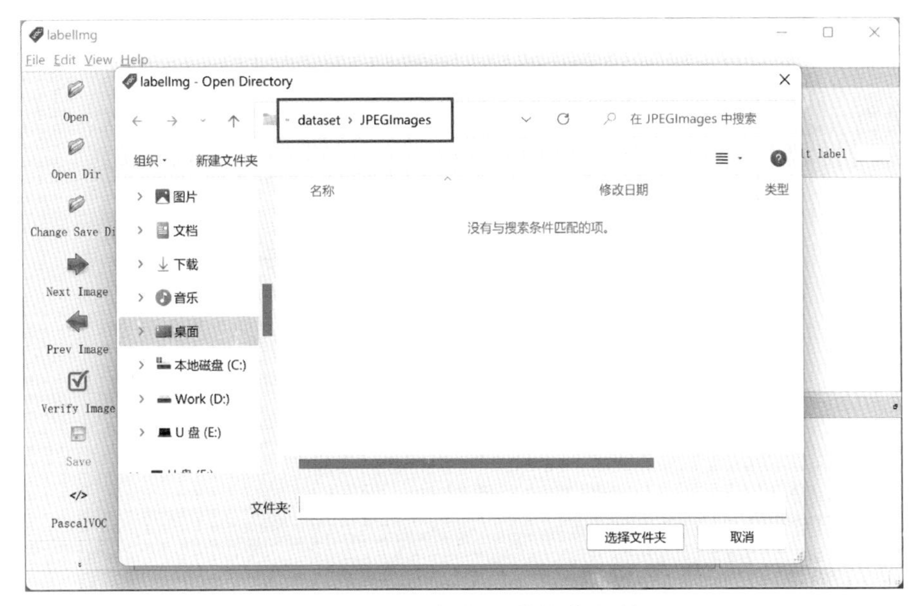

▲ 圖 3.3 選擇擷取的影像目錄

（3）labelImg 會自動彈窗，要求選擇標註檔案的儲存路徑 AILabelImg\dataset\Annotations，如圖 3.4 所示。

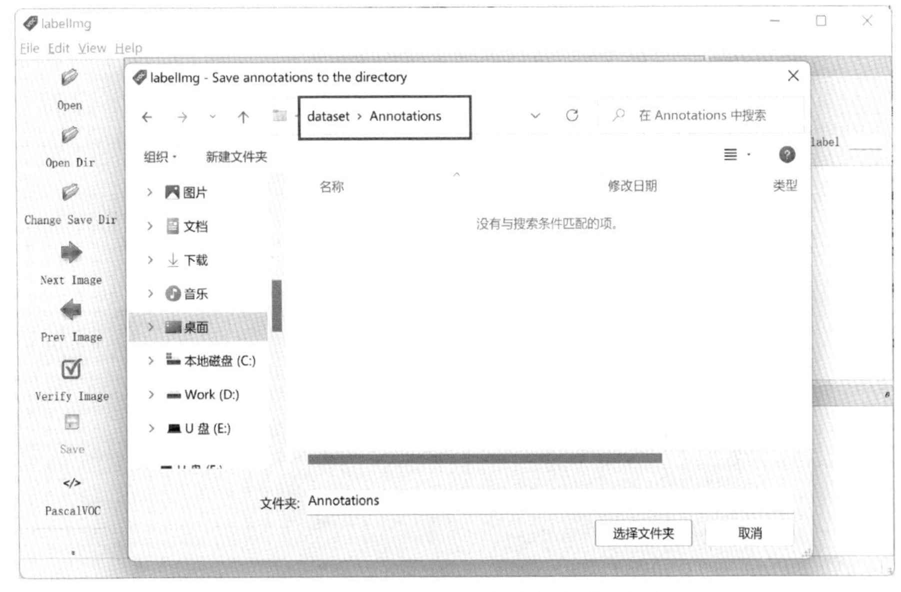

▲ 圖 3.4 選擇標註檔案的儲存路徑

（4）設定完成後，labelImg 將載入影像並在檔案清單（File List）中顯示載入的影像，按一下第一幅影像將在 labelImg 執行介面的中間顯示該影像。載入影像如圖 3.5 所示。

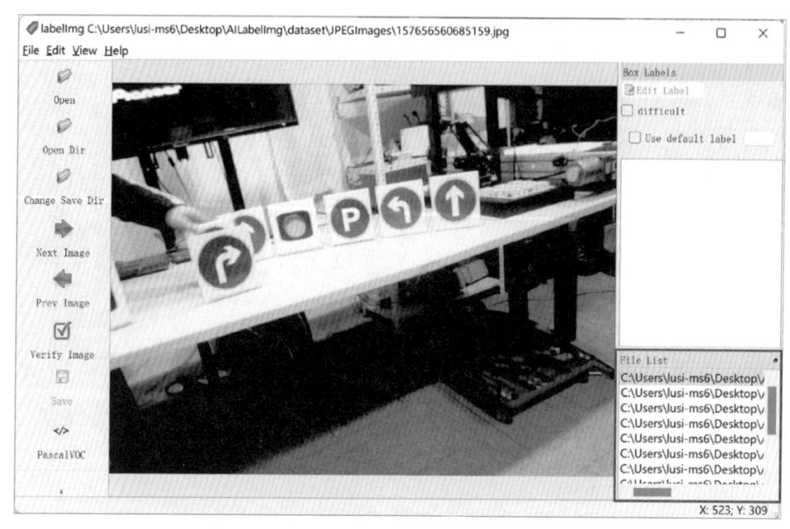

▲ 圖 3.5 載入影像

（5）在 labelImg 執行介面中選擇選單「View」→「Auto Saving」，如圖 3.6 所示，即可設定為自動儲存標註檔案。

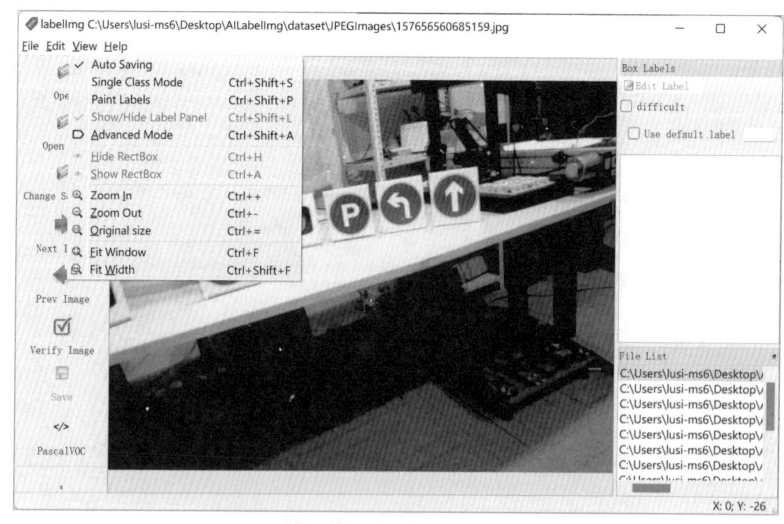

▲ 圖 3.6 選擇選單「View」→「Auto Saving」

2）影像標註

（1）影像標註：在 labelImg 執行介面的檔案清單中按一下第一幅影像，labelImg 將開始逐幅對影像進行標註。按「W」鍵可對目標物品進行矩形框標註，在需要標註的目標左上角按下滑鼠左鍵並且不要鬆開，移動到目標的右下角後鬆開滑鼠左鍵，即可對一個目標進行標註，如圖 3.7 所示。

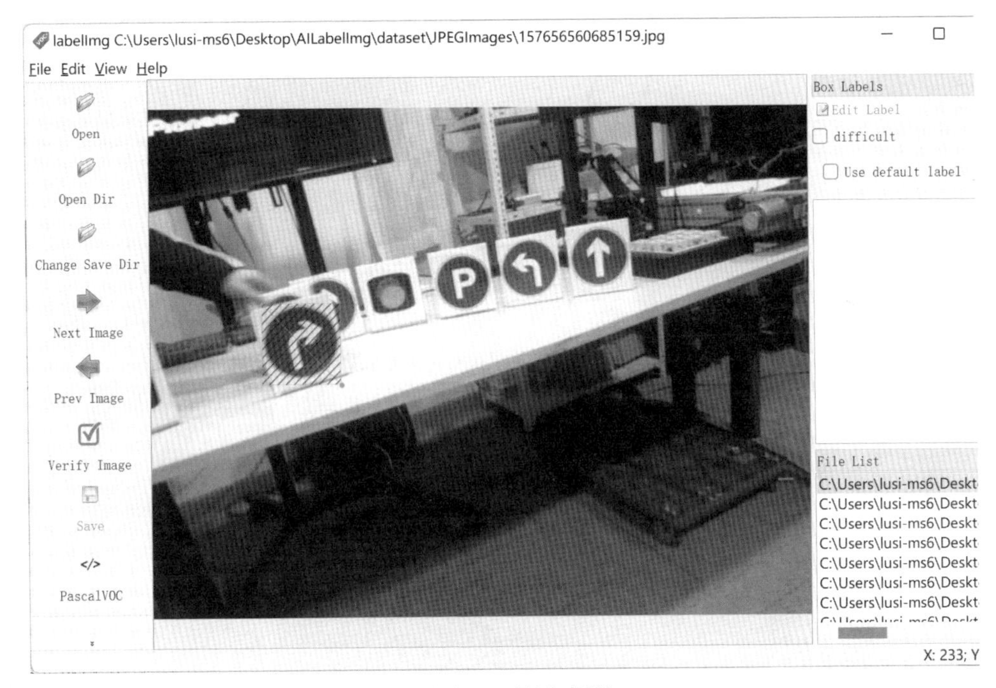

▲ 圖 3.7 影像標註

（2）設定標籤（Label）：勾選「Box Labels」中的「Edit Label」，在彈出的「labelImg」對話方塊輸入標籤名稱，如 right（再畫矩形框時就會有記錄供直接選擇），按一下「OK」按鈕即可完成影像標註，如圖 3.8 所示。

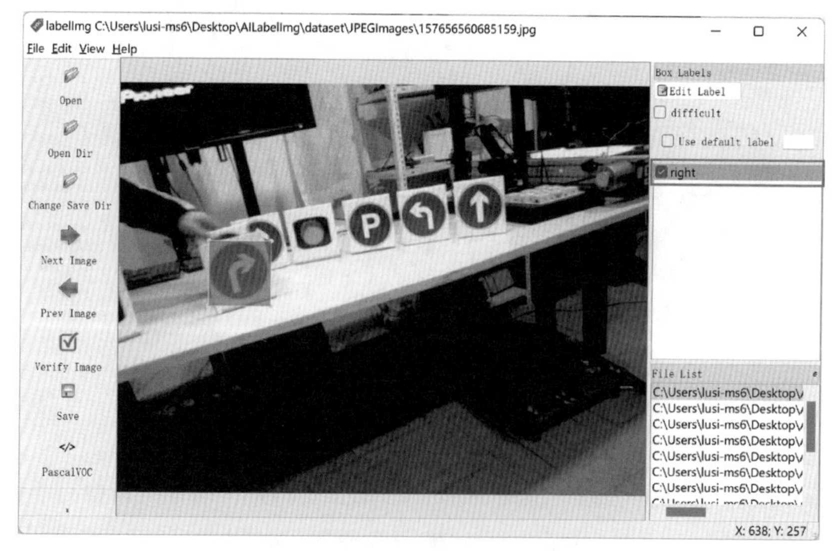

▲ 圖 3.8 設定標籤後完成影像標註

（3）預設標籤：勾選「Box Labels」中的「Use default label」，可以對同一個標籤進行批次標註，這樣在每次畫矩形框後就會自動將標註設定為預設的標籤，從而避免每次增加標籤，如圖 3.9 所示。

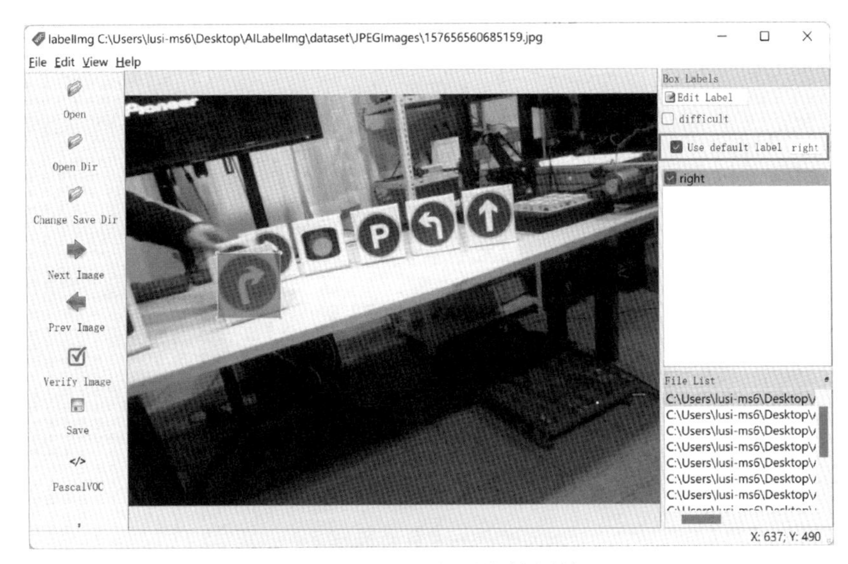

▲ 圖 3.9 設定預設的標籤

（4）連續標註：按「D」鍵切換到下一幅影像，繼續按「W」鍵可進行影像標註，如圖 3.10 所示。

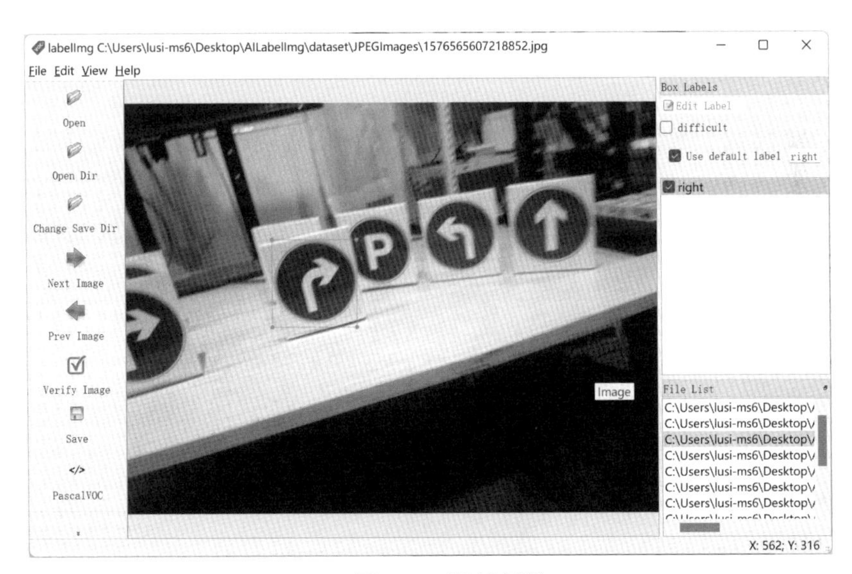

▲ 圖 3.10 連續標註

（5）標註技巧：讀者可以首先將其中一個標籤設定為預設的標籤，把所有影像內的該標籤全部標註一次；然後將其他的標籤設定為預設的標籤，繼續從頭開始將所有影像中新的標籤全部標註一次，直到影像中所有的標籤都進行了標註，如圖 3.11 所示。

▲ 圖 3.11　標註技巧

（6）開啟一個 xml 檔案，可以看到 object 屬性內包含了目標的資訊，包括目標類別、座標等資訊。程式如下：

```
<annotation>
    <folder>JPEGImages</folder>
    <filename>157656560685159.jpg</filename>
    <path>C:\Users\lusi-ms6\Desktop\AILabelImg\dataset\JPEGImages\157656560685159.
jpg</path>
    <source>
        <database>Unknown</database>
    </source>
    <size>
        <width>640</width>
        <height>480</height>
        <depth>3</depth>
```

```
</size>
<segmented>0</segmented>
<object>
    <name>right</name>
    <pose>Unspecified</pose>
    <truncated>0</truncated>
    <difficult>0</difficult>
    <bndbox>
        <xmin>109</xmin>
        <ymin>184</ymin>
        <xmax>185</xmax>
        <ymax>260</ymax>
    </bndbox>
</object>
<object>
    <name>straight</name>
    <pose>Unspecified</pose>
    <truncated>0</truncated>
    <difficult>0</difficult>
    <bndbox>
        <xmin>408</xmin>
        <ymin>112</ymin>
        <xmax>465</xmax>
        <ymax>169</ymax>
    </bndbox>
</object>
<object>
    <name>left</name>
    <pose>Unspecified</pose>
    <truncated>0</truncated>
    <difficult>0</difficult>
    <bndbox>
        <xmin>339</xmin>
        <ymin>123</ymin>
        <xmax>397</xmax>
        <ymax>181</ymax>
    </bndbox>
</object>
<object>
```

```
        <name>green</name>
        <pose>Unspecified</pose>
        <truncated>0</truncated>
        <difficult>0</difficult>
        <bndbox>
            <xmin>209</xmin>
            <ymin>153</ymin>
            <xmax>255</xmax>
            <ymax>199</ymax>
        </bndbox>
    </object>
</annotation>
```

當標籤的類別比較多時，建議標註完一個類別後再從頭開始標註下一個類別，這樣可以提高標註速度，但在提高速度時千萬不能標註錯誤，標註完成之後建議進行檢查，錯誤的標註對於模型訓練的影響非常大。

3.1.3 本節小結

本節主要介紹資料獲取與標註的一般流程，透過 CreateImg、labelImg，以及實際案例介紹了影像擷取與標註的步驟。

3.1.4 思考與擴充

（1）在影像擷取過程中，應該如何保證資料樣本的真實性和多樣性？

（2）當標籤的類別較多時，選擇何種標註策略會提高標註速度？

3.2 YOLOv3 模型的訓練與驗證

模型的訓練與驗證是確保模型能夠在實際場景中有效工作的關鍵步驟，良好的訓練與驗證能夠提高模型的性能、泛化能力，同時減少過擬合的風險。

1）訓練的作用

（1）學習資料模式：訓練能夠使模型學習輸入資料的模式和特徵。透過調整參數，模型能夠最小化預測結果與實際標籤之間的差異，從而提高模型的性能。

（2）泛化能力：透過訓練，模型不僅在訓練集上有良好的表現，還能夠泛化到未見過的資料集。泛化能力是模型在真實場景中保持實用性和有效性的關鍵。

（3）參數最佳化：在訓練階段，透過反向傳播演算法和最佳化器可以更新模型的權重和偏置，使其能夠更進一步地擬合訓練的資料，這有助模型更進一步地獲取資料的複雜關係。

（4）特徵學習：對於深度學習模型，訓練還包括對特徵的學習。透過層層堆疊的神經網路，模型可以自動學習表示資料的高級特徵。

2）驗證的作用

（1）評估性能：驗證用於評估模型在未見過的資料集上的性能，有助使用者了解模型的泛化能力，即模型在新資料集上的表現。

（2）超參數調整：驗證集通常用於調整模型的超參數，如學習率、正則化等。透過多次調整這些超參數並在驗證集上評估性能，可以找到最佳的超參數配置，從而提高模型的性能。

（3）防止過擬合：模型在訓練階段可能會出現過擬合，即過分適應訓練集中的雜訊而失去泛化能力。透過驗證集，可以檢測和防止過擬合，確保模型對新資料集的表現更為堅固。

（4）模型選擇：在深度學習中可能會使用多個模型進行比較，透過在驗證集上比較模型的性能，可以選擇最適合的模型。

（5）信任度和可靠性：透過驗證可以對模型的信任度和可靠性進行更深入的了解，從而為實際應用提供更好的支援。

本節的基礎知識如下：

- 了解主流深度學習的開發框架、物件辨識演算法原理。
- 了解 YOLO 系列模型，掌握 YOLOv3 模型的演算法原理。
- 結合交通號誌辨識範例，掌握 YOLOv3 模型及 Darknet 模型的訓練與驗證過程。

3.2.1 原理分析與開發設計

3.2.1.1 深度學習物件辨識演算法概述

物件辨識與辨識技術是一種透過電腦自動、智慧檢測周圍環境並辨識相應目標的電腦視覺技術。物件辨識與辨識技術的核心是演算法能夠對影像或視訊進行檢測，發現感興趣的目標、確定目標的具體位置並辨識目標的類別。自從將卷積神經網路和深度學習引入物件辨識領域後，物件辨識技術具備了驚人的檢測效果和優越的性能，使其成為當下人工智慧領域的熱門研究方向。深度學習利用卷積神經網路將卷積運算和人工網路結合在一起，透過模擬人腦神經結構，建構了深度卷積網路（包括卷積層、歸一化層、啟動層、池化層、全連接層等），能從巨量資料中有效提取影像的各種特徵，並對特徵進行抽象、學習。

1）卷積神經網路結構概述

卷積神經網路（Convolutional Neural Networks，CNN）是一種深度學習模型或類似於類神經網路的多層感知器，主要由資料登錄（Input）層、卷積（CONV）層、啟動（ReLU）層、池化（Pooling layer）層、全連接（FC）層組成。

（1）資料登錄層：主要對原始影像資料進行前置處理，包括取平均值、歸一化、PCA 降維等。

（2）卷積層：是卷積神經網路最重要的層，也是卷積神經網路名字的來源。卷積層由一組濾波器（Filter）組成，濾波器採用三維結構，其深度由輸入資料的深度決定，一個濾波器是由多個卷積核心堆疊形成的。這些濾波器在輸入資料上滑動進行卷積運算，從輸入資料中提取特徵。在訓練時，濾波器的權重使用隨機值進行初始化，並根據訓練集進行學習，逐步進行最佳化。

圖 3.12 所示為卷積運算的基本想法，對於一幅影像，卷積運算從影像的左上角開始，從左往右、從上往下，以一個像素或指定數量像素的間距依次滑過影像的每一個區域。

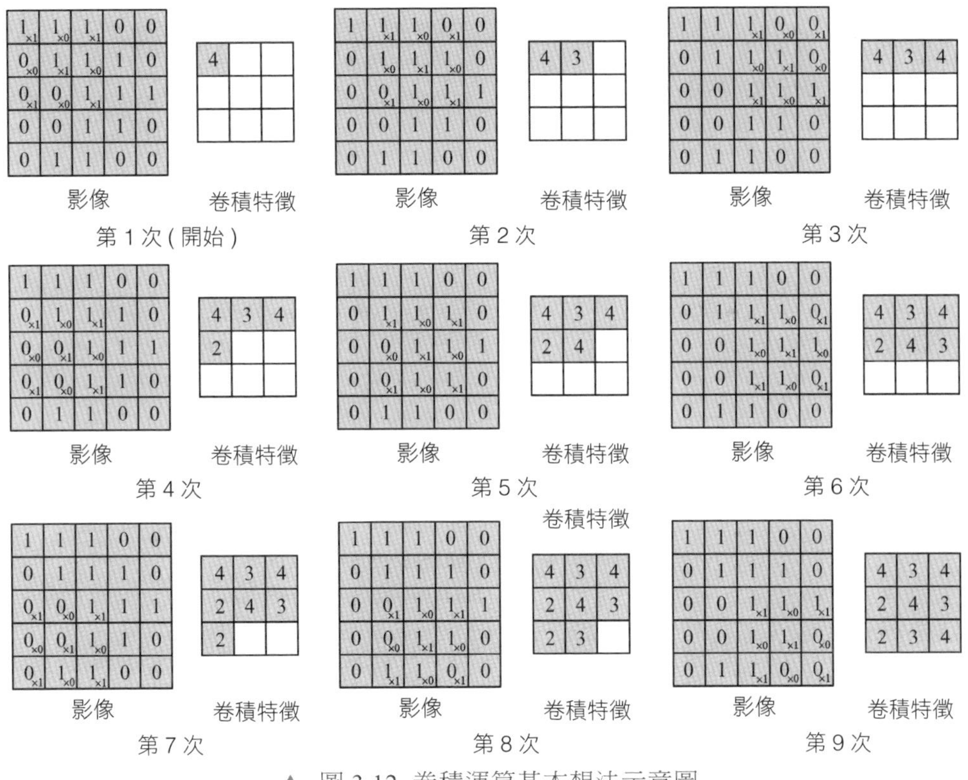

▲ 圖 3.12 卷積運算基本想法示意圖

卷積運算是指以一定間隔滑動卷積核心的視窗，將各個位置上卷積核心的元素和輸入的對應元素相乘，然後求和（有時也稱為乘累積加運算），最後將這個結果儲存到輸出的對應位置。卷積核心可以視為權重，其尺寸大小可以變化，一般取奇數。卷積核心每次滑動的像素數稱為步進值。

每一個卷積核心（Convolution Kernel）都可以當做一個特徵提取運算元，令一個特徵提取運算元在影像上不斷滑動，得出的濾波結果稱為特徵圖（Feature Map）。我們不必人工設計卷積核心，而是先使用隨機數值進行初始化來得到很多卷積核心，再透過反向傳播最佳化這些卷積核心，以期得到更好的辨識結果。卷積核心是二維的權重矩陣，濾波器（Filter）是多個卷積核心堆疊而成的三維矩陣。

卷積運算能夠更進一步地提取影像特徵，使用不同大小的卷積核心能夠提取影像的各個尺度特徵。

（3）啟動層：對卷積層的輸出結果進行非線性映射，如圖 3.13 所示。

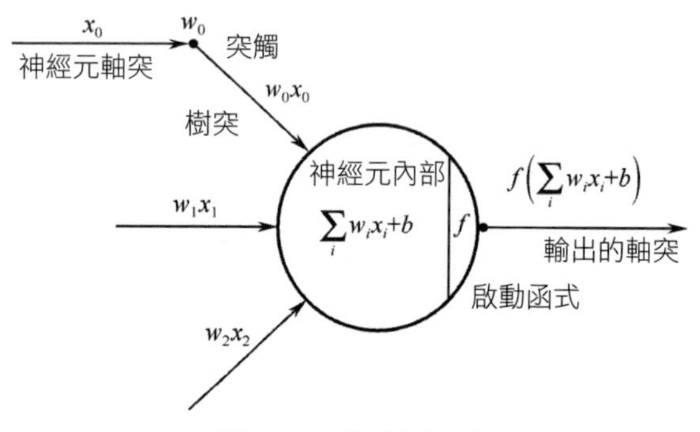

▲ 圖 3.13 啟動層示意圖

CNN 採用的啟動函式一般為 ReLU（Rectified Linear Unit，修正線性單元），該啟動函式的特點是收斂快、求梯度簡單，但較脆弱，如圖 3.14 所示。

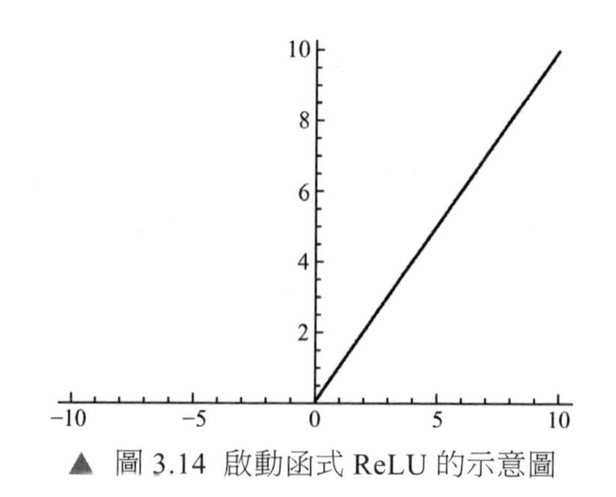

▲ 圖 3.14 啟動函式 ReLU 的示意圖

（4）池化層：也稱匯聚層，實際是一個下採樣（Down-Sample）過程，用來縮小高度和長度，減小模型規模，提高運算速度，同時提高所提取特徵的堅固性。簡單來說，池化層的目的是提取一定區域的主要特徵，並減少參數量，防止模型過擬合。如果輸入的是影像，那麼池化層的主要目的是壓縮影像。池化層在卷積層之後，二者相互交替出現，並且每個卷積層都與一個池化層相對應。池化操作也有一個類似卷積核心一樣東西在特徵圖上移動，稱為池化視窗，池化視窗也有大小，移動的時候有步進值。池化層示意圖如圖 3.15 所示。

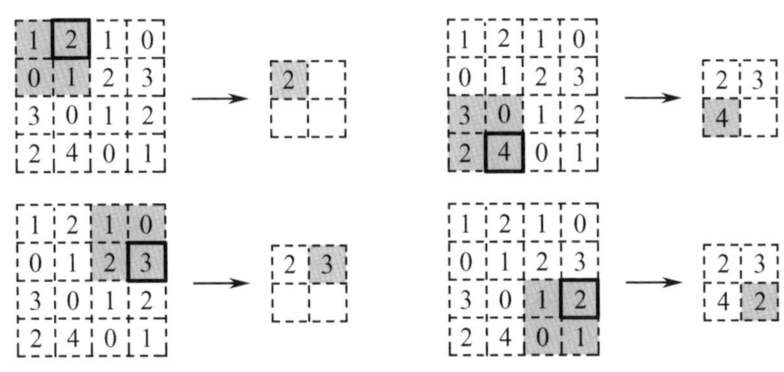

▲ 圖 3.15 池化層示意圖

（5）全連接層：通常在卷積神經網路的尾部，和傳統的神經網路神經元的連接方式是一樣的，如圖 3.16 所示。

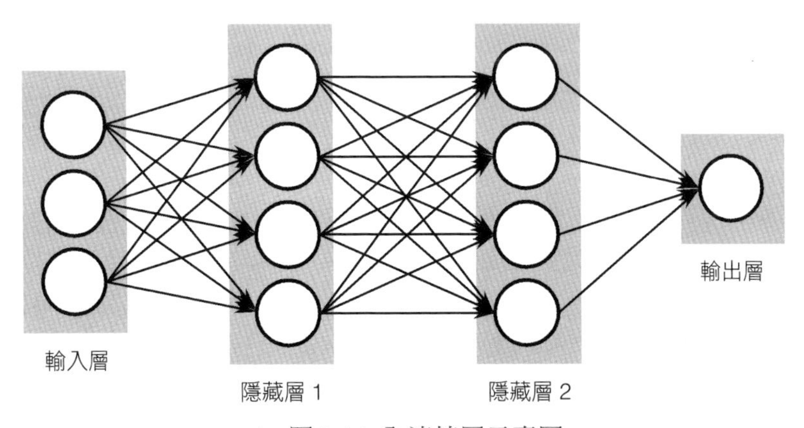

▲ 圖 3.16 全連接層示意圖

2）基於深度學習的物件辨識演算法

基於深度學習的物件辨識演算法大致可以分為兩類：基於候選區域的深度學習物件辨識演算法和基於回歸方法的深度學習物件辨識演算法，也稱為兩階段（Two-Stage）演算法和單階段（One-Stage）演算法。

（1）Two-Stage 演算法：該演算法首先在影像上產生候選區域，然後在候選區域內對目標進行分類和回歸。Two-Stage 演算法的代表演算法有 Faster R-CNN、R-FCN、FPN、Mask R-CNN 等，Two-Stage 演算法在獲得高檢測精度的同時，普遍存在檢測速度慢的缺點。

R-CNN 演算法是最早的基於深度學習的物件辨識演算法，解決了傳統演算法檢測效率較低、精度不高的問題。R-CNN 演算法首先使用選擇性搜索（Selective Search）方式生成候選框，然後利用 CNN 對生成的候選框進行特徵提取，接著使用支援向量機（Support Vector Machine，SVM）對提取到的特徵進行分類，最後透過訓練好的回歸演算法修正候選框的位置。R-CNN 演算法的原理如圖 3.17 所示。

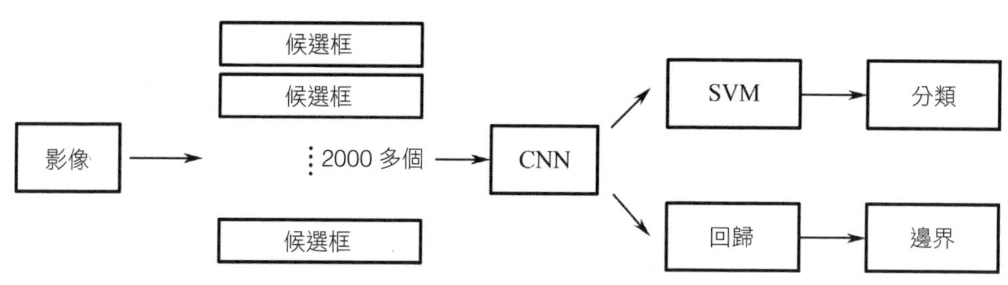

▲ 圖 3.17 R-CNN 演算法的原理

R-CNN 演算法解決了傳統演算法在選擇候選框時需要列舉所有框的問題，但透過選擇性搜索方式生成的 2000 多個候選框都需要在 CNN 中進行特徵提取，並且還需要透過 SVM 對這些特徵分類，因此計算量相當巨大，導致 R-CNN 演算法的檢測速度很慢。實際測試中，R-CNN 演算法的訓練時間需要 84 h，檢測一幅影像需要 47 s。雖然 R-CNN 演算法的檢測精度比傳統演算法有較大的提升，但在即時性方面遠遠達不到實際使用的要求。

改進的 R-CNN 演算法均在檢測速度上做了最佳化，例如在 Faster R-CNN 中，CNN 不再對每個候選框進行特徵提取，而是直接對影像進行卷積，這樣極大地減少了生成 2000 多個候選框帶來的重複計算。R-CNN 演算法需要使用 SVM 對提取到的特徵進行分類，使用回歸演算法修正候選框，Faster R-CNN 則直接用 ROI（Regions Of Interest）池化層將不同尺寸的特徵輸入映射到固定大小的特徵向量，這樣即使輸入影像的尺寸不同，也能為每個區域提取到固定的特徵，再透過 Softmax 函式進行分類。Faster R-CNN 不再使用選擇性搜索方式生成候選框，而是使用 RPN（Region Proposal Networks）生成候選框，將候選框的數量降低到了 300 個，極大地減少了過多候選框帶來的重複計算，同時共用了生成候選框的卷積網路和特徵提取網路，因此 Faster R-CNN 在檢測速度上有了進一步提升。

（2）One-Stage 演算法。One-Stage 演算法省略了候選區域，將整幅影像作為輸入，使用一個神經網路就可以直接輸出目標的位置和類別，將物件辨識問題轉為回歸問題。One-Stage 演算法的最大優勢就是其檢測速度，與 Two-Stage 演算法相比，One-Stage 演算法犧牲了一定的檢測精度。One-Stage 演算法直接利用卷積網路獲得物體的位置資訊和類別，不再需要用兩個網路進行分類與修正候選框。雖然 One-Stage 演算法的檢測精度沒有 Two-Stage 演算法高，但 One-Stage 演算法的檢測速度更快，更適合用於實際部署。One-Stage 演算法中比較有代表性的是 SSD 演算法和 YOLO。

SSD 演算法的網路結構較 R-CNN 更為簡單，僅透過卷積網路提取輸入影像的特徵資訊並進行特徵映射。SSD 演算法的網路結構如圖 3.18 所示，在網路的前半部分採用 VGG16 的卷積結構，並將其中兩個全連接層替換成卷積層；在網路的後半部分新增了 4 個不同尺度的卷積層，這 6 個不同尺度的卷積層作為輸出用於預測和分類。SSD 演算法使用了 6 個尺度的特徵圖實現了物件辨識，在淺層特徵中有豐富的位置資訊，有利於小目標的檢測。與僅使用深層特徵的 YOLOv1 相比，SSD 演算法的小物件辨識精度有一定優勢。SSD 演算法的邊界框（Bounding Box）有 8732 個，因此該演算法在不同場景下的檢測精度較為穩定，對於被遮擋的物體有更低的漏檢率。但是 SSD 演算法的網路結構較深，候選框的數量過多，使得 SSD 演算法在訓練速度和檢測速度上還有待提升。

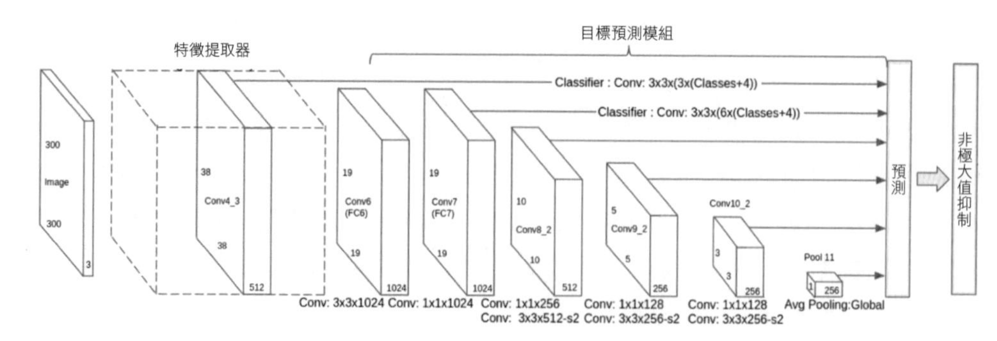

▲ 圖 3.18 SSD 演算法的網路結構

與 Two-Stage 演算法不同的是，YOLO 將物體分類與位置判定放在一起完成，檢測速度有了很大的提升，解決了 Two-Stage 演算法在檢測速度上無法滿足使用需求的問題。YOLO 基於 Darknet 框架不斷進行發展和改進，從 YOLOv1 發展到了 YOLOv5，在網路結構上不斷最佳化，不僅在檢測速度方面有較大改進，其中 YOLOv4 在 Tesla V100 上的檢測速度可達 65 FPS，還更適用於嵌入式裝置。

3.2.1.2 開發框架

1）開發框架概述

深度學習框架的選擇非常重要，選擇一個合適的框架能造成事半功倍的作用。目前，全世界最為流行的深度學習開發框架有 TensorFlow、PyTorch、Caffe、PaddlePaddle、Darknet 等。

（1）TensorFlow。TensorFlow 是 Google Brain 團隊基於 Google 在 2011 年開發的深度學習基礎框架 DistBelief 建構的。TensorFlow 使用資料流程圖進行數值計算，資料流程圖中的節點代表數學運算，邊代表在這些節點之間傳遞的多維陣列。TensorFlow 的程式設計介面支援 Python 和 C++ 語言。TensorFlow 1.0 版本開始支援 Java、Go、R 語言和 Haskell API 的 Alpha 版本，此外，TensorFlow 還可以在 Google Cloud 和 AWS 上執行。

（2）PyTorch。PyTorch 是一個 Python 優先的深度學習框架，能夠在強大的 GPU 加速基礎上實現張量和動態神經網路。PyTorch 提供了完整的使用文件、循序漸進的使用者指南，PyTorch 的開發者親自維護 PyTorch 討論區，方便使用者交流和解決問題。Facebook（現為 Meta）人工智慧研究院（FAIR）對 PyTorch 的推廣提供了大力支持，FAIR 的支援足以確保 PyTorch 獲得持續開發、更新的保障，而不會像一些個人開發的框架那樣曇花一現。如有需要，使用者也可以使用 Python 軟體套件（如 NumPy、SciPy 和 Cython）來擴充 PyTorch。

（3）Caffe。Caffe 是基於 C++ 撰寫的深度學習框架，是原始程式開放的（遵循 Licensed BSD），並提供了命令列工具，以及 MATLAB 和 Python 介面。Caffe 是深度學習研究者使用的框架，很多研究人員在上面進行了開發和最佳化。Caffe2 在工程上做了很多最佳化，如執行速度、跨平臺、可擴充性等，它可以看成 Caffe 更細粒度的重構；但在設計上，Caffe2 和 TensorFlow 更像。目前 Caffe2 的程式已開放原始碼。

（4）PaddlePaddle。PaddlePaddle（飛槳）以百度多年的深度學習技術研究和業務應用為基礎，集深度學習核心訓練和推理框架、基礎模型庫、點對點開發套件、豐富的工具元件於一體，為一功能完備開放原始碼的產業級深度學習平臺。

（5）Darknet。Darknet 是 Joseph Redmon 為了 YOLO 開發的框架，Darknet 幾乎沒有依賴函式庫，是基於 C 語言和 CUDA 的深度學習開放原始碼框架，支援 CPU 和 GPU。Darknet 跟 Caffe 頗有幾分相似之處，卻更加輕量級，非常值得學習使用。

2）Darknet 框架

Darknet 是一個用於實現深度學習演算法的開放原始碼神經網路框架，它是由 Joseph Redmon 開發的，主要用於物件辨識和影像辨識任務。Darknet 框架完全使用 C 語言撰寫，以其高效的實現和速度受到了廣泛關注，並在許多電腦視覺競賽中獲得了優異的結果。

Darknet 框架的特點包括：

（1）輕量級：Darknet 被設計成一個非常輕量級的框架，它的核心函式庫只有一個表頭檔案和一個原始檔案，非常易於使用和整合。

（2）高速度：Darknet 針對高效的計算做了最佳化，特別適合在嵌入式裝置上執行。它能夠在 CPU 和 GPU 上快速地進行計算，從而加速訓練和推理過程。

（3）支援多種演算法：Darknet 支援各種深度學習演算法，包括卷積神經網路（CNN）、全連接網路（FCN）和循環神經網路（RNN）等，可以用於影像分類、物件辨識和語義分割等多個電腦視覺任務。

（4）高度可自訂：Darknet 提供了靈活的配置選項，可以輕鬆調整網路框架、超參數和訓練設定。使用者可以根據自己的需求進行自訂最佳化和網路設計。

（5）支援多種資料型態：Darknet 支援處理不同類型的資料，包括影像、視訊和文字等。它提供了豐富的資料前置處理功能，可用於資料增強和資料清洗。

Darknet 是一個功能強大、高效的深度學習框架，適用於各種電腦視覺任務。它的速度和輕量級特點使其在資源受限的環境中表現良好，為研究人員和開發者提供了一個快速、靈活和可自訂的工具。

3.2.1.3 YOLO 系列模型

YOLO（You Only Look Once）系列模型是基於回歸的深度學習物件辨識演算法，即 One-Stage 演算法的典型代表，它不需要像 Two-Stage 演算法那樣先生成候選區域再透過 CNN 進行目標分類，因此 YOLO 系列模型的檢測速度快。

YOLO 系列模型的工作原理是：首先將影像均等地分成若干個網格，然後將每個網格的影像送入 CNN，預測每個網格是否包含目標並得到對應的邊界框和類別，最後對邊界框進行非最大值抑制處理，從而得到最終的邊界框。YOLO 系列模型的示意圖如圖 3.19 所示。

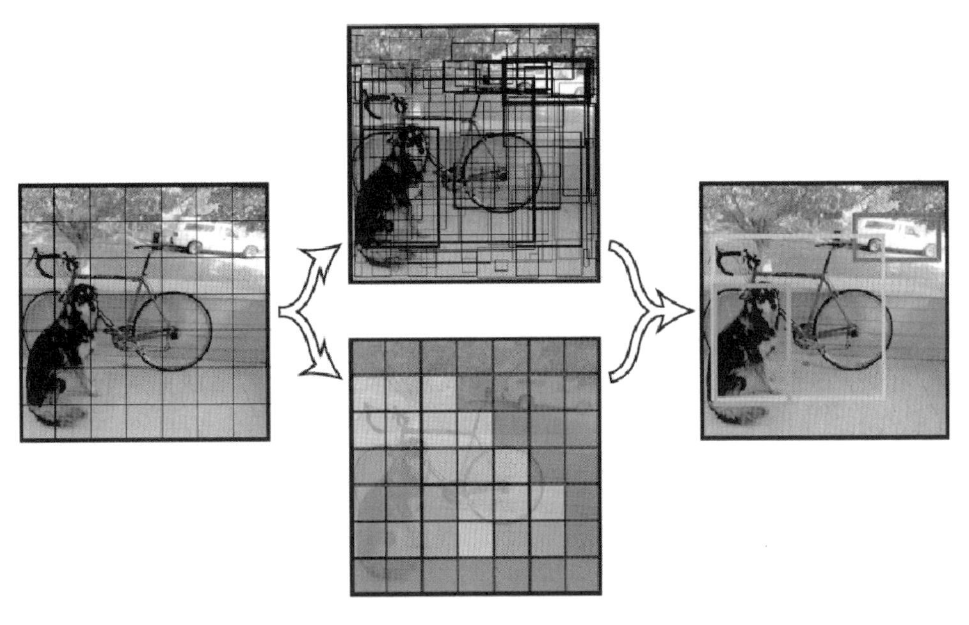

▲ 圖 3.19 YOLO 系列模型的示意圖

1）YOLOv1 模型

YOLOv1 模型的網路結構主要分為特徵提取網路、物件辨識網路、非極大值抑制（Non Maximum Suppression，NMS）處理，整體檢測過程一步實現。

YOLOv1 模型的網路結構如圖 3.20 所示。具體流程是先將輸入的影像調整成固定尺寸的影像，然後經過特徵提取網路的卷積層提取特徵，再經由物件辨識網路輸出 7×7×30 的特徵圖，並得到類別資訊和每個邊界框的置信度，最後透過 NMS 處理去除重疊程度高的邊界框，留下最合適的邊界框作為檢測結果。YOLOv1 模型的特徵提取網路最終將輸入影像分為 49 個網格，共 98 個邊界框，透過演算法去除置信度低於設定設定值的邊界框，利用 NMS 處理過濾掉重疊較多的邊界框，最終輸出邊界框的位置資訊和物體類別。

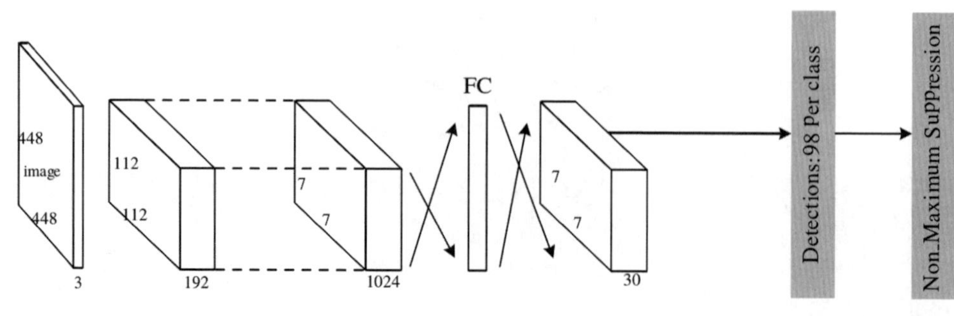

▲ 圖 3.20 YOLOv1 模型的網路結構

YOLOv1 模型的網路結構簡單、檢測速度較快，更適合部署於對即時性需求較高的場景，但由於 YOLOv1 模型的邊界框較少，當多個類別目標重疊到同一個網格時，會出現漏檢的情況；對於小目標以及目標較多的複雜場景，YOLOv1 模型容易出現定位不準的情況，因此比 R-CNN 演算法的檢測準確度低。

2）YOLOv2 模型

為了進一步提升定位的準確度和速度，同時保持分類的準確度，Joseph Redmon 和 Ali Farhadi 在 YOLOv1 的基礎上於 2017 年提出了 YOLOv2，即 YOLO 9000。YOLOv2 模型的主要流程是：首先在資料集上訓練 Darknet 網路，接著凍結網路結構的參數，移除最後一層的卷積層、全域池化層和分類層，並替換為 3 個 3×3 的卷積層、直通（Passthrough）層和 1×1 的卷積層。YOLOv2 模型的網路結構如圖 3.21 所示。

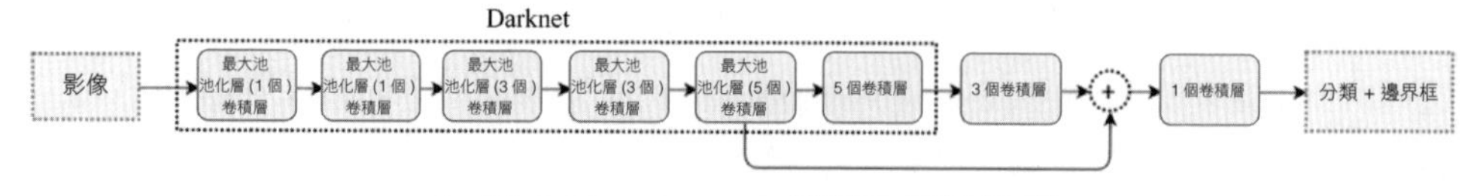

▲ 圖 3.21 YOLOv2 模型的網路結構

YOLOv2 模型使用由 22 個卷積層和 5 個最大池化層組成的特徵提取網路，可以比其他基於深度學習的檢測演算法獲得更高的檢測精度，在 PASCAL VOC2007 資料集中獲得了 76.8% 的 mAP 和 67FPS。YOLOv2 模型改進了 YOLOv1 模型定位精度不高和召回率低的問題，有效提升了檢測精度。

3）YOLOv3 模型

YOLOv3 模型是由 Joseph Redmon 和 Ali Farhadi 於 2018 年提出的，其網路結構如圖 3.22 所示。YOLOv3 模型使得物件辨識與辨識能力到達一個頂峰。相比於 Faster R-CNN 演算法，在相同條件下，YOLOv3 模型和 Faster R-CNN 演算法的檢測效果和檢測精度相差無幾，但 YOLOv3 模型的檢測速度是 Faster R-CNN 演算法的 100 倍，為即時物件辨識與辨識提供了基礎。YOLOv3 參考了 YOLOv1 模型和 YOLOv2 模型，在保持 YOLO 系列模型速度優勢的同時，提升了檢測精度，尤其對於小物體的檢測能力。YOLOv3 模型使用一個單獨神經網路處理影像，將影像劃分成多個區域並預測邊界框和每個區域的機率。

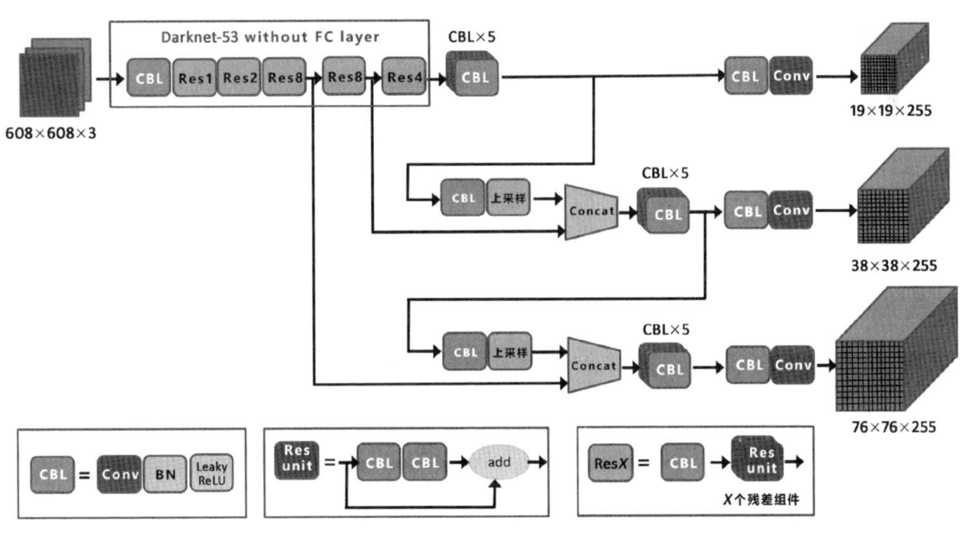

▲ 圖 3.22 YOLOv3 模型的網路結構

YOLOv3 在 Darknet-19 網路的基礎上提出了 Darknet-53 網路。Darknet-53 網路封包含 53 個卷積層，其基本組成單元 CBL 由 Conv、BN 和 Leaky ReLU 組成。YOLOv3 模型的網路結構沒有池化層，使用步幅為 2 的卷積層替代池化層進行特徵圖的上採樣，這樣可以有效阻止由於池化層導致的低層級特徵的損失，避免了池化層對梯度帶來的負面影響。為了提升檢測精度，YOLOv3 模型在深層特徵網路中引入了特徵金字塔（FPN），使用 3 個尺度的特徵層進行分類與回歸預測，對 2 個尺度的特徵層上採樣與淺層特徵進行融合。YOLOv3 模型的多尺度特徵融合使用 Concat（連接）操作，不同於殘差網路的 add（加）操作僅將特徵通道資訊相加，Concat 操作會進行張量拼接，擴充特徵圖的維度。YOLOv3 模型將 YOLOv2 模型的損失函式的計算方式改為交叉熵的損失計算方法，在類別預測和位置預測上獲得了更好的效果。YOLOv3 模型在檢測精度上與 Faster R-CNN 基本持平，但檢測速度卻是 Faster R-CNN 的 2 倍。

YOLOv3-Tiny 模型的網路結構如圖 3.23 所示。YOLOv3-Tiny 模型的網路結構在 YOLOv3 模型的基礎上壓縮了很多，沒有使用殘差層，只使用了 2 個不同尺度的輸出層（y1 和 y2）。YOLOv3-Tiny 模型的整體想法和 YOLOv3 模型是一樣的，被廣泛應用於行人、車輛檢測等，可以在很多硬體上實現。

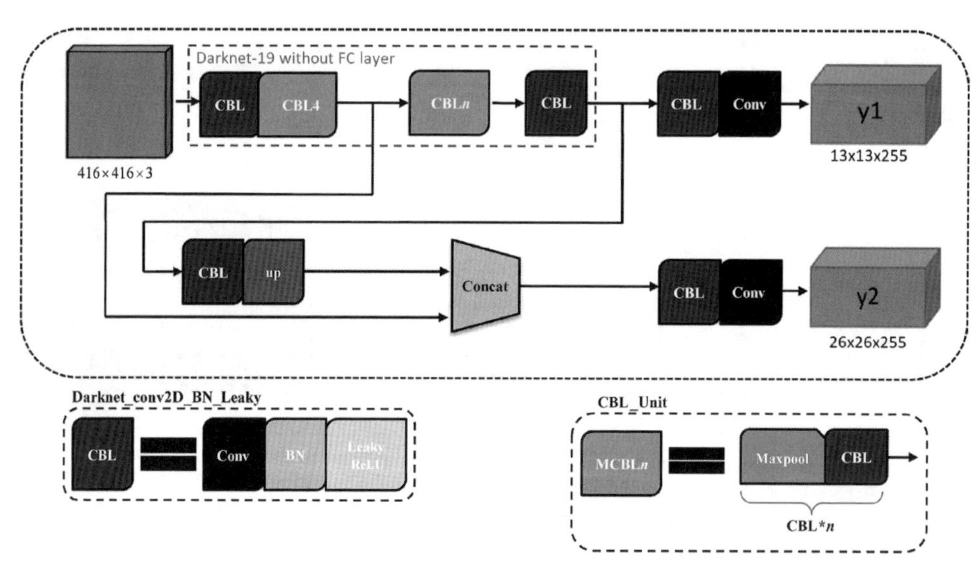

▲ 圖 3.23 YOLOv3-Tiny 模型的網路結構

4）YOLOv4 模型

YOLOv4 模型的網路結構如圖 3.24 所示。YOLOv4 模型僅需一個 GPU 就可以進行訓練和推理，引入了 CSPDarknet-53 網路〔在主幹網絡（BackBone）中〕、Leaky ReLU 啟動函式和 SPP（Spatial Pyramid Pooling）。YOLOv4 模型能夠在較深的網路結構中實現快速的訓練和推理，在檢測精度及速度上均有較大提升。SPP 使用 5×5、9×9、13×13 的最大池化方式，可以有效融合多尺度特徵，增大感受野。在 YOLOv4 模型的網路結構中，深層特徵提取網路使用了 FPN+PANet 結構，透過特徵層上採樣和下採樣，能夠提取到更加豐富的特徵資訊。

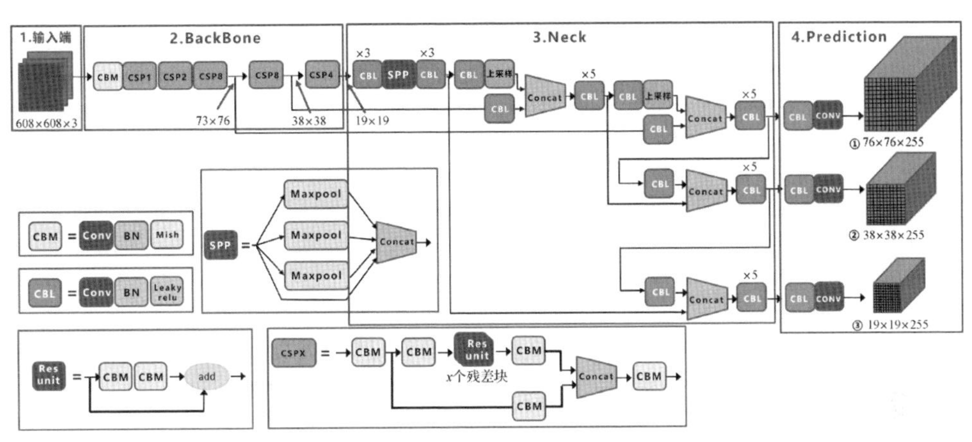

▲ 圖 3.24 YOLOv4 模型的網路結構

CSPDarknet-53 網路結構如圖 3.25 所示。

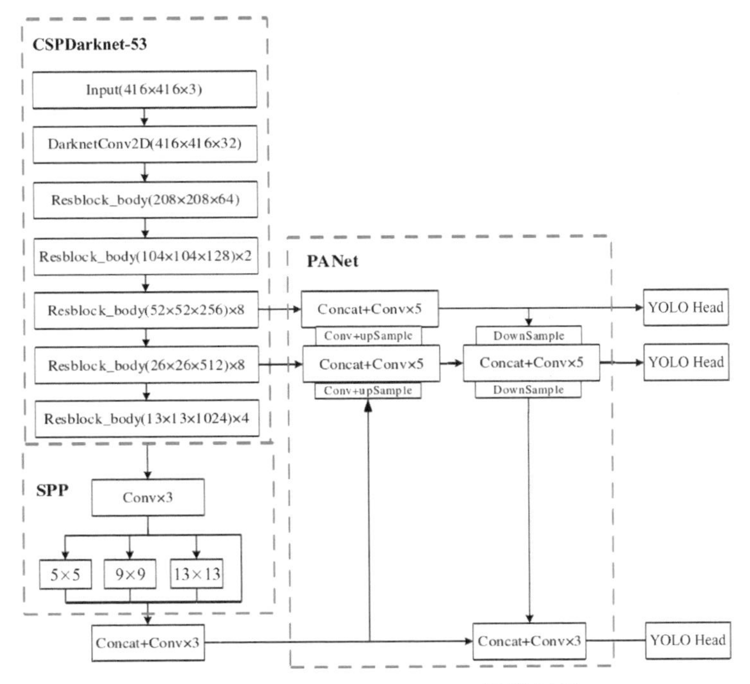

▲ 圖 3.25 CSPDarknet-53 網路結構

YOLOv4 模型使用 3 個尺度的特徵層進行分類與回歸預測，每一層的特徵資訊都與其他兩層進行了融合，有效提升了特徵提取能力。雖然 YOLOv4 模型僅靠單一 GPU 即可快速完成訓練和推理，但對於大多數算力有限的裝置而言，想要部署 YOLOv4 模型並實現即時物件辨識仍有很大的挑戰性。

3.2.1.4 YOLOv3 模型分析

YOLOv3 模型在物件辨識、物體辨識等應用中的表現非常出色。YOLOv3 模型透過不斷壓縮特徵圖的寬與高、擴張通道數的方式進行多級特徵提取，然後將提取到的特徵以上採樣的方式獲得不同大小的特徵層，並傳入特徵金字塔結構中，與上一層特徵進行堆疊，重複三次堆疊過程，得到三種不同尺度的特徵層，最終用於物件辨識。在得到特徵層後需要對目標進行預測，以確定預測框，最終實現檢測的目的。這種多尺度特徵提取的思想，使 YOLOv3 演算法的檢測效果獲得了明顯的提升，模型的整體適應性也因此獲得了提高。在 One-Stage 演算法發展過程中，YOLOv3 模型的提出，對比前幾代網路雖然檢測速度略有下降，但在檢測精度與準確度方面有了很大的提升。

YOLOv3 模型包含 107 層，0 ～ 74 層為卷積層和 Res 層，其目的是提取影像的上層特徵；75 ～ 106 層是三個 YOLO 分支（y1、y2、y3），使得模型具備檢測、分類和回歸的功能。由此可見，YOLOv3 模型相對來言還是比較複雜的。為了兼顧 YOLOv3 在物件辨識、物體辨識的速度，YOLOv3-Tiny 是在 YOLOv3 模型基礎上的簡化版，其目的兼顧準確率的同時適應訓練、推理速度要求比較高的業務場景。YOLOv3-Tiny 在 YOLOv3 的基礎上去掉了一些特徵層，只保留了 2 個獨立預測分支。

由於 Darknet-53 在執行時期存在計算參數多、成本高等缺點，因此在 YOLOv3 模型將殘差塊中的傳統卷積方式改進為深度可分離卷積，並加入注意力機制重新建構殘差單元塊，組成新的特徵提取網路。

原始的殘差塊採用的卷積核心大小為 3×3，本節首先交替使用 3×3 與 5×5 的卷積核心，透過交替使用不同大小的卷積核心，減少了最終建構出的總殘差網路的層數。本節其次將這兩種大小的卷積核心進行卷積的方法替換為深度可分離卷積，達到了減少模型整體參數量與模型執行時期的計算量的目的。伴隨著網路執行過程中計算量的減少，難免存在特徵提取不精細的問題，為了解決該問題，YOLOv3 模型將注意力機制加入殘差塊的深度可分離卷積後，其內部結構為一次全域平均池化，連接兩個 1×1 卷積核心，將通道數大小調整為一個相對較大、另一個相對較小，這樣調整的目的是對特徵進行壓縮與擴充，豐富整體網路的特徵表示內容。本節最後採用乘積運算來合併兩種卷積核心的結果，完成整體注意力機制結構的架設，並採用 1×1 的卷積核心對注意力機制所提取出的特徵進行卷積降維，並與最初輸入到殘差塊的原始特徵進行疊加，作為最終輸出結果；同時將 YOLOv3 中 BatchNorm 層後原本採用的 Leaky ReLU 啟動函式全部替換為 Swish 啟動函式，即：

$$f(x) = x \cdot \text{sigmoid}(\beta x) = \frac{x}{1 + \mathrm{e}^{-\beta x}} \tag{3-1}$$

式中，β 表示抑制參數變化的較小值，當 $\beta=0$ 時，$f(x)=0.5x$，Swish 函式就是一個一維的線性函式；當 β 趨於正無窮時，$f(x)=\max(0,x)$，Swish 函式就和 ReLU 函式相同。正是因為 β 參數的引入，Swish 函式整體可以看成一個平滑函式，且相較於其他損失函式來說，Swish 函式更適合神經網路對更深層次進行訓練。改進後的殘差塊的結構如圖 3.26 所示。

YOLOv3 模型的主幹特徵提取網路如圖 3.27 所示。主幹特徵提取網路對原始的特徵提取網路進行改進，堆疊了 16 層新殘差塊，建構了新的主幹特徵提取網路。與 YOLOv3 模型中的 23 層殘差網路相比較，新的主幹特徵提取網路減少了 7 層。為了與原始 YOLOv3 模型相對應，將新的主幹特徵提取網路中的 Block3 的輸出作為第一個有效特徵層，此時的特徵圖尺寸是原始輸入影像的 1/3。由於主幹特徵提取網路的深度決定了所提取特徵的有效程度，將 Block4 與 Block5 合併，並將 Block5 的輸出結果定義為第二個有效特徵層。同理合併 Block6 與 Block7，將 Block7 的輸出作為第三個有效特徵層。

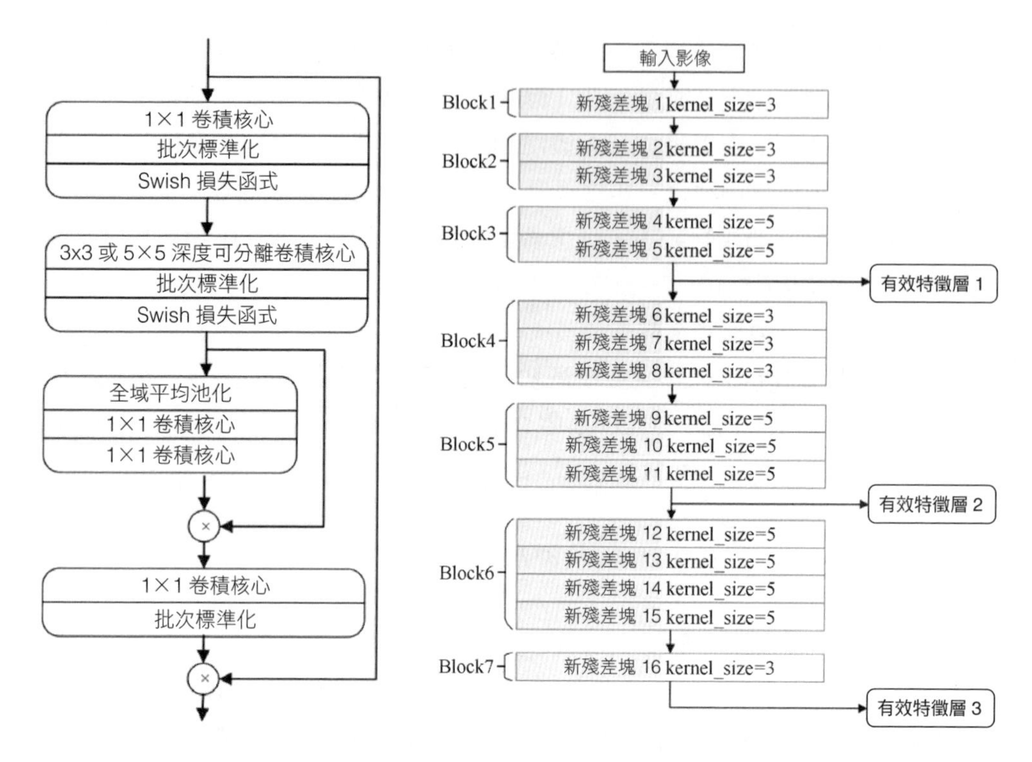

▲ 圖 3.26 改進後的殘差塊機構　▲ 圖 3.27 YOLOv3 模型的主幹特徵提取網路

基於上述的理論分析，對 YOLOv3 模型的主幹特徵提取網路進行強化處理，繼續運用原始的特徵金字塔結構對特徵進行多尺度劃分，透過兩次上採樣操作對不同深度層次提取到的不同尺度特徵進行融合，並進行最終的目標預測，再透過多個 1×1 卷積核心對最終結果進行降維，使網路可以準確預測被檢測影像中的多尺度目標。改進後的 YOLOv3 模型的網路結構如圖 3.28 所示。

YOLOv3 模型的損失函式將預測框定位的誤差、置信度誤差以及分類結果誤差進行統一整合，預測框定位的損失函式，採用的是均方誤差（Mean Square Error，MSE）函式計算，分類與置信度的損失函式，採用的是交叉熵函式。整體損失函式為：

$$\text{Loss=cooError+claError+conError} \tag{3-2}$$

式中，cooError 表示資料標註的座標框與測試後的檢測框座標之間的預測框定位損失；claError 表示分類損失；conError 表示置信度損失。

預測框定位損失為：

$$\begin{aligned}\text{cooError} = &\sum_{i=1}^{s^2}\sum_{j=1}^{B} I_{ij}^{\text{obj}}[(x_i^j - \hat{x}_i^j)^2 + (y_i^j - y_i^j)^2] + \\ &\sum_{i=1}^{s^2}\sum_{j=1}^{B} I_{ij}^{\text{obj}}[(\sqrt{w_i^j} - \sqrt{\hat{w}_i^j})^2 + (\sqrt{h_i^j} - \sqrt{\hat{h}_i^j})^2]\end{aligned} \tag{3-3}$$

分類損失為：

$$\text{claError} = -\sum_{i=1}^{s^2} I_{ij}^{\text{obj}} \cdot \sum_{c\in\text{classes}} [\hat{P}_i^j \log(P_i^j) + (1-P_i^j)\log(1-P_i^j)] \tag{3-4}$$

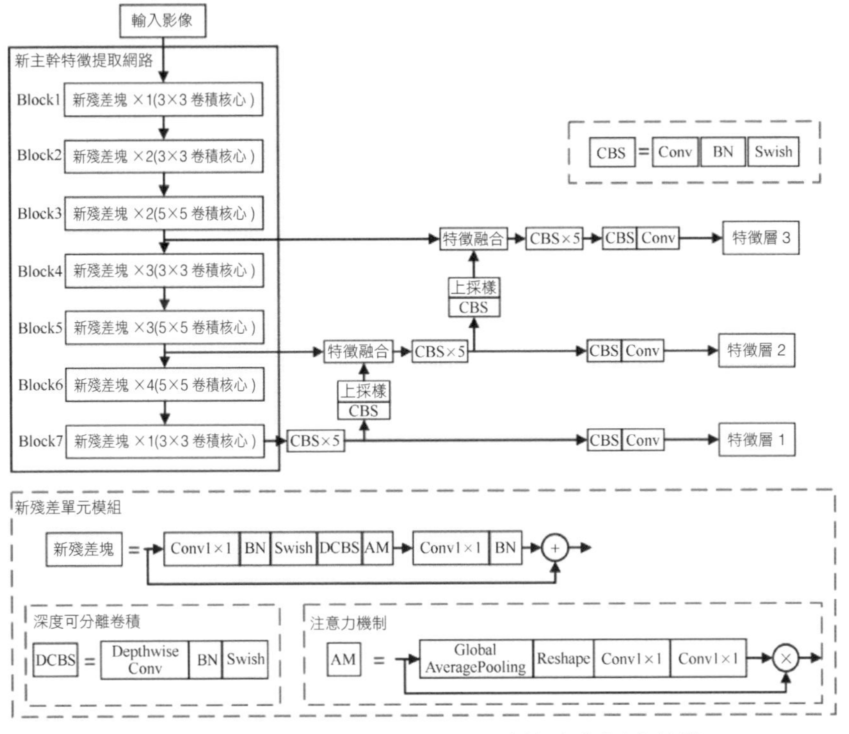

▲ 圖 3.28 改進後的 YOLOv3 演算法的網路結構

置信度損失為：

$$\text{conError} = -\sum_{i=1}^{s^2}\sum_{j=1}^{B} I_{ij}^{\text{obj}}[\hat{C}_i^j \log(C_i^j)\cdot(1-C_i^j)\log(1-C_i^j)] - \lambda_{\text{noobj}}\sum_{i=1}^{s^2}\sum_{j=1}^{B} I_{ij}^{\text{noobj}}[\hat{C}_i^j \log(C_i^j)\cdot(1-C_i^j)\log(1-C_i^j)] \quad (3\text{-}5)$$

式中，S^2 表示儲存格數量；B 表示錨框（Anchor Box）；I_{ij}^{obj} 表示第 i 個網格中的第 j 個候選視窗中是否存在需要被檢測出的目標，如果存在需要被檢測的目標，則 $I_{ij}^{\text{obj}}=1$，否則 $I_{ij}^{\text{obj}}=0$；I_{ij}^{noobj} 表示第 i 個網格的第 j 個檢測框中不存在需要被檢測的目標；x、y、w、h 依次為目標真實所在位置的中心點水平座標、垂直座標、檢測框的寬度和高度（以像素值衡量）；參數置信 $\hat{C}_{ij}^{\text{obj}}$ 表示真實值，由檢測框中有沒有被檢測的目標決定，若檢測框中有被檢測目標，則 $\hat{C}_{ij}^{\text{obj}}=1$，否則 $\hat{C}_{ij}^{\text{obj}}=0$。

YOLOv3 模型透過對預測框與真實框的 L1 範數或 L2 範數進行計算，獲得了位置回歸損失，但在測評時採用 IoU 損失來判斷預測框是否選中目標，二者並不是完全等價的；在 L2 範數或 L1 範數相等的情況下，IoU 的值並不會因為預測框與真實框距離相同就固定不變。在上一輪損失函式計算到的預測框與真實框間誤差數值大於 1 的情況下，MSE 損失在下一輪則會進行計算並傳回更大的誤差，以及更高的權重，整體的模型性能就會受到影響。當遇到預測框與真實框之間沒有任何重疊的情況時，IoU 的值為 0，傳回的損失也為 0，無法進行下一次最佳化。

為了避免 IoU 作為損失函式時所存在的缺點對損失計算造成大偏差的影響，YOLOv3 模型引入了 GIoU 的概念。IoU 與 GIoU 的損失計算原理如式（3-6）和式（3-7）所示。

$$R_{\text{GIoU}}=R_{\text{IoU}}-\frac{|C|(A\cup B)|}{|C|} \quad (3\text{-}6)$$

$$R_{\text{IoU}}=\frac{|A\cap B|}{|A\cup B|} \quad (3\text{-}7)$$

當訓練過程中資料集存在負樣本數量較多，與正樣本的比例不均衡的情況時，依舊會在很大的程度上影響訓練過程中傳回的權重。本專案的物件辨識網路結構中置信度損失以 Focal Loss 計算誤差，透過在傳回的損失中增加檢測到的各個分類的權重與各分類樣本檢測難度的權重排程因數，達到緩解樣本中不同分類數量以及檢測難度不平衡等問題。Focal Loss 的定義如公式（3-8）和式（3-9）所示。

$$L_{\mathrm{FL}(p_t)} = -\alpha_t(1-p_t)^{\gamma}\ln(p_t) \tag{3-8}$$

$$p_t = \begin{cases} p, & y=1 \\ 1-p, & \text{其他} \end{cases} \tag{3-9}$$

式中，α_t 為分類權重因數，用來協調正負樣本之間的比例；γ 為各分類樣本檢測難度權重排程因數，用來控制網路迭代樣本權重降低的速度。

改進後的 YOLOv3 模型在預測框損失中採用 GIoU Loss 函式進行預測，在置信度損失中採用 Focal Loss 函式進行預測，達到了更精準計算預測框、解決資料分類分佈不均衡帶來的問題的目的。

3.2.1.5 開發設計

1）資料格式的轉換

voc2yolo.py 檔案實現了資料格式的轉換，該檔案的功能是將影像標註工具生成的 voc 格式資料轉為 yolo 格式，程式如下：

```
import xml.etree.ElementTree as ET
import pickle
import os
from os import listdir, getcwd
from glob import glob
import random
from os.path import join

classes = ['red','green','left','straight','right']
```

```
dataset_path = "./dataset/"
txt_label_path = dataset_path + '/labels'
test_ratio = 0.1

if not os.path.exists(txt_label_path): os.mkdir(txt_label_path)

def convert(size, box):
    dw = 1. / size[0]
    dh = 1. / size[1]
    x = (box[0] + box[1]) / 2.0
    y = (box[2] + box[3]) / 2.0
    w = box[1] - box[0]
    h = box[3] - box[2]
    x = x * dw
    w = w * dw
    y = y * dh
    h = h * dh
    return (x, y, w, h)
    #return (int(x), int(y), int(w), int(h))

def convert_annotation(image_id):
    # 這裡改為 .xml 資料夾的路徑
    in_file = open(os.path.join(dataset_path, 'Annotations/%s.xml' % (image_id)))
    # 這裡是生成每幅影像對應的 .txt 檔案的路徑
    out_file = open(os.path.join(txt_label_path, '%s.txt' % (image_id)), 'w')
    tree = ET.parse(in_file)
    root = tree.getroot()
    size = root.find('size')
    w = int(size.find('width').text)
    h = int(size.find('height').text)  #

    for obj in root.iter('object'):
        cls = obj.find('name').text
        if cls not in classes:
            continue
        cls_id = classes.index(cls)
        xmlbox = obj.find('bndbox')
        b = (float(xmlbox.find('xmin').text), float(xmlbox.find('xmax').text),
```

```
float(xmlbox.find('ymin').text), float(xmlbox.find('ymax').text))
        bb = convert((w, h), b)
        #list_file.write(str(cls_id) + " " + " ".join([str(a) for a in bb]) + '\n')
        #list_file.write(" " + " ".join([str(a) for a in bb]) + " " + str(cls_id))
        out_file.write(str(cls_id) + " " + " ".join([str(a) for a in bb]) + '\n')

    #list_file.write('\n')

anno_files = glob(os.path.join(dataset_path, 'Annotations', '*.xml'))
anno_files = [item.split(os.sep)[-1].split('.')[0] for item in anno_files]
print("files:", anno_files[:10])
random.shuffle(anno_files)
test_num = int(len(anno_files) * test_ratio)
image_ids_val = anno_files[:test_num]
image_ids_train = anno_files[test_num:]
list_file_train = open('./dataset/object_train.txt', 'w')
list_file_val = open('./dataset/object_val.txt', 'w')
for image_id in image_ids_train:
    # 這裡改為樣本影像所在資料夾的路徑
    list_file_train.write(os.path.join(dataset_path, 'JPEGImages', '%s.jpg\n' % (image_
id)))
    convert_annotation(image_id)
list_file_train.close()
for image_id in image_ids_val:
    # 這裡改為樣本影像所在資料夾的路徑
    list_file_val.write(os.path.join(dataset_path, 'JPEGImages', '%s.jpg\n' % (image_
id)))
    convert_annotation(image_id)
list_file_val.close()
```

2）錨點（Anchor）座標的計算

train_anchors.py 檔案實現了錨點座標的計算，程式如下：

```
import xml.etree.ElementTree as ET
from glob import glob
import os

classes = ['red','green','left','straight','right']
```

```
dataset_path = './dataset'

def convert_annotation(_image_file, _list_file):
    xml_file = _image_file.split(os.sep)[-1].replace('.jpg', '.xml').
replace('JPEGImages', 'labels')
    xml_file = os.path.join(dataset_path, 'Annotations', xml_file)
    print(xml_file)
    #in_file = open(xml_file, 'r', encoding='UTF-8')
    tree = ET.parse(xml_file)
    root = tree.getroot()

    for obj in root.iter('object'):
        difficult = obj.find('difficult').text
        cls = obj.find('name').text
        if cls not in classes or int(difficult) == 1:
            continue
        cls_id = classes.index(cls)
        xmlbox = obj.find('bndbox')
        b = (int(xmlbox.find('xmin').text), int(xmlbox.find('ymin').text), int(xmlbox.
find('xmax').text), int(xmlbox.find('ymax').text))
        _list_file.write(" " + ",".join([str(a) for a in b]) + ',' + str(cls_id))

image_files = glob(os.path.join(dataset_path, 'JPEGImages', '*.jpg'))
list_file = open('train_anchors.txt', 'w')
for image_file in image_files:
    list_file.write(image_file)
    convert_annotation(image_file, list_file)
    list_file.write('\n')
list_file.close()
```

3）聚類演算法的實現

kmeans.py 檔案實現了聚類演算法，該檔案的功能和 train_anchors.py 檔案的功能分開是為了使用者採用不同的聚類演算法，程式如下：

```
import numpy as np

class YOLO_Kmeans:
```

```
def __init__(self, cluster_number, filename):
    self.cluster_number = cluster_number
    self.filename = filename

def iou(self, boxes, clusters):  #1 box -> k clusters
    n = boxes.shape[0]
    k = self.cluster_number

    box_area = boxes[:, 0] * boxes[:, 1]
    box_area = box_area.repeat(k)
    box_area = np.reshape(box_area, (n, k))

    cluster_area = clusters[:, 0] * clusters[:, 1]
    cluster_area = np.tile(cluster_area, [1, n])
    cluster_area = np.reshape(cluster_area, (n, k))

    box_w_matrix = np.reshape(boxes[:, 0].repeat(k), (n, k))
    cluster_w_matrix = np.reshape(np.tile(clusters[:, 0], (1, n)), (n, k))
    min_w_matrix = np.minimum(cluster_w_matrix, box_w_matrix)

    box_h_matrix = np.reshape(boxes[:, 1].repeat(k), (n, k))
    cluster_h_matrix = np.reshape(np.tile(clusters[:, 1], (1, n)), (n, k))
    min_h_matrix = np.minimum(cluster_h_matrix, box_h_matrix)
    inter_area = np.multiply(min_w_matrix, min_h_matrix)

    result = inter_area / (box_area + cluster_area - inter_area)
    return result

def avg_iou(self, boxes, clusters):
    accuracy = np.mean([np.max(self.iou(boxes, clusters), axis=1)])
    return accuracy

def kmeans(self, boxes, k, dist=np.median):
    box_number = boxes.shape[0]
    distances = np.empty((box_number, k))
    last_nearest = np.zeros((box_number,))
    np.random.seed()
    clusters = boxes[np.random.choice(
```

```
            box_number, k, replace=False)]  #init k clusters
        while True:

            distances = 1 - self.iou(boxes, clusters)

            current_nearest = np.argmin(distances, axis=1)
            if (last_nearest == current_nearest).all():
                break  #clusters won't change
            for cluster in range(k):
                clusters[cluster] = dist(  #update clusters
                    boxes[current_nearest == cluster], axis=0)

            last_nearest = current_nearest

        return clusters

    def result2txt(self, data):
        f = open("yolo_anchors.txt", 'w')
        row = np.shape(data)[0]
        for i in range(row):
            if i == 0:
                x_y = "%d,%d" % (data[i][0], data[i][1])
            else:
                x_y = ", %d,%d" % (data[i][0], data[i][1])
            f.write(x_y)
        f.close()

    def txt2boxes(self):
        f = open(self.filename, 'r')
        dataSet = []
        for line in f:
            infos = line.strip().split(" ")
            length = len(infos)
            for i in range(1, length):
                print(i,length,infos[0],len(infos[-1]))
                width = int(float(infos[i].split(",")[2])) - int(float(infos[i].
split(",")[0]))
                height = int(float(infos[i].split(",")[3])) - int(float(infos[i].
split(",")[1]))
```

```
                dataSet.append([width, height])
            result = np.array(dataSet)
            f.close()
            return result

    def txt2clusters(self):
        all_boxes = self.txt2boxes()
        result = self.kmeans(all_boxes, k=self.cluster_number)
        result = result[np.lexsort(result.T[0, None])]
        self.result2txt(result)
        print("K anchors:\n {}".format(result))
        print("Accuracy: {:.2f}%".format(
            self.avg_iou(all_boxes, result) * 100))

if __name__ == "__main__":
    cluster_number = 6
    filename = "train_anchors.txt"
    kmeans = YOLO_Kmeans(cluster_number, filename)
    kmeans.txt2clusters()
```

3.2.2 開發步驟與驗證

3.2.2.1 專案部署

1）硬體部署

詳見 2.1.2.1 節。

2）專案部署

（1）執行 MobaXterm 工具，透過 SSH 登入深度學習伺服器。

（2）在 SSH 終端執行以下命令，建立開發專案目錄。

```
$ mkdir -p ~/aiedge-exp
```

（3）將本專案的專案程式上傳到 ~/aicam-exp 目錄下，並採用 unzip 命令進行解壓縮。

```
$ cd ~/aiedge-exp
$ unzip object_detection_darknet.zip
```

3.2.2.2 資料處理

透過 voc2yolo 工具將 voc 格式的資料轉為 yolo 格式的資料，並分開訓練集和驗證集。

（1）將製作好的資料集 traffic_dataset_v1.3.zip 透過 SSH 上傳到 ~/aiedge-exp 目錄下。

說明：模型的精度依賴於資料集，本專案採用已標註好的交通號誌資料集進行模型訓練。

（2）解壓縮資料集，並將資料集複製到 darknet 專案目錄下：

```
$ cd ~/aiedge-exp
$ unzip traffic_dataset_v1.3.zip
$ cd object_detection_darknet/
$ cp -a ../traffic_dataset_v1.3/dataset ./
```

（3）根據交通號誌的類別修改 voc2yolo.py 檔案內的目標類別，透過 SSH 將修改後的檔案上傳到邊緣計算閘道。修改後的 voc2yolo.py 檔案內容如下：

```
classes = ['red','green','left','straight','right']
```

（4）在 SSH 終端輸入以下命令進行資料格式的轉換和資料集的切分，在 dataset 資料夾中生成 yolo 格式的資料集 labels，以及訓練集 object_train.txt 和驗證集 object_val.txt。

```
$ cd ~/aiedge-exp/object_detection_darknet
$ conda activate py36_tf114_torch15_cpu_cv345     //Ubuntu 20.04 作業系統下需要切換環境
$ python3 voc2yolo.py
files: ['1638757925381593', '16387579716174064', '1640220985345892',
'1640220986479758', '16387579476261072', '16387579697302487', '1640221005206378',
'16387579900259972', '1638757972922645', '159238212456671337387']
```

3.2.2.3 參數配置

建立交通號誌辨識模型的設定檔。

（1）在 dataset 資料夾下建立交通號誌資料集的標籤檔案 traffic.names，內容如下：

```
red
green
left
straight
right
```

（2）在 SSH 終端輸入以下命令複製 yolov3-tiny-pill.cfg/yolov3-tiny-pill.data 檔案，並命名為 yolov3-tiny-traffic.cfg/yolov3-tiny-traffic.data。

```
$ cd ~/aiedge-exp/object_detection_darknet
$ cp cfg/yolov3-tiny-pill.cfg cfg/yolov3-tiny-traffic.cfg
$ cp cfg/yolov3-tiny-pill.data cfg/yolov3-tiny-traffic.data
$ ls cfg/
yolov3-tiny-pill.cfg  yolov3-tiny-pill.data  yolov3-tiny-traffic.cfg  yolov3-tiny-traffic.
data
```

（3）根據交通號誌的類別可知 classes 為 5，修改 yolov3-tiny-traffic.data 檔案內相關參數：

```
classes= 5
train  = ./dataset/object_train.txt
valid  = ./dataset/object_val.txt
names = ./dataset/traffic.names
backup = ./backup/
```

（4）生成錨點。根據交通號誌的類別修改 train_anchors.py 檔案中與 classes 相關的參數，透過 SSH 將修改好的檔案上傳到邊緣計算閘道。修改後的 train_anchors.py 檔案如下：

```
classes = ['red','green','left','straight','right']
```

在 SSH 終端輸入以下命令執行 train_anchors.py，生成用於錨點聚類的資料檔案 train_ anchors.txt：

```
$ cd ~/aiedge-exp/object_detection_darknet
$ conda activate py36_tf114_torch15_cpu_cv345    //Ubuntu 20.04 作業系統下需要切換環境
$ python3 train_anchors.py
./dataset/Annotations/1638757980448458.xml
./dataset/Annotations/16387579456090543.xml
./dataset/Annotations/16387579667065635.xml
./dataset/Annotations/164022096542211.xml
./dataset/Annotations/16387579403672035.xml
……
```

在 SSH 終端輸入以下命令執行 kmeans.py，生成錨點的資料檔案 yolo_ anchors.txt：

```
$ cd ~/aiedge-exp/object_detection_darknet
$ conda activate py36_tf114_torch15_cpu_cv345    //Ubuntu 20.04 作業系統下需要切換環境
$ python3 kmeans.py
1 4 ./dataset/JPEGImages/1638757980448458.jpg 17
2 4 ./dataset/JPEGImages/1638757980448458.jpg 17
3 4 ./dataset/JPEGImages/1638757980448458.jpg 17
1 2 ./dataset/JPEGImages/16387579456090543.jpg 15
……
K anchors:
 [[ 43  51]
 [ 62  79]
 [ 86  95]
 [104 109]
 [125 132]
 [163 175]]
Accuracy: 84.59%
```

從 yolo_anchors.txt 檔案獲取的錨點為（43,51）、（62,79）、（86,95）、（104,109）、（125,132）、（163,175）。

（5）修改交通號誌辨識設定檔 cfg/yolov3-tiny-traffic.cfg。根據交通號誌的類別（classes=5）修改對應的卷積數〔filters=(classes+5)×3=30〕，搜索 cfg/yolov3-tiny-traffic.cfg 檔案內的 yolo 關鍵字（有兩處），修改 filters、anchors、classes。修改後的交通號誌辨識模型的設定檔如圖 3.29 所示。

```
[convolutional]
size=1
stride=1
pad=1
filters=30
activation=linear

[yolo]
mask = 3,4,5
anchors = 43,51, 62,79, 86,95, 104,109, 125,132, 163,175
classes=5
num=6
jitter=.3
ignore_thresh = .7
truth_thresh = 1
random=1
```

```
[convolutional]
size=1
stride=1
pad=1
filters=30
activation=linear

[yolo]
mask = 0,1,2
anchors = 43,51, 62,79, 86,95, 104,109, 125,132, 163,175
classes=5
num=6
jitter=.3
ignore_thresh = .7
truth_thresh = 1
random=1
```

▲ 圖 3.29 修改後的交通號誌辨識模型的設定檔（一）

由於左轉和右轉交通號誌是垂直鏡像的關係，所以設定檔內要增加選項 flip=0，如圖 3.30 所示。

```
[net]
# Testing
batch=16
subdivisions=1
# Training
# batch=64
# subdivisions=2
width=416
height=416
channels=3
momentum=0.9
decay=0.0005
angle=0
flip=0
saturation = 1.5
exposure = 1.5
hue=.1
```

▲ 圖 3.30 修改後的交通號誌辨識模型的設定檔（二）

至此，交通號誌辨識模型的參數檔案修改完畢。

3.2.2.4 模型訓練

透過 Darknet 框架可以完成交通號誌辨識模型的訓練，可選擇普通的電腦或整合了支援模型訓練 GPU 的電腦進行模型訓練。

（1）在 SSH 終端輸入以下命令進入專案目錄，改變 Darknet 的執行許可權。

```
$ cd ~/aiedge-exp/object_detection_darknet
$ chmod 755 darknet-*
```

（2）透過以下兩種方式進行模型訓練。

方式一：透過普通的電腦進行模型訓練（至少需要 300 h 以上）。

```
$ cd ~/aiedge-exp/object_detection_darknet
$ conda activate py36_tf114_torch15_cpu_cv345     //Ubuntu 20.04 作業系統下需要切換環境
$ ./darknet-cpu detector train cfg/yolov3-tiny-traffic.data cfg/yolov3-tiny-traffic.cfg 
yolov3-tiny.conv.15 15
  GPU isn't used
```

```
 OpenCV isn't used - data augmentation will be slow
yolov3-tiny-traffic
mini_batch = 16, batch = 16, time_steps = 1, train = 1
   layer      filters    size/strd(dil)     input                output
   0 conv       16       3 x 3/ 1     416 x 416 x   3 ->  416 x 416 x 16 0.150 BF
   1 max                 2x 2/ 2      416 x 416 x  16 ->  208 x 208 x 16 0.003 BF
   2 conv       32       3 x 3/ 1     208 x 208 x  16 ->  208 x 208 x 32 0.399 BF
   3 max                 2x 2/ 2      208 x 208 x  32 ->  104 x 104 x 32 0.001 BF
   4 conv       64       3 x 3/ 1     104 x 104 x  32 ->  104 x 104 x 64 0.399 BF
   5 max                 2x 2/ 2      104 x 104 x  64 ->  52 x 52 x 64 0.001 BF
   6 conv      128       3 x 3/ 1      52 x  52 x  64 ->  52 x 52 x 128 0.399 BF
   7 max                 2x 2/ 2       52 x  52 x 128 ->  26 x 26 x 128 0.000 BF
   8 conv      256       3 x 3/ 1      26 x  26 x 128 ->  26 x 26 x 256 0.399 BF
   9 max                 2x 2/ 2       26 x  26 x 256 ->  13 x 13 x 256 0.000 BF
  10 conv      512       3 x 3/ 1      13 x  13 x 256 ->  13 x 13 x 512 0.399 BF
  11 max                 2x 2/ 1       13 x  13 x 512 ->  13 x 13 x 512 0.000 BF
  12 conv     1024       3 x 3/ 1      13 x  13 x 512 ->  13 x 13 x1024 1.595 BF
  13 conv      256       1 x 1/ 1      13 x  13 x1024 ->  13 x 13 x 256 0.089 BF
  14 conv      512       3 x 3/ 1      13 x  13 x 256 ->  13 x 13 x 512 0.399 BF
  15 conv       30       1 x 1/ 1      13 x  13 x 512 ->  13 x 13 x 30 0.005 BF
  16 yolo
[yolo] params: iou loss: mse (2), iou_norm: 0.75, obj_norm: 1.00, cls_norm: 1.00,
delta_norm: 1.00, scale_x_y: 1.00
  17 route  13                                 ->     13 x 13 x 256
  18 conv      128       1 x 1/ 1      13 x 13 x 256 ->   13 x 13 x 128 0.011 BF
  19 upsample            2x            13 x 13 x 128 ->   26 x 26 x 128
  20 route  19 8                               ->     26 x 26 x 384
  21 conv      256       3 x 3/ 1      26 x 26 x 384 ->   26 x 26 x 256 1.196 BF
  22 conv       30       1 x 1/ 1      26 x 26 x 256 ->   26 x 26 x  30 0.010 BF
  23 yolo
[yolo] params: iou loss: mse (2), iou_norm: 0.75, obj_norm: 1.00, cls_norm: 1.00,
delta_norm: 1.00, scale_x_y: 1.00
Total BFLOPS 5.454
avg_outputs = 325691
Loading weights from yolov3-tiny.conv.15...
 seen 64, trained: 0 K-images (0 Kilo-batches_64)
Done! Loaded 15 layers from weights-file
Learning Rate: 0.001, Momentum: 0.9, Decay: 0.0005
   Detection layer: 16 - type = 28
```

```
    Detection layer: 23 - type = 28
Resizing, random_coef = 1.40

608 x 608
Create 64 permanent cpu-threads
Loaded: 0.069905 seconds
v3 (mse loss, Normalizer: (iou: 0.75, obj: 1.00, cls: 1.00) Region 16 Avg (IOU:
0.462607), count: 19, class_loss = 377.545441, iou_loss = 1.748474, total_loss =
379.293915
v3 (mse loss, Normalizer: (iou: 0.75, obj: 1.00, cls: 1.00) Region 23 Avg (IOU:
0.409510), count: 23, class_loss = 1121.026611, iou_loss = 3.438599, total_loss =
1124.465210, total_bbox = 42, rewritten_bbox = 0.000000 %

 1: 751.879578, 751.879578 avg loss, 0.000000 rate, 103.777236 seconds, 16 images,
-1.000000 hours left
Loaded: 0.000072 seconds
v3 (mse loss, Normalizer: (iou: 0.75, obj: 1.00, cls: 1.00) Region 16 Avg (IOU:
0.341734), count: 25, class_loss = 376.413269, iou_loss = 4.249542, total_loss =
380.662811
v3 (mse loss, Normalizer: (iou: 0.75, obj: 1.00, cls: 1.00) Region 23 Avg (IOU:
0.372197), count: 22, class_loss = 1121.051880, iou_loss = 4.764771, total_loss =
1125.816650
......
25009: 0.062177, 0.086965 avg loss, 0.000010 rate, 0.128478 seconds, 400144 images,
0.002451 hours left
Loaded: 0.000057 seconds
v3 (mse loss, Normalizer: (iou: 0.75, obj: 1.00, cls: 1.00) Region 16 Avg (IOU:
0.937889), count: 26, class_loss = 0.053363, iou_loss = 0.020515, total_loss = 0.073878
v3 (mse loss, Normalizer: (iou: 0.75, obj: 1.00, cls: 1.00) Region 23 Avg (IOU:
0.914718), count: 20, class_loss = 0.097458, iou_loss = 0.049408, total_loss = 0.146867
total_bbox = 1091785, rewritten_bbox = 0.001282 %

25010: 0.110372, 0.089306 avg loss, 0.000010 rate, 0.119459 seconds, 400160 images,
0.002427 hours left
Saving weights to ./backup//yolov3-tiny-traffic_final.weights
If you want to train from the beginning, then use flag in the end of training command:
-clear
```

方式二：透過整合了支援模型訓練 GPU 的電腦進行模型訓練（如顯示卡 RTX3080，需要 30 ～ 60 min）。

```
$ cd ~/aiedge-exp/object_detection_darknet
$ conda activate py36_tf114_torch15_cpu_cv345    //Ubuntu 20.04 作業系統下需要切換環境
$ ./darknet-gpu detector train cfg/yolov3-tiny-traffic.data cfg/yolov3-tiny-traffic.cfg
yolov3-tiny.conv.15 15
    CUDA-version: 11030 (11040), cuDNN: 8.2.1, GPU count: 1
    OpenCV isn't used - data augmentation will be slow
yolov3-tiny-traffic
    0 : compute_capability = 860, cudnn_half = 0, GPU: NVIDIA GeForce RTX 3080
net.optimized_memory = 0
mini_batch = 16, batch = 16, time_steps = 1, train = 1
    layer   filters  size/strd(dil)      input                output
    0 Create CUDA-stream - 0
    Create cudnn-handle 0

    conv         16        3 x 3/ 1         416 x 416 x 3 ->     416 x 416 x 16 0.150 BF
    1 max                  2x 2/ 2          416 x 416 x 16 ->    208 x 208 x 16 0.003 BF
    2 conv       32        3 x 3/ 1         208 x 208 x 16 ->    208 x 208 x 32 0.399 BF
    3 max                  2x 2/ 2          208 x 208 x 32 ->    104 x 104 x 32 0.001 BF
    4 conv       64        3 x 3/ 1         104 x 104 x 32 ->    104 x 104 x 64 0.399 BF
    5 max                  2x 2/ 2          104 x 104 x 64 ->    52 x  52 x 64 0.001 BF
    6 conv       128       3 x 3/ 1         52 x 52 x 64 ->      52 x 52 x 128 0.399 BF
    7 max                  2x 2/ 2          52 x 52 x 128 ->     26 x 26 x 128 0.000 BF
    8 conv       256       3 x 3/ 1         26 x 26 x 128 ->     26 x 26 x 256 0.399 BF
    9 max                  2x 2/ 2          26 x 26 x 256 ->     13 x 13 x 256 0.000 BF
    10 conv      512       3 x 3/ 1         13 x 13 x 256 ->     13 x 13 x 512 0.399 BF
    11 max                 2x 2/ 1          13 x 13 x 512 ->     13 x 13 x 512 0.000 BF
    12 conv      1024      3 x 3/ 1         13 x 13 x 512 ->     13 x 13 x1024 1.595 BF
    13 conv      256       1 x 1/ 1         13 x 13 x1024 ->     13 x 13 x 256 0.089 BF
    14 conv      512       3 x 3/ 1         13 x 13 x 256 ->     13 x 13 x 512 0.399 BF
    15 conv      30        1 x 1/ 1         13 x 13 x 512 ->     13 x 13 x  30 0.005 BF
    16 yolo
[yolo] params: iou loss: mse (2), iou_norm: 0.75, obj_norm: 1.00, cls_norm: 1.00,
delta_norm:
                  1.00, scale_x_y: 1.00
    17 route     13                                   ->       13 x 13 x 256
    18 conv      128       1 x 1/ 1         13 x 13 x 256 ->     13 x 13 x 128 0.011 BF
```

```
    19 upsample          2x      13 x 13 x 128 ->    26 x 26 x 128
    20 route     19 8                           ->        26 x 26 x 384
    21 conv      256      3 x 3/ 1          26 x 26 x 384 ->    26 x 26 x 256 1.196 BF
    22 conv      30       1 x 1/ 1          26 x 26 x 256 ->    26 x 26 x  30 0.010 BF
    23 yolo
[yolo] params: iou loss: mse (2), iou_norm: 0.75, obj_norm: 1.00, cls_norm: 1.00,
delta_norm:
                   1.00, scale_x_y: 1.00
Total BFLOPS 5.454
avg_outputs = 325691
Allocate additional workspace_size = 132.12 MB
Loading weights from yolov3-tiny.conv.15...
seen 64, trained: 0 K-images (0 Kilo-batches_64)
Done! Loaded 15 layers from weights-file
Learning Rate: 0.001, Momentum: 0.9, Decay: 0.0005
    Detection layer: 16 - type = 28
    Detection layer: 23 - type = 28
Resizing, random_coef = 1.40

    608 x 608
    Create 64 permanent cpu-threads
    try to allocate additional workspace_size = 163.97 MB
    CUDA allocate done!
Loaded: 0.000046 seconds
v3 (mse loss, Normalizer: (iou: 0.75, obj: 1.00, cls: 1.00) Region 16 Avg (IOU:
0.364088), count: 18, class_loss = 274.552612, iou_loss = 3.491913, total_loss =
278.044525
v3 (mse loss, Normalizer: (iou: 0.75, obj: 1.00, cls: 1.00) Region 23 Avg (IOU:
0.252610), count: 23, class_loss = 1262.276123, iou_loss = 6.435059, total_loss =
1268.711182
total_bbox = 41, rewritten_bbox = 0.000000 %

1: 773.377869, 773.377869 avg loss, 0.000000 rate, 0.265851 seconds, 16 images,
-1.000000 hours left
Loaded: 0.000054 seconds
v3 (mse loss, Normalizer: (iou: 0.75, obj: 1.00, cls: 1.00) Region 16 Avg (IOU:
0.318203), count: 22, class_loss = 277.137756, iou_loss = 4.733521, total_loss =
281.871277
```

```
v3 (mse loss, Normalizer: (iou: 0.75, obj: 1.00, cls: 1.00) Region 23 Avg (IOU:
0.347696), count: 25, class_loss = 1261.224731, iou_loss = 6.360596, total_loss
1267.585327
total_bbox = 88, rewritten_bbox = 0.000000 %

2: 774.728271, 773.512939 avg loss, 0.000000 rate, 0.117972 seconds, 32 images,
1.847303 hours left
Loaded: 0.000066 seconds
v3 (mse loss, Normalizer: (iou: 0.75, obj: 1.00, cls: 1.00) Region 16 Avg (IOU:
0.325834), count: 18, class_loss = 275.294678, iou_loss = 4.547729, total_loss =
279.842407
v3 (mse loss, Normalizer: (iou: 0.75, obj: 1.00, cls: 1.00) Region 23 Avg (IOU:
0.346565), count: 23, class_loss = 1260.499756, iou_loss = 4.489014, total_loss =
1264.988770
 total_bbox = 129, rewritten_bbox = 0.000000 %
......
25009: 0.062177, 0.086965 avg loss, 0.000010 rate, 0.128478 seconds, 400144 images,
0.002451 hours left
Loaded: 0.000057 seconds
v3 (mse loss, Normalizer: (iou: 0.75, obj: 1.00, cls: 1.00) Region 16 Avg (IOU:
0.937889), count: 26, class_loss = 0.053363, iou_loss = 0.020515, total_loss = 0.073878
v3 (mse loss, Normalizer: (iou: 0.75, obj: 1.00, cls: 1.00) Region 23 Avg (IOU:
0.914718), count: 20, class_loss = 0.097458, iou_loss = 0.049408, total_loss = 0.146867
total_bbox = 1091785, rewritten_bbox = 0.001282 %

25010: 0.110372, 0.089306 avg loss, 0.000010 rate, 0.119459 seconds, 400160 images,
0.002427 hours left
Saving weights to ./backup//yolov3-tiny-traffic_final.weights
If you want to train from the beginning, then use flag in the end of training command:
-clear 3)
```

訓練完成後，可以在 backup 目錄下看到訓練完成的模型檔案 yolov3-tiny-traffic_final.weights，如圖 3.31 所示。

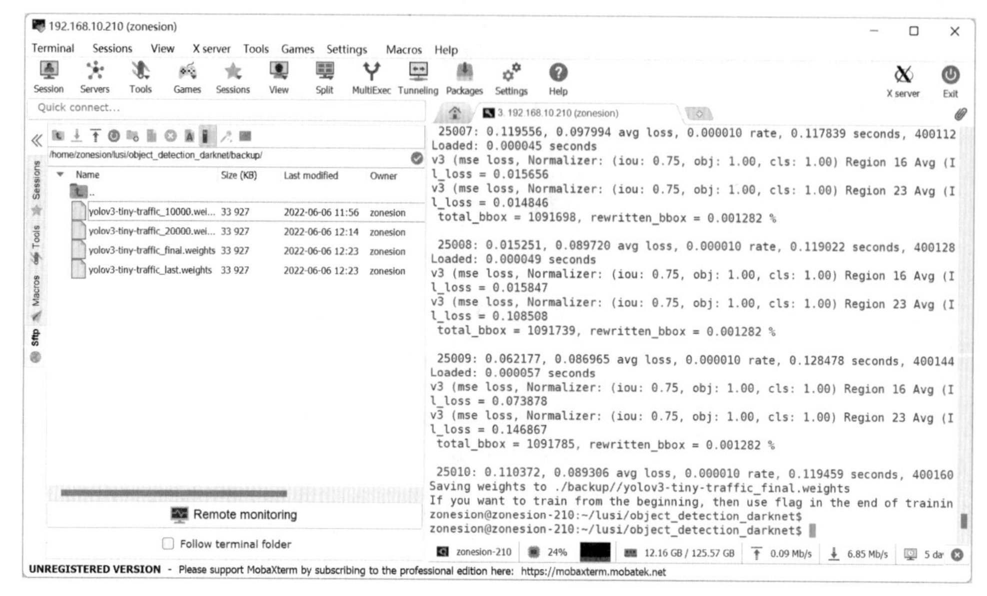

▲ 圖 3.31 訓練完成的模型檔案

3.2.2.5 模型驗證

透過前面的步驟可完成交通號誌辨識模型的訓練，得到 yolov3-tiny-traffic_final.weights。如果沒有條件訓練出最終的模型，也可以解壓本專案提供的最終模型檔案，將其複製到 backup 目錄下。

（1）在資料集中選擇一幅測試樣圖，這裡選擇的是 dataset/JPEGImages/10050.jpg。

（2）在 SSH 終端輸入以下命令對測試樣圖進行測試，執行命令與結果如下，成功地將目標（交通號誌）辨識為「right」。

```
$ cd ~/aiedge-exp/object_detection_darknet
$ conda activate py36_tf114_torch15_cpu_cv345     //Ubuntu 20.04 作業系統下需要切換環境
$ ./darknet-cpu detector test cfg/yolov3-tiny-traffic.data cfg/yolov3-tiny-traffic.cfg
                  backup/yolov3-tiny-traffic_final.weights dataset/JPEGImages/10050.jpg
GPU isn't used
OpenCV isn't used - data augmentation will be slow
mini_batch = 1, batch = 1, time_steps = 1, train = 0
```

```
   layer     filters   size/strd(dil)    input              output
   0 conv    16        3 x 3/ 1          416 x 416 x 3 ->   416 x 416 x 16 0.150 BF
   1 max               2x 2/ 2           416 x 416 x 16 ->  208 x 208 x 16 0.003 BF
   2 conv    32        3 x 3/ 1          208 x 208 x 16 ->  208 x 208 x 32 0.399 BF
   3 max               2x 2/ 2           208 x 208 x 32 ->  104 x 104 x 32 0.001 BF
   4 conv    64        3 x 3/ 1          104 x 104 x 32 ->  104 x 104 x 64 0.399 BF
   5 max               2x 2/ 2           104 x 104 x 64 ->  52 x 52 x  64 0.001 BF
   6 conv    128       3 x 3/ 1          52 x 52 x 64 ->    52 x 52 x 128 0.399 BF
   7 max               2x 2/ 2           52 x 52 x 128 ->   26 x 26 x 128 0.000 BF
   8 conv    256       3 x 3/ 1          26 x 26 x 128 ->   26 x 26 x 256 0.399 BF
   9 max               2x 2/ 2           26 x 26 x 256 ->   13 x 13 x 256 0.000 BF
   10 conv   512       3 x 3/ 1          13 x 13 x 256 ->   13 x 13 x 512 0.399 BF
   11 max              2x 2/ 1           13 x 13 x 512 ->   13 x 13 x 512 0.000 BF
   12 conv   1024      3 x 3/ 1          13 x 13 x 512 ->   13 x 13 x1024 1.595 BF
   13 conv   256       1 x 1/ 1          13 x 13 x1024 ->   13 x 13 x 256 0.089 BF
   14 conv   512       3 x 3/ 1          13 x 13 x 256 ->   13 x 13 x 512 0.399 BF
   15 conv   30        1 x 1/ 1          13 x 13 x 512 ->   13 x 13 x 30 0.005 BF
   16 yolo
[yolo] params: iou loss: mse (2), iou_norm: 0.75, obj_norm: 1.00, cls_norm: 1.00,
delta_norm: 1.00, scale_x_y: 1.00
   17 route  13                                     ->    13 x 13 x 256
   18 conv   128       1 x 1/ 1          13 x 13 x 256 ->   13 x 13 x 128 0.011 BF
   19 upsample         2x                13 x 13 x 128 ->   26 x 26 x 128
   20 route  198                                    ->    26 x 26 x 384
   21 conv   256       3 x 3/ 1          26 x 26 x 384 ->   26 x 26 x 256 1.196 BF
   22 conv   30        1 x 1/ 1          26 x 26 x 256 ->   26 x 26 x 30 0.010 BF
   23 yolo
[yolo] params: iou loss: mse (2), iou_norm: 0.75, obj_norm: 1.00, cls_norm: 1.00,
delta_norm: 1.00, scale_x_y: 1.00
Total BFLOPS 5.454
avg_outputs = 325691
Loading weights from backup/yolov3-tiny-traffic_final.weights...
   seen 64, trained: 400 K-images (6 Kilo-batches_64)
Done! Loaded 24 layers from weights-file
   Detection layer: 16 - type = 28
   Detection layer: 23 - type = 28
dataset/JPEGImages/10050.jpg: Predicted in 826.494000 milli-seconds.
right: 100%
Not compiled with OpenCV, saving to predictions.png instead
```

執行完成後會在專案目錄生成 predictions.jpg 檔案，如圖 3.32 所示。

▲ 圖 3.32 生成 predictions.jpg 檔案

3.2.3 本節小結

本節首先介紹了 R-CNN 演算法，然後介紹了兩種基於深度學習的物件辨識演算法類型，以及深度學習主流開發框架，接著詳細地對 YOLO 系列模型進行了剖析，最後結合案例介紹了 YOLOv3 模型及 Darknet 框架的專案部署、開發步驟、模型訓練及驗證。

3.2.4 思考與擴充

（1）在物件辨識過程中，YOLO 系列模型與 Faster R-CNN 演算法的想法有何不同？各自的優劣勢表現在哪些方面？

（2）為什麼 Darknet 框架在電腦視覺領域應用很廣泛？簡述 YOLOv3 模型使用的 Darknet-53 網路特點。

3.3 YOLOv5 模型的訓練與驗證

本節的基礎知識如下：

- 掌握模型訓練與驗證作用。
- 掌握 YOLOv5 模型的原理。
- 結合口罩檢測案例，掌握基於 PyTorch 框架和 YOLOv5 模型的專案部署、模型訓練與模型驗證的步驟。

3.3.1 原理分析與開發設計

3.3.1.1 PyTorch 框架

PyTorch 是一個開放原始碼的深度學習框架，該框架由 Facebook 人工智慧研究院（FAIR）的 Torch7 團隊開發，它的底層基於 Torch，但實現與運用全部是使用 Python 完成的。PyTorch 主要用於人工智慧領域的科學研究與應用程式開發。

PyTorch 最主要的功能有兩個：一是擁有 GPU 張量，該張量可以透過 GPU 加速，可滿足在短時間內處理巨量資料的要求；二是支援動態神經網路，可逐層對神經網路進行修改，並且神經網路具備自動求導的功能。

PyTorch 框架的後續版本增加了很多的新特性，如無縫行動裝置部署、量化模型（用於加速推理）、前端改進（如對張量進行命名和建立更乾淨的程式）等，PyTorch 官方同時還開放原始碼了很多工具和函式庫，使得 PyTorch 框架的眾多功能與 TensorFlow 趨同，同時保持了原有特性，競爭力得到極大的增強

3.3.1.2 YOLOv5 模型

YOLOv5 模型是與 YOLOv4 模型幾乎同時提出的，兩者極為相似，YOLOv5 模型有 YOLOv4 的所有優點和特性，並擁有其不具備的輕量性優勢。YOLOv5 透過對模型容量、規模大小進行縮放，提出了 s、m、l、x 四個版本，用於追求極致輕量化或高精度，並結合演算法思想不斷迭代更新這四個版本。

YOLOv5 模型的網路結構如圖 3.33 所示，主要包括輸入、BackBone（主幹網絡）、特徵融合和輸出等模組。

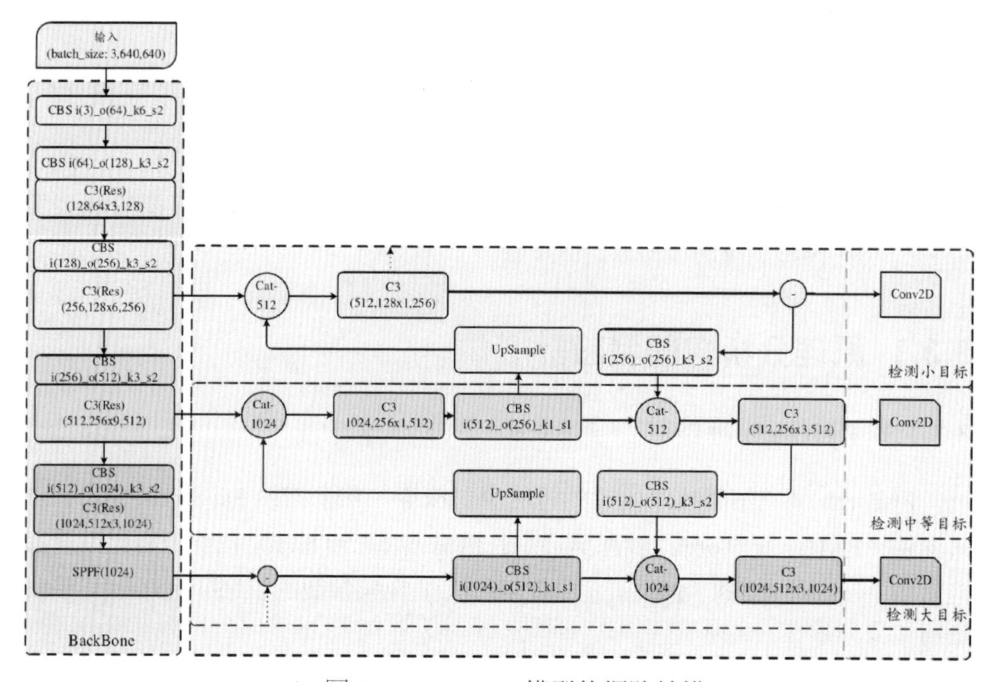

▲ 圖 3.33 YOLOv5 模型的網路結構

1）輸入模組

輸入模組包含一系列影像增強方法，飽和度、色調、亮度的隨機變化增強，用於解決影像失真的種種問題，以適應晝夜及各種惡劣天氣；對影像進行隨機旋轉、平移、翻轉、剪貼、透視等操作，用於解決影像採樣帶來的問題；資料增強 Mosaic 和 Mixup 可對影像組進行組合、拼接，增強影像樣本的背景資訊種類，提高樣本檢出率與精度；資料增強 Copy-Paste 透過對目標實例進行跨影像遷移，可均衡化各類別實例數量。Mosaic 是指將 4 幅不同的影像拼接在一起形成一幅新影像；Mixup 是指將 2 幅不同的影像按照一定的比例進行混合，生成一幅新影像；Copy-Paste 是指從來源影像中剪貼物件塊並貼上到目標影像，從而獲得組合數量的合成訓練資料，顯著提高檢測 / 分割性能。

2）BackBone（主幹網絡）

主幹網絡頭部多次利用下採樣卷積和 C3 模組進行特徵提取，下採樣卷積是步進值為 2 的卷積，C3 模組為 CSP 化的殘差模組，並支援無殘差模式，網路深度縮放功能只針對 C3 模組內部，可提高特徵提取效率、解決深度神經網路的梯度消失問題。CSP 化可以在不降低殘差模組特徵提取的效率的同時加速 40%。主幹網絡尾部增加了 SPPF 池化模組，以特徵共用形式降低了 SPP 模組的計算量。主幹網絡使用輕量化的 SiLU（Sigmoid Linear Unit）啟動函式，計算量低於 YOLOv4 模型中的 Mish 啟動函式，同時精度優於 Leaky ReLU、Mish 等啟動函式。C3 模組和 SPPF 池化模組如圖 3.34 所示。

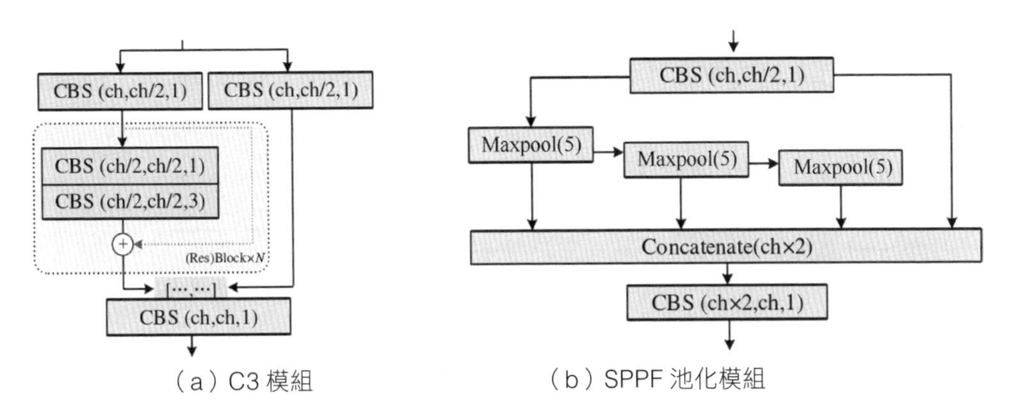

（a）C3 模組　　（b）SPPF 池化模組

▲ 圖 3.34 C3 模組和 SPPF 池化模組

在 CSPNet 和網路縮放的基礎上，YOLOv5 模型改進了主幹網絡，提供了主幹網絡深度、寬度縮放功能，如圖 3.35 所示。YOLOv5 模型的四個版本（s、m、l、x）的深度和寬度縮放比例分別為（0.33, 0.5）、（0.66, 0.75）、（1.0, 1.0）、（1.33, 1.25），可兼顧各類資料集和硬體裝置。

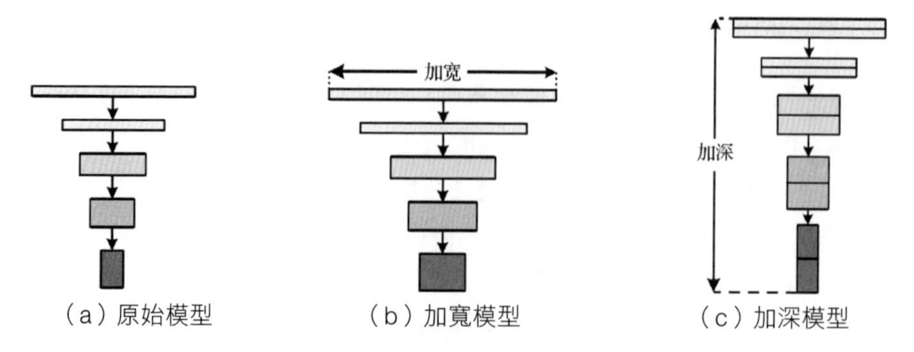

▲ 圖 3.35 主幹網絡的深度和寬度縮放

3）特徵融合

從 YOLOv3 模型開始，YOLO 系列模型採用特徵金字塔（FPN）、PANet 特徵融合的方式提高中小目標的定位和分類精度。YOLOv5 模型則採用輕量型的 PANet 特徵融合，如圖 3.36 所示，圖中，〔…，…〕表示 Concat。

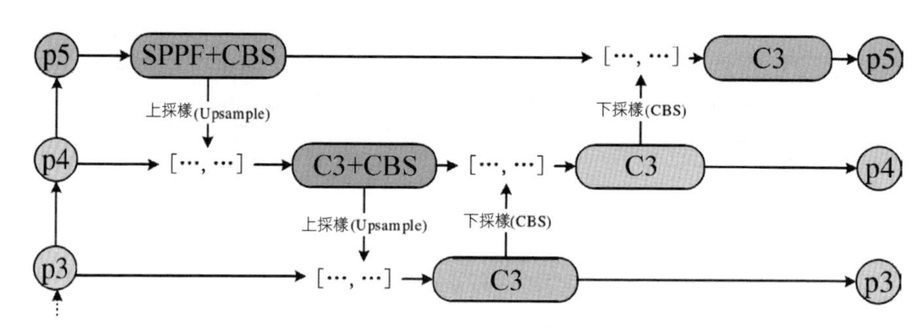

▲ 圖 3.36 輕量型 PANet 特徵融合

FPN 採用由深至淺的單向迭代式融合，具有多尺度串聯和特徵混合的作用。FPN 首先對主幹網絡的多尺度特徵進行降維，一方面可以降低特徵融合整體的計算量，造成加速的作用，另一方面可以對主幹網絡多尺度輸出特徵進行折半處理，造成特徵選擇、降低維度的效果，可同時平衡本尺度檢測任務和其他尺度檢測任務，緩解不同尺度任務的回傳梯度在主幹網絡上的矛盾。然後對特徵組進行迭代式串聯混合，有助本尺度特徵獲取更深層次的語義資訊，提高本尺度的物件偵測精度。PANet 採用由淺至深的單向迭代式融合，起著資訊流通的作用，可提高本尺度的目標定位精度和細粒度辨識精度。

4）解碼輸出及最佳化損失

YOLOv5 模型的輸出和 YOLOv4 模型一樣，但 YOLOv5 模型採用的是 Focal Loss 損失函式，該損失函式有利於均衡正負樣本、解決影像中目標稀少的問題。

與 YOLOv4 模型對比，YOLOv5 模型的 BackBone 沒有太大的變化，這對於現有的一些 GPU 裝置及其相應的最佳化演算法更加高效。在 Neck 部分，YOLOv5 模型首先將 SPP 換成了 SPPF，後者效率更高；其次另外一個不同點就是 CSP-PAN，在 YOLOv4 模型中，Neck 的 PAN 沒有引入 CSP，但在 YOLOv5 模型中，PAN 中加入了 CSP。CSP 將原輸入分成兩個分支，分別進行卷積操作使得通道數減半，一個分支先進行 Bottleneck × *N* 操作，再透過連接兩個分支，使得 Bottlenneck CSP 的輸入與輸出的大小是一樣的，這樣可以讓模型學習到更多的特徵。

YOLOv5s 模型是 YOLOv5 模型中深度最小、特徵圖的寬度最小的模型。YOLOv5s 模型的網路結構如圖 3.37 所示。

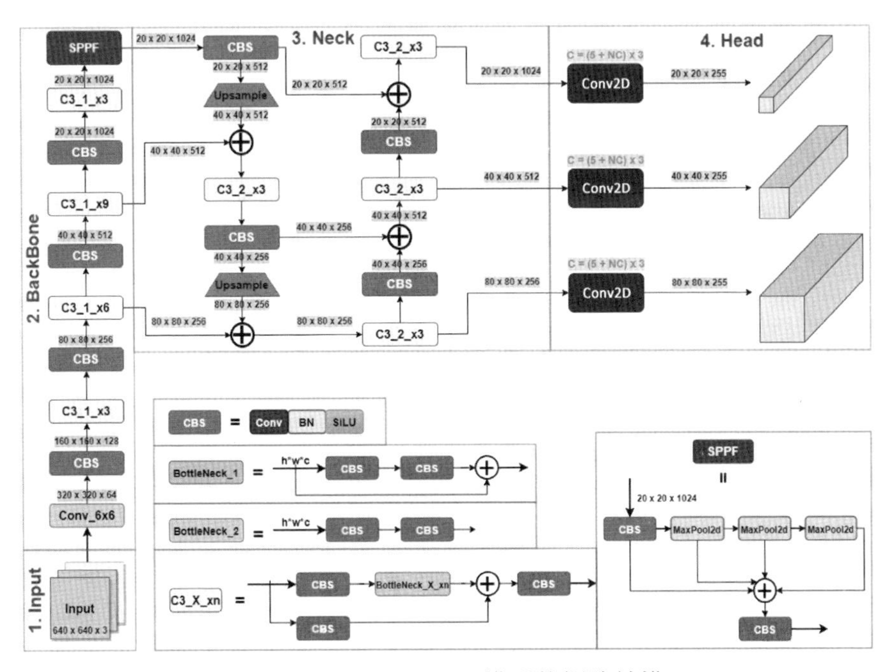

▲ 圖 3.37 YOLOv5s 模型的網路結構

3.3.1.3 開發設計

在 YOLOv5 模型專案中，voc2yolo.py 檔案為標註資料格式轉換的，YOLOv5 模型自動計算錨點座標，不需要 YOLOv3 模型專案中的 train_anchors.py 和 kmeans.py 檔案。另外，train.py 是訓練模型檔案，detect.py 是訓練完成後的測試模型檔案，export.py 是模型輸出檔案，可以輸出為 onnx 格式的結果。

1）資料格式轉換

資料格式的轉換是在 voc2yolo.py 檔案中實現的，該檔案的功能是將影像標註工具生成的 voc 格式資料轉為 yolo 格式的資料，程式如下：

```
import xml.etree.ElementTree as ET
import pickle
import os
from os import listdir, getcwd
from glob import glob
import random
from os.path import join

classes = ['face','face_mask']
dataset_path = "./dataset/"
txt_label_path = dataset_path + '/labels'
test_ratio = 0.1

if not os.path.exists(txt_label_path):
    os.mkdir(txt_label_path)

def convert(size, box):
    dw = 1. / size[0]
    dh = 1. / size[1]
    x = (box[0] + box[1]) / 2.0
    y = (box[2] + box[3]) / 2.0
    w = box[1] - box[0]
    h = box[3] - box[2]
    x = x * dw
    w = w * dw
    y = y * dh
```

```
    h = h * dh
    return (x, y, w, h)
    #return (int(x), int(y), int(w), int(h))

def convert_annotation(image_id):
    # 這裡改為 .xml 資料夾的路徑
    in_file = open(os.path.join(dataset_path, 'Annotations/%s.xml' % (image_id)))
    # 這裡是生成每幅影像對應的 .txt 檔案的路徑
    out_file = open(os.path.join(txt_label_path, '%s.txt' % (image_id)), 'w')
    tree = ET.parse(in_file)
    root = tree.getroot()
    size = root.find('size')
    w = int(size.find('width').text)
    h = int(size.find('height').text)  #

    for obj in root.iter('object'):
        cls = obj.find('name').text
        if cls not in classes:
            continue
        cls_id = classes.index(cls)
        xmlbox = obj.find('bndbox')
        b = (float(xmlbox.find('xmin').text), float(xmlbox.find('xmax').text),
float(xmlbox.find('ymin').text), float(xmlbox.find('ymax').text))
        bb = convert((w, h), b)
        #list_file.write(str(cls_id) + " " + " ".join([str(a) for a in bb]) + '\n')
        #list_file.write(" " + " ".join([str(a) for a in bb]) + " " + str(cls_id))
        out_file.write(str(cls_id) + " " + " ".join([str(a) for a in bb]) + '\n')

    #list_file.write('\n')

anno_files = glob(os.path.join(dataset_path, 'Annotations', '*.xml'))
anno_files = [item.split(os.sep)[-1].split('.')[0] for item in anno_files]
print("Totally convert %d files." % len(anno_files))
random.shuffle(anno_files)
test_num = int(len(anno_files) * test_ratio)
image_ids_val = anno_files[:test_num]
image_ids_train = anno_files[test_num:]
list_file_train = open('./object_train.txt', 'w')
list_file_val = open('./object_val.txt', 'w')
```

```
for image_id in image_ids_train:
    # 這裡改為樣本影像所在資料夾的路徑
    list_file_train.write(os.path.join(dataset_path, 'images', '%s.jpg\n' % (image_
id)))
    convert_annotation(image_id)
list_file_train.close()
for image_id in image_ids_val:
    # 這裡改為樣本影像所在資料夾的路徑
    list_file_val.write(os.path.join(dataset_path, 'images', '%s.jpg\n' % (image_id)))
    convert_annotation(image_id)
# 修改 JPEGImages 目錄為 images，調配 YOLOv5 模型訓練
os.rename(dataset_path+'/JPEGImages',dataset_path+'/images')
list_file_val.close()
```

2）訓練模型

train.py 是訓練模型檔案，程式如下：

```
import argparse
import math
import os
import random
import sys
import time
from copy import deepcopy
from datetime import datetime
from pathlib import Path

import numpy as np
import torch
import torch.distributed as dist
import torch.nn as nn
import yaml
from torch.cuda import amp
from torch.nn.parallel import DistributedDataParallel as DDP
from torch.optim import SGD, Adam, AdamW, lr_scheduler
from tqdm import tqdm

FILE = Path(__file__).resolve()
```

```
ROOT = FILE.parents[0]  #YOLOv5 root directory
if str(ROOT) not in sys.path:
    sys.path.append(str(ROOT))  #add ROOT to PATH
ROOT = Path(os.path.relpath(ROOT, Path.cwd()))  #relative

import val  #for end-of-epoch mAP
from models.experimental import attempt_load
from models.yolo import Model
from utils.autoanchor import check_anchors
from utils.autobatch import check_train_batch_size
from utils.callbacks import Callbacks
from utils.datasets import create_dataloader
from utils.downloads import attempt_download
from utils.general import (LOGGER, check_dataset, check_file, check_git_status,
                    check_img_size, check_requirements, check_suffix, check_yaml,
                    colorstr, get_latest_run, increment_path, init_seeds, intersect_
                    dicts, labels_to_class_weights, labels_to_image_weights, methods,
                    one_cycle, print_args, print_mutation, strip_optimizer)
from utils.loggers import Loggers
from utils.loggers.wandb.wandb_utils import check_wandb_resume
from utils.loss import ComputeLoss
from utils.metrics import fitness
from utils.plots import plot_evolve, plot_labels
from utils.torch_utils import EarlyStopping, ModelEMA, de_parallel, select_device,
                               torch_distributed_zero_first

LOCAL_RANK = int(os.getenv('LOCAL_RANK', -1))  #https://pytorch.org/docs/stable/
elastic/run.html
RANK = int(os.getenv('RANK', -1))
WORLD_SIZE = int(os.getenv('WORLD_SIZE', 1))

def train(hyp,  #path/to/hyp.yaml or hyp dictionary
          opt,
          device,
          callbacks
       ):
    save_dir, epochs, batch_size, weights, single_cls, evolve, data, cfg, resume,
noval, nosave,
            workers, freeze = Path(opt.save_dir), opt.epochs, opt.batch_size, opt.
```

```
        weights, opt.single_cls, opt.evolve, opt.data, opt.cfg, opt.resume, opt.
        noval, opt.nosave, opt.workers, opt.freeze

    #Directories
    w = save_dir / 'weights'  #weights dir
    (w.parent if evolve else w).mkdir(parents=True, exist_ok=True)  #make dir
    last, best = w / 'last.pt', w / 'best.pt'

    #Hyperparameters
    if isinstance(hyp, str):
        with open(hyp, errors='ignore') as f:
            hyp = yaml.safe_load(f)  #load hyps dict
    LOGGER.info(colorstr('hyperparameters: ') + ', '.join(f'{k}={v}' for k, v in hyp.
items()))

    #Save run settings
    if not evolve:
        with open(save_dir / 'hyp.yaml', 'w') as f:
            yaml.safe_dump(hyp, f, sort_keys=False)
        with open(save_dir / 'opt.yaml', 'w') as f:
            yaml.safe_dump(vars(opt), f, sort_keys=False)

    #Loggers
    data_dict = None
    if RANK in [-1, 0]:
        loggers = Loggers(save_dir, weights, opt, hyp, LOGGER)  #loggers instance
        if loggers.wandb:
            data_dict = loggers.wandb.data_dict
            if resume:
                weights, epochs, hyp, batch_size = opt.weights, opt.epochs, opt.hyp,
opt.batch_size
        #Register actions
        for k in methods(loggers):
            callbacks.register_action(k, callback=getattr(loggers, k))

    #Config
    plots = not evolve  #create plots
    cuda = device.type != 'cpu'
    init_seeds(1 + RANK)
```

```
    with torch_distributed_zero_first(LOCAL_RANK):
        data_dict = data_dict or check_dataset(data)  #check if None
    train_path, val_path = data_dict['train'], data_dict['val']
    nc = 1 if single_cls else int(data_dict['nc'])  #number of classes
    names = ['item'] if single_cls and len(data_dict['names']) != 1 else data_
dict['names']  #class names
    assert len(names) == nc, f'{len(names)} names found for nc={nc} dataset in {data}'
#check
    is_coco = isinstance(val_path, str) and val_path.endswith('coco/val2017.txt')
#COCO dataset

    #Model
    check_suffix(weights, '.pt')  #check weights
    pretrained = weights.endswith('.pt')
    if pretrained:
        with torch_distributed_zero_first(LOCAL_RANK):
            weights = attempt_download(weights)      #download if not found locally
            #load checkpoint to CPU to avoid CUDA memory leak
            #load checkpoint to CPU to avoid CUDA memory leak
            ckpt = torch.load(weights, map_location='cpu')
            model = Model(cfg or ckpt['model'].yaml, ch=3, nc=nc,
                          anchors=hyp.get('anchors')).to(device)  #create
        exclude = ['anchor'] if (cfg or hyp.get('anchors')) and not resume else []
#exclude keys
        csd = ckpt['model'].float().state_dict()  #checkpoint state_dict as FP32
        csd = intersect_dicts(csd, model.state_dict(), exclude=exclude)  #intersect
        model.load_state_dict(csd, strict=False)  #load
        LOGGER.info(f'Transferred {len(csd)}/{len(model.state_dict())} items from
{weights}')
    else:
        model = Model(cfg, ch=3, nc=nc, anchors=hyp.get('anchors')).to(device)  #create

    #Freeze
    freeze = [f'model.{x}.' for x in (freeze if len(freeze) > 1 else
range(freeze[0]))]  #layers to freeze
    for k, v in model.named_parameters():
        v.requires_grad = True  #train all layers
        if any(x in k for x in freeze):
            LOGGER.info(f'freezing {k}')
```

```
            v.requires_grad = False

    #Image size
    gs = max(int(model.stride.max()), 32)  #grid size (max stride)
    imgsz = check_img_size(opt.imgsz, gs, floor=gs * 2)  #verify imgsz is gs-multiple

    #Batch size
    if RANK == -1 and batch_size == -1:  #single-GPU only, estimate best batch size
        batch_size = check_train_batch_size(model, imgsz)
        loggers.on_params_update({"batch_size": batch_size})

    #Optimizer
    nbs = 64  #nominal batch size
    accumulate = max(round(nbs / batch_size), 1)  #accumulate loss before optimizing
    hyp['weight_decay'] *= batch_size * accumulate / nbs  #scale weight_decay
    LOGGER.info(f"Scaled weight_decay = {hyp['weight_decay']}")

    g0, g1, g2 = [], [], []  #optimizer parameter groups
    for v in model.modules():
        if hasattr(v, 'bias') and isinstance(v.bias, nn.Parameter):  #bias
            g2.append(v.bias)
        if isinstance(v, nn.BatchNorm2d):  #weight (no decay)
            g0.append(v.weight)
        elif hasattr(v, 'weight') and isinstance(v.weight, nn.Parameter):  #weight
(with decay)
            g1.append(v.weight)

    if opt.optimizer == 'Adam':
        optimizer = Adam(g0, lr=hyp['lr0'], betas=(hyp['momentum'], 0.999)) #adjust
beta1 to momentum
    elif opt.optimizer == 'AdamW':
        optimizer = AdamW(g0, lr=hyp['lr0'], betas=(hyp['momentum'], 0.999)) #adjust
beta1 to momentum
    else:
        optimizer = SGD(g0, lr=hyp['lr0'], momentum=hyp['momentum'], nesterov=True)

    #add g1 with weight_decay
    optimizer.add_param_group({'params': g1, 'weight_decay': hyp['weight_decay']})
    optimizer.add_param_group({'params': g2})  #add g2 (biases)
```

```
    LOGGER.info(f"{colorstr('optimizer:')} {type(optimizer).__name__} with parameter
groups "
                f"{len(g0)} weight (no decay), {len(g1)} weight, {len(g2)} bias")
    del g0, g1, g2

    #Scheduler
    if opt.cos_lr:
        lf = one_cycle(1, hyp['lrf'], epochs)  #cosine 1->hyp['lrf']
    else:
        lf = lambda x: (1 - x / epochs) * (1.0 - hyp['lrf']) + hyp['lrf']  #linear
    #plot_lr_scheduler(optimizer, scheduler, epochs)
    scheduler = lr_scheduler.LambdaLR(optimizer, lr_lambda=lf)

    #EMA
    ema = ModelEMA(model) if RANK in [-1, 0] else None

    #Resume
    start_epoch, best_fitness = 0, 0.0
    if pretrained:
        #Optimizer
        if ckpt['optimizer'] is not None:
            optimizer.load_state_dict(ckpt['optimizer'])
            best_fitness = ckpt['best_fitness']

        #EMA
        if ema and ckpt.get('ema'):
            ema.ema.load_state_dict(ckpt['ema'].float().state_dict())
            ema.updates = ckpt['updates']

        #Epochs
        start_epoch = ckpt['epoch'] + 1
        if resume:
            assert start_epoch > 0, f'{weights} training to {epochs} epochs is
finished, nothing to resume.'
        if epochs < start_epoch:
            LOGGER.info(f"{weights} has been trained for {ckpt['epoch']} epochs.
                        Fine-tuning for {epochs} more epochs.")
            epochs += ckpt['epoch']  #finetune additional epochs
```

```
        del ckpt, csd

    #DP mode
    if cuda and RANK == -1 and torch.cuda.device_count() > 1:
        LOGGER.warning('WARNING: DP not recommended, use torch.
                       distributed.run for best DDP Multi-GPU results.\n'
                       'See Multi-GPU Tutorial at https://github.com/ultralytics/
yolov5/issues/475 to get started.')
        model = torch.nn.DataParallel(model)

    #SyncBatchNorm
    if opt.sync_bn and cuda and RANK != -1:
        model = torch.nn.SyncBatchNorm.convert_sync_batchnorm(model).to(device)
        LOGGER.info('Using SyncBatchNorm()')

    #Trainloader
    train_loader, dataset = create_dataloader(train_path, imgsz, batch_size //WORLD_
SIZE,
                      gs, single_cls, hyp=hyp, augment=True, cache=None if opt.cache
==
                      'val' else opt.cache, rect=opt.rect, rank=LOCAL_RANK,
workers=workers,
                      image_weights=opt.image_weights, quad=opt.quad,
                      prefix=colorstr('train: '), shuffle=True)
    mlc = int(np.concatenate(dataset.labels, 0)[:, 0].max())  #max label class
    nb = len(train_loader)  #number of batches
    assert mlc < nc, f'Label class {mlc} exceeds nc={nc} in {data}. Possible class
labels are 0-{nc - 1}'

    #Process 0
    if RANK in [-1, 0]:
        val_loader = create_dataloader(val_path, imgsz, batch_size //WORLD_SIZE * 2, gs,
single_cls,
                                hyp=hyp, cache=None if noval else opt.cache,
                                rect=True, rank=-1, workers=workers * 2, pad=0.5,
                                prefix=colorstr('val: '))[0]

        if not resume:
```

```
        labels = np.concatenate(dataset.labels, 0)
        #c = torch.tensor(labels[:, 0])  #classes
        #cf = torch.bincount(c.long(), minlength=nc) + 1.  #frequency
        #model._initialize_biases(cf.to(device))
        if plots:
            plot_labels(labels, names, save_dir)

        #Anchors
        if not opt.noautoanchor:
            check_anchors(dataset, model=model, thr=hyp['anchor_t'], imgsz=imgsz)
        model.half().float()  #pre-reduce anchor precision

    callbacks.run('on_pretrain_routine_end')

#DDP mode
if cuda and RANK != -1:
    model = DDP(model, device_ids=[LOCAL_RANK], output_device=LOCAL_RANK)

#Model attributes
nl = de_parallel(model).model[-1].nl  #number of detection layers (to scale hyps)
hyp['box'] *= 3 / nl  #scale to layers
hyp['cls'] *= nc / 80 * 3 / nl  #scale to classes and layers
hyp['obj'] *= (imgsz / 640) ** 2 * 3 / nl  #scale to image size and layers
hyp['label_smoothing'] = opt.label_smoothing
model.nc = nc  #attach number of classes to model
model.hyp = hyp  #attach hyperparameters to model
model.class_weights = labels_to_class_weights(dataset.labels, nc).to(device) * nc
model.names = names

#Start training
t0 = time.time()
nw = max(round(hyp['warmup_epochs'] * nb), 1000)  #number of warmup iterations,
        max(3 epochs, 1k iterations)
#nw = min(nw, (epochs - start_epoch) / 2 * nb)  #limit warmup to < 1/2 of training
last_opt_step = -1
maps = np.zeros(nc)  #mAP per class
results = (0, 0, 0, 0, 0, 0, 0)  #P, R, mAP@.5, mAP@.5-.95, val_loss(box, obj, cls)
scheduler.last_epoch = start_epoch - 1  #do not move
scaler = amp.GradScaler(enabled=cuda)
```

```
    stopper = EarlyStopping(patience=opt.patience)
    compute_loss = ComputeLoss(model)  #init loss class
    LOGGER.info(f'Image sizes {imgsz} train, {imgsz} val\n'
                f'Using {train_loader.num_workers * WORLD_SIZE} dataloader workers\n'
                f"Logging results to {colorstr('bold', save_dir)}\n"
                f'Starting training for {epochs} epochs...')
    for epoch in range(start_epoch, epochs):  #epoch ------------------------------------------
        model.train()

        #Update image weights (optional, single-GPU only)
        if opt.image_weights:
            cw = model.class_weights.cpu().numpy() * (1 - maps) ** 2 / nc  #class weights
            iw = labels_to_image_weights(dataset.labels, nc=nc, class_weights=cw)  #image
weights
            dataset.indices = random.choices(range(dataset.n), weights=iw, k=dataset.n)
#rand weighted idx

        #Update mosaic border (optional)
        #b = int(random.uniform(0.25 * imgsz, 0.75 * imgsz + gs) //gs * gs)
        #dataset.mosaic_border = [b - imgsz, -b]  #height, width borders

        mloss = torch.zeros(3, device=device)  #mean losses
        if RANK != -1:
            train_loader.sampler.set_epoch(epoch)
        pbar = enumerate(train_loader)
        LOGGER.info(('\n' + '%10s' * 7) % ('Epoch', 'gpu_mem', 'box', 'obj', 'cls',
'labels', 'img_size'))
        if RANK in [-1, 0]:
            pbar = tqdm(pbar, total=nb, bar_format='{l_bar}{bar:10}{r_bar}{bar:-10b}')
#progress bar
        optimizer.zero_grad()
        for i, (imgs, targets, paths, _) in pbar:  #batch ------------------------------------
            ni = i + nb * epoch  #number integrated batches (since train start)
            imgs = imgs.to(device, non_blocking=True).float() / 255  #uint8 to float32, 0-255
to 0.0-1.0

            #Warmup
            if ni <= nw:
                xi = [0, nw]  #x interp
```

```
                #compute_loss.gr = np.interp(ni, xi, [0.0, 1.0])  #iou loss ratio (obj_loss
= 1.0 or iou)
                accumulate = max(1, np.interp(ni, xi, [1, nbs / batch_size]).round())
                for j, x in enumerate(optimizer.param_groups):
                    #bias lr falls from 0.1 to lr0, all other lrs rise from 0.0 to lr0
                    x['lr'] = np.interp(ni, xi, [hyp['warmup_bias_lr'] if j == 2 else 0.0,
                                  x['initial_lr'] * lf(epoch)])
                    if 'momentum' in x:
                        x['momentum'] = np.interp(ni, xi, [hyp['warmup_momentum'],
hyp['momentum']])

            #Multi-scale
            if opt.multi_scale:
                sz = random.randrange(imgsz * 0.5, imgsz * 1.5 + gs) //gs * gs  #size
                sf = sz / max(imgs.shape[2:])  #scale factor
                if sf != 1:
                    #new shape (stretched to gs-multiple)
                    ns = [math.ceil(x * sf / gs) * gs for x in imgs.shape[2:]]
                    imgs = nn.functional.interpolate(imgs, size=ns, mode='bilinear', align_
corners=False)

            #Forward
            with amp.autocast(enabled=cuda):
                pred = model(imgs)  #forward
                loss, loss_items = compute_loss(pred, targets.to(device))  #loss scaled by
batch_size
                if RANK != -1:
                    loss *= WORLD_SIZE  #gradient averaged between devices in DDP mode
                if opt.quad:
                    loss *= 4.

            #Backward
            scaler.scale(loss).backward()

            #Optimize
            if ni - last_opt_step >= accumulate:
                scaler.step(optimizer)  #optimizer.step
                scaler.update()
                optimizer.zero_grad()
```

```
                if ema:
                    ema.update(model)
                last_opt_step = ni

            #Log
            if RANK in [-1, 0]:
                mloss = (mloss * i + loss_items) / (i + 1)  #update mean losses
                mem = f'{torch.cuda.memory_reserved() / 1E9 if torch.cuda.is_available()
                                                     else 0:.3g}G'  #(GB)
                pbar.set_description(('%10s' * 2 + '%10.4g' * 5) % (
                    f'{epoch}/{epochs - 1}', mem, *mloss, targets.shape[0], imgs.shape[-1]))
                callbacks.run('on_train_batch_end', ni, model, imgs, targets, paths, plots,
opt.sync_bn)
                if callbacks.stop_training:
                    return
            #end batch ----------------------------------------------------------------------

        #Scheduler
        lr = [x['lr'] for x in optimizer.param_groups]  #for loggers
        scheduler.step()

        if RANK in [-1, 0]:
            #mAP
            callbacks.run('on_train_epoch_end', epoch=epoch)
            ema.update_attr(model, include=['yaml', 'nc', 'hyp', 'names', 'stride', 'class_
weights'])
            final_epoch = (epoch + 1 == epochs) or stopper.possible_stop
            if not noval or final_epoch:  #Calculate mAP
                results, maps, _ = val.run(data_dict,
                                           batch_size=batch_size //WORLD_SIZE * 2,
                                           imgsz=imgsz,
                                           model=ema.ema,
                                           single_cls=single_cls,
                                           dataloader=val_loader,
                                           save_dir=save_dir,
                                           plots=False,
                                           callbacks=callbacks,
                                           compute_loss=compute_loss)
```

```
        #Update best mAP
        #weighted combination of [P, R, mAP@.5, mAP@.5-.95]
        fi = fitness(np.array(results).reshape(1, -1))
        if fi > best_fitness:
            best_fitness = fi
        log_vals = list(mloss) + list(results) + lr
        callbacks.run('on_fit_epoch_end', log_vals, epoch, best_fitness, fi)

        #Save model
        if (not nosave) or (final_epoch and not evolve):  #if save
            ckpt = {'epoch': epoch,
                    'best_fitness': best_fitness,
                    'model': deepcopy(de_parallel(model)).half(),
                    'ema': deepcopy(ema.ema).half(),
                    'updates': ema.updates,
                    'optimizer': optimizer.state_dict(),
                    'wandb_id': loggers.wandb.wandb_run.id if loggers.wandb else None,
                    'date': datetime.now().isoformat()}

            #Save last, best and delete
            torch.save(ckpt, last)
            if best_fitness == fi:
                torch.save(ckpt, best)
            if (epoch > 0) and (opt.save_period > 0) and (epoch % opt.save_period == 0):
                torch.save(ckpt, w / f'epoch{epoch}.pt')
            del ckpt
            callbacks.run('on_model_save', last, epoch, final_epoch, best_fitness, fi)

        #Stop Single-GPU
        if RANK == -1 and stopper(epoch=epoch, fitness=fi):
            break

        #Stop DDP TODO: known issues shttps://github.com/ultralytics/yolov5/pull/4576
        #stop = stopper(epoch=epoch, fitness=fi)
        #if RANK == 0:
        #    dist.broadcast_object_list([stop], 0)  #broadcast 'stop' to all ranks

    #Stop DPP
    #with torch_distributed_zero_first(RANK):
```

```
        #if stop:
        #    break  #must break all DDP ranks

        #end epoch ----------------------------------------------------------------------------
    #end training ------------------------------------------------------------------------------
    if RANK in [-1, 0]:
        LOGGER.info(f'\n{epoch - start_epoch + 1} epochs completed in {(time.time() -
                                                t0) / 3600:.3f} hours.')
        for f in last, best:
            if f.exists():
                strip_optimizer(f)  #strip optimizers
                if f is best:
                    LOGGER.info(f'\nValidating {f}...')
                    results, _, _ = val.run(data_dict, batch_size=batch_size //WORLD_SIZE * 2,
                            imgsz=imgsz, model=attempt_load(f, device).half(),
                            iou_thres=0.65 if is_coco else 0.60,  #best pycocotools results
at 0.65
                            single_cls=single_cls, dataloader=val_loader, save_dir=save_dir,
                            save_json=is_coco, verbose=True, plots=True, callbacks=callbacks,
                            compute_loss=compute_loss)  #val best model with plots
                    if is_coco:
                        callbacks.run('on_fit_epoch_end', list(mloss) + list(results) +
                                lr, epoch, best_fitness, fi)

        callbacks.run('on_train_end', last, best, plots, epoch, results)
        LOGGER.info(f"Results saved to {colorstr('bold', save_dir)}")

    torch.cuda.empty_cache()
    return results
```

3）訓練測試

detect.py 是用於訓練完成測試模型檔案，程式如下：

```
import argparse
import os
import sys
from pathlib import Path
```

```
import cv2
import torch
import torch.backends.cudnn as cudnn

FILE = Path(__file__).resolve()
ROOT = FILE.parents[0]  #YOLOv5 root directory
if str(ROOT) not in sys.path: sys.path.append(str(ROOT))  #add ROOT to PATH
ROOT = Path(os.path.relpath(ROOT, Path.cwd()))  #relative

from models.common import DetectMultiBackend
from utils.datasets import IMG_FORMATS, VID_FORMATS, LoadImages, LoadStreams
from utils.general import (LOGGER, check_file, check_img_size, check_imshow, check_
requirements, colorstr, increment_path, non_max_suppression, print_args, scale_coords,
strip_optimizer, xyxy2xywh)
from utils.plots import Annotator, colors, save_one_box
from utils.torch_utils import select_device, time_sync

@torch.no_grad()
def run(weights=ROOT / 'yolov5s.pt',          #model.pt path(s)
        source=ROOT / 'data/images',          #file/dir/URL/glob, 0 for webcam
        data=ROOT / 'data/coco128.yaml',      #dataset.yaml path
        imgsz=(640, 640),                     #inference size (height, width)
        conf_thres=0.25,                      #confidence threshold
        iou_thres=0.45,                       #NMS IOU threshold
        max_det=1000,                         #maximum detections per image
        device='',                            #cuda device, i.e. 0 or 0,1,2,3 or cpu
        view_img=False,                       #show results
        save_txt=False,                       #save results to *.txt
        save_conf=False,                      #save confidences in --save-txt labels
        save_crop=False,                      #save cropped prediction boxes
        nosave=False,                         #do not save images/videos
        classes=None,                         #filter by class: --class 0, or --class 0 2 3
        agnostic_nms=False,                   #class-agnostic NMS
        augment=False,                        #augmented inference
        visualize=False,                      #visualize features
        update=False,                         #update all models
        project=ROOT / 'runs/detect',         #save results to project/name
        name='exp',                           #save results to project/name
```

```
        exist_ok=False,                     #existing project/name ok, do not increment
        line_thickness=3,                   #bounding box thickness (pixels)
        hide_labels=False,                  #hide labels
        hide_conf=False,                    #hide confidences
        half=False,                         #use FP16 half-precision inference
        dnn=False,                          #use OpenCV DNN for ONNX inference
):
    source = str(source)
    save_img = not nosave and not source.endswith('.txt')  #save inference images
    is_file = Path(source).suffix[1:] in (IMG_FORMATS + VID_FORMATS)
    is_url = source.lower().startswith(('rtsp://', 'rtmp://', 'http://', 'https://'))
    webcam = source.isnumeric() or source.endswith('.txt') or (is_url and not is_file)
    if is_url and is_file:
        source = check_file(source)  #download

    #Directories
    save_dir = increment_path(Path(project) / name, exist_ok=exist_ok)  #increment run
    (save_dir / 'labels' if save_txt else save_dir).mkdir(parents=True, exist_ok=True)  #make dir

    #Load model
    device = select_device(device)
    model = DetectMultiBackend(weights, device=device, dnn=dnn, data=data)
    stride, names, pt, jit, onnx, engine = model.stride, model.names, model.pt,
                                  model.jit, model.onnx, model.engine
    imgsz = check_img_size(imgsz, s=stride)  #check image size

    #Half
    #FP16 supported on limited backends with CUDA
    half &= (pt or jit or onnx or engine) and device.type != 'cpu'
    if pt or jit:
        model.model.half() if half else model.model.float()

    #Dataloader
    if webcam:
        view_img = check_imshow()
        cudnn.benchmark = True     #set True to speed up constant image size inference
        dataset = LoadStreams(source, img_size=imgsz, stride=stride, auto=pt)
        bs = len(dataset)                             #batch_size
    else:
```

```
        dataset = LoadImages(source, img_size=imgsz, stride=stride, auto=pt)
        bs = 1                                  #batch_size
    vid_path, vid_writer = [None] * bs, [None] * bs

    #Run inference
    model.warmup(imgsz=(1 if pt else bs, 3, *imgsz), half=half)  #warmup
    dt, seen = [0.0, 0.0, 0.0], 0
    for path, im, im0s, vid_cap, s in dataset:
        t1 = time_sync()
        im = torch.from_numpy(im).to(device)
        im = im.half() if half else im.float()  #uint8 to fp16/32
        im /= 255  #0 - 255 to 0.0 - 1.0
        if len(im.shape) == 3:
            im = im[None]  #expand for batch dim
        t2 = time_sync()
        dt[0] += t2 - t1

        #Inference
        visualize = increment_path(save_dir / Path(path).stem, mkdir=True) if visualize
else False
        pred = model(im, augment=augment, visualize=visualize)
        t3 = time_sync()
        dt[1] += t3 - t2

        #NMS
        pred = non_max_suppression(pred, conf_thres, iou_thres, classes, agnostic_nms, max_
det=max_det)
        dt[2] += time_sync() - t3

        #Second-stage classifier (optional)
        #pred = utils.general.apply_classifier(pred, classifier_model, im, im0s)

        #Process predictions
        for i, det in enumerate(pred):  #per image
            seen += 1
            if webcam:  #batch_size >= 1
                p, im0, frame = path[i], im0s[i].copy(), dataset.count
                s += f'{i}: '
            else:
```

```
            p, im0, frame = path, im0s.copy(), getattr(dataset, 'frame', 0)

        p = Path(p)  #to Path
        save_path = str(save_dir / p.name)                #im.jpg
        txt_path = str(save_dir / 'labels' / p.stem) + ('' if dataset.mode == 'image'
else f'_{frame}') #im.txt
        s += '%gx%g ' % im.shape[2:]                      #print string
        gn = torch.tensor(im0.shape)[[1, 0, 1, 0]]        #normalization gain whwh
        imc = im0.copy() if save_crop else im0            #for save_crop
        annotator = Annotator(im0, line_width=line_thickness, example=str(names))
        if len(det):
            #Rescale boxes from img_size to im0 size
            det[:, :4] = scale_coords(im.shape[2:], det[:, :4], im0.shape).round()

            #Print results
            for c in det[:, -1].unique():
                n = (det[:, -1] == c).sum()  #detections per class
                s += f"{n} {names[int(c)]}{'s' * (n > 1)}, "  #add to string

            #Write results
            for *xyxy, conf, cls in reversed(det):
                if save_txt:  #Write to file
                    xywh = (xyxy2xywh(torch.tensor(xyxy).view(1, 4)) / gn).
                           view(-1).tolist()  #normalized xywh
                    line = (cls, *xywh, conf) if save_conf else (cls, *xywh)  #label
format
                    with open(txt_path + '.txt', 'a') as f:
                        f.write(('%g ' * len(line)).rstrip() % line + '\n')

                if save_img or save_crop or view_img:  #Add bbox to image
                    c = int(cls)  #integer class
                    label = None if hide_labels else (names[c] if hide_conf
                                else f'{names[c]} {conf:.2f}')
                    annotator.box_label(xyxy, label, color=colors(c, True))
                    if save_crop:
                        save_one_box(xyxy, imc, file=save_dir / 'crops' / names[c] /
                                     f'{p.stem}.jpg', BGR=True)

        #Stream results
```

```
            im0 = annotator.result()
            if view_img:
                cv2.imshow(str(p), im0)
                cv2.waitKey(1)  #1 millisecond

            #Save results (image with detections)
            if save_img:
                if dataset.mode == 'image':
                    cv2.imwrite(save_path, im0)
                else:                                       #'video' or 'stream'
                    if vid_path[i] != save_path:            #new video
                        vid_path[i] = save_path
                        if isinstance(vid_writer[i], cv2.VideoWriter):
                            vid_writer[i].release()  #release previous video writer
                        if vid_cap:                                  #video
                            fps = vid_cap.get(cv2.CAP_PROP_FPS)
                            w = int(vid_cap.get(cv2.CAP_PROP_FRAME_WIDTH))
                            h = int(vid_cap.get(cv2.CAP_PROP_FRAME_HEIGHT))
                        else:  #stream
                            fps, w, h = 30, im0.shape[1], im0.shape[0]
                        #force *.mp4 suffix on results videos
                        save_path = str(Path(save_path).with_suffix('.mp4'))
                        vid_writer[i] = cv2.VideoWriter(save_path,
                                    cv2.VideoWriter_fourcc(*'mp4v'), fps, (w, h))
                    vid_writer[i].write(im0)

        #Print time (inference-only)
        LOGGER.info(f'{s}Done. ({t3 - t2:.3f}s)')

    #Print results
    t = tuple(x / seen * 1E3 for x in dt)  #speeds per image
    LOGGER.info(f'Speed: %.1fms pre-process, %.1fms inference,
                    %.1fms NMS per image at shape {(1, 3, *imgsz)}' % t)
    if save_txt or save_img:
        s = f"\n{len(list(save_dir.glob('labels/*.txt')))} labels saved to {save_dir /
'labels'}" if save_txt else ''
        LOGGER.info(f"Results saved to {colorstr('bold', save_dir)}{s}")
    if update:
        strip_optimizer(weights)  #update model (to fix SourceChangeWarning)
```

4）模型輸出

export.py 是用於輸出模型檔案，可以輸出 onnx 格式的結果，程式如下：

```
import argparse
import json
import os
import platform
import subprocess
import sys
import time
import warnings
from pathlib import Path

import pandas as pd
import torch
import torch.nn as nn
from torch.utils.mobile_optimizer import optimize_for_mobile

FILE = Path(__file__).resolve()
ROOT = FILE.parents[0]                                  #YOLOv5 root directory
if str(ROOT) not in sys.path:
    sys.path.append(str(ROOT))                          #add ROOT to PATH
ROOT = Path(os.path.relpath(ROOT, Path.cwd()))          #relative

from models.common import Conv
from models.experimental import attempt_load
from models.yolo import Detect
from utils.activations import SiLU
from utils.datasets import LoadImages
from utils.general import (LOGGER, check_dataset, check_img_size, check_requirements,
                    check_version, colorstr, file_size, print_args, url2file)
from utils.torch_utils import select_device

def export_formats():
    #YOLOv5 export formats
    x = [['PyTorch', '-', '.pt'],
        ['TorchScript', 'torchscript', '.torchscript'],
        ['ONNX', 'onnx', '.onnx'],
```

```
        ['OpenVINO', 'openvino', '_openvino_model'],
        ['TensorRT', 'engine', '.engine'],
        ['CoreML', 'coreml', '.mlmodel'],
        ['TensorFlow SavedModel', 'saved_model', '_saved_model'],
        ['TensorFlow GraphDef', 'pb', '.pb'],
        ['TensorFlow Lite', 'tflite', '.tflite'],
        ['TensorFlow Edge TPU', 'edgetpu', '_edgetpu.tflite'],
        ['TensorFlow.js', 'tfjs', '_web_model']]
    return pd.DataFrame(x, columns=['Format', 'Argument', 'Suffix'])

def export_torchscript(model, im, file, optimize, prefix=colorstr('TorchScript:')):
    #YOLOv5 TorchScript model export
    try:
        LOGGER.info(f'\n{prefix} starting export with torch {torch.__version__}...')
        f = file.with_suffix('.torchscript')

        ts = torch.jit.trace(model, im, strict=False)
        d = {"shape": im.shape, "stride": int(max(model.stride)), "names": model.names}
        extra_files = {'config.txt': json.dumps(d)}  #torch._C.ExtraFilesMap()
        if optimize:  #https://pytorch.org/tutorials/recipes/mobile_interpreter.html
            optimize_for_mobile(ts)._save_for_lite_interpreter(str(f), _extra_files=extra_
files)
        else:
            ts.save(str(f), _extra_files=extra_files)

        LOGGER.info(f'{prefix} export success, saved as {f} ({file_size(f):.1f} MB)')
        return f
    except Exception as e:
        LOGGER.info(f'{prefix} export failure: {e}')

def export_onnx(model, im, file, opset, train, dynamic, simplify, prefix=colorstr('ONNX:')):
    #YOLOv5 ONNX export
    try:
        check_requirements(('onnx',))
        import onnx

        LOGGER.info(f'\n{prefix} starting export with onnx {onnx.__version__}...')
        f = file.with_suffix('.onnx')
```

```
        torch.onnx.export(model, im, f, verbose=False, opset_version=opset,
                          training=torch.onnx.TrainingMode.TRAINING if train
                          else torch.onnx.TrainingMode.EVAL,
                          do_constant_folding=not train,
                          input_names=['images'], output_names=['output'],
                          dynamic_axes={'images': {0: 'batch', 2: 'height', 3: 'width'},
#shape(1,3,640,640)
                                        'output': {0: 'batch', 1: 'anchors'}  #shape(1,25200,85)
                                        } if dynamic else None)

        #Checks
        model_onnx = onnx.load(f)                                    #load onnx model
        onnx.checker.check_model(model_onnx)                         #check onnx model
        #LOGGER.info(onnx.helper.printable_graph(model_onnx.graph))  #print

        #Simplify
        if simplify:
            try:
                check_requirements(('onnx-simplifier',))
                import onnxsim

                LOGGER.info(f'{prefix} simplifying with onnx-simplifier {onnxsim.__
version__}...')
                model_onnx, check = onnxsim.simplify(
                    model_onnx,
                    dynamic_input_shape=dynamic,
                    input_shapes={'images': list(im.shape)} if dynamic else None)
                assert check, 'assert check failed'
                onnx.save(model_onnx, f)
            except Exception as e:
                LOGGER.info(f'{prefix} simplifier failure: {e}')
        LOGGER.info(f'{prefix} export success, saved as {f} ({file_size(f):.1f} MB)')
        return f
    except Exception as e:
        LOGGER.info(f'{prefix} export failure: {e}')

def export_openvino(model, im, file, prefix=colorstr('OpenVINO:')):
    #YOLOv5 OpenVINO export
    try:
```

```
        #requires openvino-dev:https://pypi.org/project/openvino-dev/
        check_requirements(('openvino-dev',))
        import openvino.inference_engine as ie

        LOGGER.info(f'\n{prefix} starting export with openvino {ie.__version__}...')
        f = str(file).replace('.pt', '_openvino_model' + os.sep)

        cmd = f"mo --input_model {file.with_suffix('.onnx')} --output_dir {f}"
        subprocess.check_output(cmd, shell=True)

        LOGGER.info(f'{prefix} export success, saved as {f} ({file_size(f):.1f} MB)')
        return f
    except Exception as e:
        LOGGER.info(f'\n{prefix} export failure: {e}')

def export_coreml(model, im, file, prefix=colorstr('CoreML:')):
    #YOLOv5 CoreML export
    try:
        check_requirements(('coremltools',))
        import coremltools as ct

        LOGGER.info(f'\n{prefix} starting export with coremltools {ct.__version__}...')
        f = file.with_suffix('.mlmodel')

        ts = torch.jit.trace(model, im, strict=False)  #TorchScript model
        ct_model = ct.convert(ts, inputs=[ct.ImageType('image', shape=im.shape,
                              scale=1 / 255, bias=[0, 0, 0])])
        ct_model.save(f)

        LOGGER.info(f'{prefix} export success, saved as {f} ({file_size(f):.1f} MB)')
        return ct_model, f
    except Exception as e:
        LOGGER.info(f'\n{prefix} export failure: {e}')
        return None, None

def export_engine(model, im, file, train, half, simplify, workspace=4,
              verbose=False, prefix=colorstr('TensorRT:')):
    #YOLOv5 TensorRT export https://developer.nvidia.com/tensorrt
    try:
```

```
check_requirements(('tensorrt',))
import tensorrt as trt

#TensorRT 7 handling https://github.com/ultralytics/yolov5/issues/6012
if trt.__version__[0] == '7':
    grid = model.model[-1].anchor_grid
    model.model[-1].anchor_grid = [a[..., :1, :1, :] for a in grid]
    export_onnx(model, im, file, 12, train, False, simplify)  #opset 12
    model.model[-1].anchor_grid = grid
else:  #TensorRT >= 8
    check_version(trt.__version__, '8.0.0', hard=True)  #require tensorrt>=8.0.0
    export_onnx(model, im, file, 13, train, False, simplify)  #opset 13
onnx = file.with_suffix('.onnx')

LOGGER.info(f'\n{prefix} starting export with TensorRT {trt.__version__}...')
assert im.device.type != 'cpu', 'export running on CPU but must be on GPU,
                    i.e. `python export.py --device 0`'
assert onnx.exists(), f'failed to export ONNX file: {onnx}'
f = file.with_suffix('.engine')  #TensorRT engine file
logger = trt.Logger(trt.Logger.INFO)
if verbose:
    logger.min_severity = trt.Logger.Severity.VERBOSE

builder = trt.Builder(logger)
config = builder.create_builder_config()
config.max_workspace_size = workspace * 1 << 30

flag = (1 << int(trt.NetworkDefinitionCreationFlag.EXPLICIT_BATCH))
network = builder.create_network(flag)
parser = trt.OnnxParser(network, logger)
if not parser.parse_from_file(str(onnx)):
    raise RuntimeError(f'failed to load ONNX file: {onnx}')

inputs = [network.get_input(i) for i in range(network.num_inputs)]
outputs = [network.get_output(i) for i in range(network.num_outputs)]
LOGGER.info(f'{prefix} Network Description:')
for inp in inputs:
    LOGGER.info(f'{prefix}\tinput "{inp.name}" with shape {inp.shape} and dtype {inp.
dtype}')
```

```
        for out in outputs:
            LOGGER.info(f'{prefix}\toutput "{out.name}" with shape {out.shape} and dtype {out.
dtype}')

        half &= builder.platform_has_fast_fp16
        LOGGER.info(f'{prefix} building FP{16 if half else 32} engine in {f}')
        if half:
            config.set_flag(trt.BuilderFlag.FP16)
        with builder.build_engine(network, config) as engine, open(f, 'wb') as t:
            t.write(engine.serialize())
        LOGGER.info(f'{prefix} export success, saved as {f} ({file_size(f):.1f} MB)')
        return f
    except Exception as e:
        LOGGER.info(f'\n{prefix} export failure: {e}')

def export_saved_model(model, im, file, dynamic, tf_nms=False, agnostic_nms=False,
                    topk_per_class=100, topk_all=100, iou_thres=0.45,
                    conf_thres=0.25, keras=False, prefix=colorstr('TensorFlow SavedModel:')):
    #YOLOv5 TensorFlow SavedModel export
    try:
        import tensorflow as tf
        from tensorflow.python.framework.convert_to_constants import convert_variables_to_
constants_v2

        from models.tf import TFDetect, TFModel

        LOGGER.info(f'\n{prefix} starting export with tensorflow {tf.__version__}...')
        f = str(file).replace('.pt', '_saved_model')
        batch_size, ch, *imgsz = list(im.shape)  #BCHW

        tf_model = TFModel(cfg=model.yaml, model=model, nc=model.nc, imgsz=imgsz)
        im = tf.zeros((batch_size, *imgsz, 3))  #BHWC order for TensorFlow
        _ = tf_model.predict(im, tf_nms, agnostic_nms, topk_per_class, topk_all, iou_thres,
conf_thres)
        inputs = tf.keras.Input(shape=(*imgsz, 3), batch_size=None if dynamic else batch_
size)
        outputs = tf_model.predict(inputs, tf_nms, agnostic_nms, topk_per_class,
                                   topk_all, iou_thres, conf_thres)
        keras_model = tf.keras.Model(inputs=inputs, outputs=outputs)
```

```
        keras_model.trainable = False
        keras_model.summary()
        if keras:
            keras_model.save(f, save_format='tf')
        else:
            m = tf.function(lambda x: keras_model(x))  #full model
            spec = tf.TensorSpec(keras_model.inputs[0].shape, keras_model.inputs[0].dtype)
            m = m.get_concrete_function(spec)
            frozen_func = convert_variables_to_constants_v2(m)
            tfm = tf.Module()
            tfm.__call__ = tf.function(lambda x: frozen_func(x), [spec])
            tfm.__call__(im)
            tf.saved_model.save(
                tfm,
                f,
                options=tf.saved_model.SaveOptions(experimental_custom_gradients=False) if
                check_version(tf.__version__, '2.6') else tf.saved_model.SaveOptions())
        LOGGER.info(f'{prefix} export success, saved as {f} ({file_size(f):.1f} MB)')
        return keras_model, f
    except Exception as e:
        LOGGER.info(f'\n{prefix} export failure: {e}')
        return None, None

def export_pb(keras_model, im, file, prefix=colorstr('TensorFlow GraphDef:')):
    #YOLOv5 TensorFlow GraphDef *.pb export https://github.com/leimao/Frozen_Graph_
TensorFlow
    try:
        import tensorflow as tf
        from tensorflow.python.framework.convert_to_constants import convert_variables_to_
constants_v2

        LOGGER.info(f'\n{prefix} starting export with tensorflow {tf.__version__}...')
        f = file.with_suffix('.pb')

        m = tf.function(lambda x: keras_model(x))  #full model
        m = m.get_concrete_function(tf.TensorSpec(keras_model.inputs[0].shape,
                                    keras_model.inputs[0].dtype))
        frozen_func = convert_variables_to_constants_v2(m)
        frozen_func.graph.as_graph_def()
```

```
        tf.io.write_graph(graph_or_graph_def=frozen_func.graph, logdir=str(f.parent),
                      name=f.name, as_text=False)

        LOGGER.info(f'{prefix} export success, saved as {f} ({file_size(f):.1f} MB)')
        return f
    except Exception as e:
        LOGGER.info(f'\n{prefix} export failure: {e}')

def export_tflite(keras_model, im, file, int8, data, ncalib, prefix=colorstr('TensorFlow
Lite:')):
    #YOLOv5 TensorFlow Lite export
    try:
        import tensorflow as tf

        LOGGER.info(f'\n{prefix} starting export with tensorflow {tf.__version__}...')
        batch_size, ch, *imgsz = list(im.shape)  #BCHW
        f = str(file).replace('.pt', '-fp16.tflite')

        converter = tf.lite.TFLiteConverter.from_keras_model(keras_model)
        converter.target_spec.supported_ops = [tf.lite.OpsSet.TFLITE_BUILTINS]
        converter.target_spec.supported_types = [tf.float16]
        converter.optimizations = [tf.lite.Optimize.DEFAULT]
        if int8:
            from models.tf import representative_dataset_gen
            dataset = LoadImages(check_dataset(data)['train'], img_size=imgsz, auto=False)
            converter.representative_dataset = lambda: representative_dataset_gen(dataset,
ncalib)
            converter.target_spec.supported_ops = [tf.lite.OpsSet.TFLITE_BUILTINS_INT8]
            converter.target_spec.supported_types = []
            converter.inference_input_type = tf.uint8
            converter.inference_output_type = tf.uint8
            converter.experimental_new_quantizer = False
            f = str(file).replace('.pt', '-int8.tflite')

        tflite_model = converter.convert()
        open(f, "wb").write(tflite_model)
        LOGGER.info(f'{prefix} export success, saved as {f} ({file_size(f):.1f} MB)')
        return f
    except Exception as e:
```

```
        LOGGER.info(f'\n{prefix} export failure: {e}')

def export_edgetpu(keras_model, im, file, prefix=colorstr('Edge TPU:')):
    #YOLOv5 Edge TPU export https://coral.ai/docs/edgetpu/models-intro/
    try:
        cmd = 'edgetpu_compiler --version'
        help_url = 'https://coral.ai/docs/edgetpu/compiler/'
        assert platform.system() == 'Linux', f'export only supported on Linux. See {help_
url}'
        if subprocess.run(cmd + ' >/dev/null', shell=True).returncode != 0:
            LOGGER.info(f'\n{prefix} export requires Edge TPU compiler. Attempting
                                                    install from {help_url}')
            #sudo installed on system
            sudo = subprocess.run('sudo --version >/dev/null', shell=True).returncode == 0
            for c in ['curl https://packages.cloud.google.com/apt/doc/apt-key.gpg | sudo
apt-key add -',
                    'echo "deb https://packages.cloud.google.com/apt coral-edgetpu-stable
main" |
                    sudo tee /etc/apt/sources.list.d/coral-edgetpu.list', 'sudo apt-get
update',
                    'sudo apt-get install edgetpu-compiler']:
                subprocess.run(c if sudo else c.replace('sudo ', ''), shell=True,
check=True)
        ver = subprocess.run(cmd, shell=True, capture_output=True, check=True).stdout.
decode().split()[-1]

        LOGGER.info(f'\n{prefix} starting export with Edge TPU compiler {ver}...')
        f = str(file).replace('.pt', '-int8_edgetpu.tflite')  #Edge TPU model
        f_tfl = str(file).replace('.pt', '-int8.tflite')  #TFLite model

        cmd = f"edgetpu_compiler -s {f_tfl}"
        subprocess.run(cmd, shell=True, check=True)

        LOGGER.info(f'{prefix} export success, saved as {f} ({file_size(f):.1f} MB)')
        return f
    except Exception as e:
        LOGGER.info(f'\n{prefix} export failure: {e}')

def export_tfjs(keras_model, im, file, prefix=colorstr('TensorFlow.js:')):
```

```
    #YOLOv5 TensorFlow.js export
    try:
        check_requirements(('tensorflowjs',))
        import re

        import tensorflowjs as tfjs

        LOGGER.info(f'\n{prefix} starting export with tensorflowjs {tfjs.__version__}...')
        f = str(file).replace('.pt', '_web_model')  #js dir
        f_pb = file.with_suffix('.pb')  #*.pb path
        f_json = f + '/model.json'  #*.json path

        cmd = f'tensorflowjs_converter --input_format=tf_frozen_model ' \
              f'--output_node_names="Identity,Identity_1,Identity_2,Identity_3" {f_pb} {f}'
        subprocess.run(cmd, shell=True)

        json = open(f_json).read()
        with open(f_json, 'w') as j:  #sort JSON Identity_* in ascending order
            subst = re.sub(
                r'{"outputs": {"Identity.?.?": {"name": "Identity.?.?"}, '
                r'"Identity.?.?": {"name": "Identity.?.?"}, '
                r'"Identity.?.?": {"name": "Identity.?.?"}, '
                r'"Identity.?.?": {"name": "Identity.?.?"}}}',
                r'{"outputs": {"Identity": {"name": "Identity"}, '
                r'"Identity_1": {"name": "Identity_1"}, '
                r'"Identity_2": {"name": "Identity_2"}, '
                r'"Identity_3": {"name": "Identity_3"}}}',
                json)
            j.write(subst)

        LOGGER.info(f'{prefix} export success, saved as {f} ({file_size(f):.1f} MB)')
        return f
    except Exception as e:
        LOGGER.info(f'\n{prefix} export failure: {e}')

@torch.no_grad()
def run(data=ROOT / 'data/coco128.yaml',                       #'dataset.yaml path'
        weights=ROOT / 'yolov5s.pt',                           #weights path
        imgsz=(640, 640),                                      #image (height, width)
```

```
        batch_size=1,                                 #batch size
        device='cpu',                                 #cuda device, i.e. 0 or 0,1,2,3 or cpu
        include=('torchscript', 'onnx'),              #include formats
        half=False,                                   #FP16 half-precision export
        inplace=False,                                #set YOLOv5 Detect() inplace=True
        train=False,                                  #model.train() mode
        optimize=False,                               #TorchScript: optimize for mobile
        int8=False,                                   #CoreML/TF INT8 quantization
        dynamic=False,                                #ONNX/TF: dynamic axes
        simplify=False,                               #ONNX: simplify model
        opset=12,                                     #ONNX: opset version
        verbose=False,                                #TensorRT: verbose log
        workspace=4,                                  #TensorRT: workspace size (GB)
        nms=False,                                    #TF: add NMS to model
        agnostic_nms=False,                           #TF: add agnostic NMS to model
        topk_per_class=100,                           #TF.js NMS: topk per class to keep
        topk_all=100,                                 #TF.js NMS: topk for all classes to keep
        iou_thres=0.45,                               #TF.js NMS: IoU threshold
        conf_thres=0.25                               #TF.js NMS: confidence threshold
        ):
    t = time.time()
    include = [x.lower() for x in include]  #to lowercase
    formats = tuple(export_formats()['Argument'][1:])  #--include arguments
    flags = [x in include for x in formats]
    assert sum(flags) == len(include), f'ERROR: Invalid --include {include},
                             valid --include arguments are {formats}'
    jit, onnx, xml, engine, coreml, saved_model, pb, tflite, edgetpu, tfjs = flags  #export
booleans
    file = Path(url2file(weights) if str(weights).startswith(('http:/', 'https:/')) else
weights) #PyTorch weights

    #Load PyTorch model
    device = select_device(device)
    assert not (device.type == 'cpu' and half), '--half only compatible with GPU export, i.e.
use --device 0'
    model = attempt_load(weights, map_location=device, inplace=True, fuse=True)  #load FP32
model
    nc, names = model.nc, model.names  #number of classes, class names
```

```
    #Checks
    imgsz *= 2 if len(imgsz) == 1 else 1  #expand
    opset = 12 if ('openvino' in include) else opset  #OpenVINO requires opset <= 12
    assert nc == len(names), f'Model class count {nc} != len(names) {len(names)}'

    #Input
    gs = int(max(model.stride))  #grid size (max stride)
    imgsz = [check_img_size(x, gs) for x in imgsz]  #verify img_size are gs-multiples
    im = torch.zeros(batch_size, 3, *imgsz).to(device)  #image size(1,3,320,192) BCHW
iDetection

    #Update model
    if half:
        im, model = im.half(), model.half()  #to FP16
    model.train() if train else model.eval()  #training mode = no Detect() layer grid
construction
    for k, m in model.named_modules():
        if isinstance(m, Conv):  #assign export-friendly activations
            if isinstance(m.act, nn.SiLU):
                m.act = SiLU()
        elif isinstance(m, Detect):
            m.inplace = inplace
            m.onnx_dynamic = dynamic
            if hasattr(m, 'forward_export'):
                m.forward = m.forward_export  #assign custom forward (optional)

    for _ in range(2):
        y = model(im)  #dry runs
    shape = tuple(y[0].shape)  #model output shape
    LOGGER.info(f"\n{colorstr('PyTorch:')} starting from {file} with output shape {shape}
                                  ({file_size(file):.1f} MB)")

    #Exports
    f = [''] * 10  #exported filenames
    warnings.filterwarnings(action='ignore', category=torch.jit.TracerWarning)  #suppress
TracerWarning
    if jit:
        f[0] = export_torchscript(model, im, file, optimize)
    if engine:  #TensorRT required before ONNX
```

```
        f[1] = export_engine(model, im, file, train, half, simplify, workspace, verbose)
    if onnx or xml:  #OpenVINO requires ONNX
        f[2] = export_onnx(model, im, file, opset, train, dynamic, simplify)
    if xml:  #OpenVINO
        f[3] = export_openvino(model, im, file)
    if coreml:
        _, f[4] = export_coreml(model, im, file)

    #TensorFlow Exports
    if any((saved_model, pb, tflite, edgetpu, tfjs)):
        if int8 or edgetpu:  #TFLite --int8 bug https://github.com/ultralytics/yolov5/
issues/5707
            check_requirements(('flatbuffers==1.12',))  #required before `import tensorflow`
        assert not (tflite and tfjs), 'TFLite and TF.js models must be exported separately,
                              please pass only one type.'
        model, f[5] = export_saved_model(model, im, file, dynamic, tf_nms=nms or agnostic_
nms or tfjs,
                              agnostic_nms=agnostic_nms or tfjs,
                              topk_per_class=topk_per_class, topk_all=topk_all,
                              conf_thres=conf_thres, iou_thres=iou_thres)  #keras model
        if pb or tfjs:  #pb prerequisite to tfjs
            f[6] = export_pb(model, im, file)
        if tflite or edgetpu:
            f[7] = export_tflite(model, im, file, int8=int8 or edgetpu, data=data, ncalib=100)
        if edgetpu:
            f[8] = export_edgetpu(model, im, file)
        if tfjs:
            f[9] = export_tfjs(model, im, file)

    #Finish
    f = [str(x) for x in f if x]  #filter out '' and None
    if any(f):
        LOGGER.info(f'\nExport complete ({time.time() - t:.2f}s)'
                    f"\nResults saved to {colorstr('bold', file.parent.resolve())}"
                    f"\nDetect: python detect.py --weights {f[-1]}"
                    f"\nPyTorch Hub: model = torch.hub.load('ultralytics/yolov5', 'custom',
'{f[-1]}')"
                    f"\nValidate: python val.py --weights {f[-1]}"
                    f"\nVisualize: https://netron.app")
    return f  #return list of exported files/dirs.
```

3.3.2 開發步驟與驗證

3.3.2.1 專案部署

本專案基於 YOLOv5 物件辨識演算法進行口罩檢測模型的訓練，開發專案套件 object_ detection_yolov5-6.1 有 NCNN（官方原版）和 RKNN（瑞芯最佳化版本）版本，訓練的步驟完全一樣，下面以 NCNN 版本的開發專案套件為例介紹。

1）硬體部署

詳見 2.1.2.1 節。

2）專案部署

（1）執行 MobaXterm 工具，透過 SSH 登入到深度學習伺服器。

（2）在 SSH 終端執行以下命令，建立開發專案目錄。

```
$ mkdir -p ~/aiedge-exp
```

（3）將本開發專案程式上傳到 ~/aicam-exp 目錄下，並採用 unzip 命令按照 3.2.2.1 節的方法進行解壓縮。

3.3.2.2 資料處理

透過 voc2yolo 工具將 VOC 格式的資料集轉為 yolo 格式的資料集，並切分訓練集和驗證集。voc2yolo 工具位於 object_detection_yolov5-6.1 資料夾內（由 object_detection_yolov5- 6.1.zip 解壓得到）。

（1）將製作的資料集 mask_dataset_v1.0.zip 透過 SSH 上傳到 ~/aiedge-exp 目錄下（模型的精度依賴於資料集，本專案採用已標註好的資料集進行模型訓練）。

（2）解壓縮資料集，並將資料集複製到 object_detection_yolov5-6.1 專案目錄下：

```
$ cd ~/aiedge-exp
$ unzip mask_dataset_v1.0.zip
$ cd object_detection_yolov5-6.1
$ cp -a ../mask_dataset_v1.0/dataset .
```

（3）根據是否戴口罩的類別修改 voc2yolo.py 內的目標類別，透過的 SSH 將修改好的檔案上傳到邊緣計算閘道。修改後的 voc2yolo.py 如下：

```
classes = ['face','face_mask']
```

（4）在 SSH 終端輸入以下命令進行資料集轉換和資料集的切分，完成後將在當前資料夾內生成 yolo 格式的資料集 labels，以及訓練集 object_train.txt 和驗證集 object_val.txt。

```
$ cd ~/aiedge-exp/object_detection_yolov5-6.1
$ conda activate py36_tf25_torch110_cuda113_cv345
$ python3 voc2yolo.py
Totally convert 2707 files.
```

3.3.2.3 參數配置

建立口罩檢測的設定檔。

（1）在 SSH 終端輸入以下命令，將口罩檢測模型的訓練設定檔 yolov5s-mask.yaml 複製到 data 資料夾：

```
$ cd ~/aiedge-exp/object_detection_yolov5-6.1
$ cp data/coco128.yaml data/yolov5s-mask.yaml
```

（2）根據口罩檢測模型的資料集，修改 data/yolov5s-mask.yaml 檔案內的相關參數：

```
path: ./                               #dataset root dir
train: ./object_train.txt              #train images (relative to 'path')
val: ./object_val.txt                  #val images (relative to 'path')
#Classes
nc: 2                                  #number of classes
names: ['face', 'face_mask']           #class names
```

（3）修改口罩檢測模型的訓練超參數設定檔 data/hyp.scratch.yaml，禁止影像鏡像翻轉，這樣在訓練時可以防止 YOLOv5 模型因自動進行影像上下左右翻轉而導致模型精度差，如圖 3.38 所示。

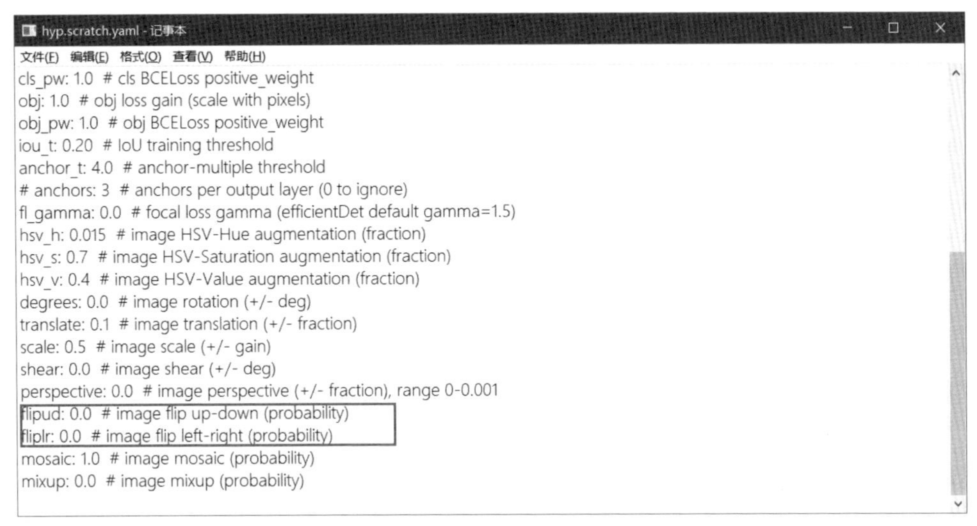
hyp.scratch.yaml - 记事本

文件(F) 编辑(E) 格式(O) 查看(V) 帮助(H)

```
cls_pw: 1.0  # cls BCELoss positive_weight
obj: 1.0  # obj loss gain (scale with pixels)
obj_pw: 1.0  # obj BCELoss positive_weight
iou_t: 0.20  # IoU training threshold
anchor_t: 4.0  # anchor-multiple threshold
# anchors: 3  # anchors per output layer (0 to ignore)
fl_gamma: 0.0  # focal loss gamma (efficientDet default gamma=1.5)
hsv_h: 0.015  # image HSV-Hue augmentation (fraction)
hsv_s: 0.7  # image HSV-Saturation augmentation (fraction)
hsv_v: 0.4  # image HSV-Value augmentation (fraction)
degrees: 0.0  # image rotation (+/- deg)
translate: 0.1  # image translation (+/- fraction)
scale: 0.5  # image scale (+/- gain)
shear: 0.0  # image shear (+/- deg)
perspective: 0.0  # image perspective (+/- fraction), range 0-0.001
flipud: 0.0  # image flip up-down (probability)
fliplr: 0.0  # image flip left-right (probability)
mosaic: 1.0  # image mosaic (probability)
mixup: 0.0  # image mixup (probability)
```

▲ 圖 3.38 修改口罩檢測模型的訓練超參數設定檔

注意：在修改口罩檢測模型的訓練超參數設定檔時，不能修改「flipud: 0.0 #image flip up-down (probability)」和「fliplr: 0.0 #image flip left-right (probability)」，否則會使模型的訓練效果變得極差，導致訓練出的模型的檢測置信度過低。

3.3.2.4 模型訓練

透過 object_detection_yolov5-6.1 框架完成口罩檢測模型的訓練（使用 RTX3080 顯示卡需要 30 ～ 60 min）。

（1）在 SSH 終端輸入以下命令進入專案目錄，啟動 conda 環境。

```
$ cd ~/aiedge-exp/object_detection_yolov5-6.1/
$ conda activate py36_tf25_torch110_cuda113_cv345
```

（2）執行命令進行模型訓練，其中，「--name yolov5s-mask」表示儲存模型的專案目錄名稱為 yolov5s-mask，該目錄位於 aiedge-exp/object_detection_yolov5-6.1/runs/train/ 目錄下。模型訓練的程式如下：

```
$ python3 train.py --img-size 640 --batch-size 20 --epochs 30
              --data ./data/yolov5s-mask.yaml --weights ./models/yolov5s.pt
              --cfg ./models/yolov5s.yaml --name yolov5s-mask
github: skipping check (not a git repository)
YOLOv5 torch 1.10.1 CUDA:0 (NVIDIA GeForce RTX 3080, 10009.625MB)

Namespace(adam=False, artifact_alias='latest', batch_size=20, bbox_interval=-1, bucket='',
        cache_images=False, cfg='./models/hub/yolov5s6.yaml', data='./data/yolov5_mask.
yaml',
        device='', entity=None, epochs=30, evolve=False, exist_ok=False, global_rank=-1,
        hyp='data/hyp.scratch.yaml', image_weights=False, img_size=[640, 640], linear_
lr=False,
        local_rank=-1, multi_scale=False, name='yolov5s-mask', noautoanchor=False,
        nosave=False, notest=False, project='runs/train', quad=False, rect=False,
        resume=False, save_dir='runs/train/yolov5s-mask2', save_period=-1, single_
cls=False,
        sync_bn=False, total_batch_size=20, upload_dataset=False, weights='./models/
yolov5s6.pt',
        workers=8, world_size=1)
tensorboard: Start with 'tensorboard --logdir runs/train', view at http://localhost:6006/
hyperparameters: lr0=0.01, lrf=0.2, momentum=0.937, weight_decay=0.0005, warmup_epochs=3.0,
             warmup_momentum=0.8, warmup_bias_lr=0.1, box=0.05, cls=0.5, cls_pw=1.0,
             obj=1.0, obj_pw=1.0, iou_t=0.2, anchor_t=4.0, fl_gamma=0.0, hsv_h=0.015,
             hsv_s=0.7, hsv_v=0.4, degrees=0.0, translate=0.1, scale=0.5, shear=0.0,
             perspective=0.0, flipud=0.0, fliplr=0.0, mosaic=1.0, mixup=0.0
wandb: Install Weights & Biases for YOLOv5 logging with 'pip install wandb' (recommended)
Overriding model.yaml nc=80 with nc=2

              from  n    params  module                                  arguments
0               -1  1      3520  models.common.Conv                      [3, 32, 6, 2, 2]
1               -1  1     18560  models.common.Conv                      [32, 64, 3, 2]
2               -1  1     18816  models.common.C3                        [64, 64, 1]
3               -1  1     73984  models.common.Conv                      [64, 128, 3, 2]
4               -1  2    115712  models.common.C3                        [128, 128, 2]
5               -1  1    295424  models.common.Conv                      [128, 256, 3, 2]
6               -1  3    625152  models.common.C3                        [256, 256, 3]
7               -1  1   1180672  models.common.Conv                      [256, 512, 3, 2]
8               -1  1   1182720  models.common.C3                        [512, 512, 1]
9               -1  1    656896  models.common.SPPF                      [512, 512, 5]
```

```
10                -1  1    131584  models.common.Conv                      [512, 256, 1, 1]
11                -1  1         0  torch.nn.modules.upsampling.Upsample   [None, 2, 'nearest']
12           [-1, 6]  1         0  models.common.Concat                    [1]
13                -1  1    361984  models.common.C3                        [512, 256, 1, False]
14                -1  1     33024  models.common.Conv                      [256, 128, 1, 1]
15                -1  1         0  torch.nn.modules.upsampling.Upsample   [None, 2, 'nearest']
16           [-1, 4]  1         0  models.common.Concat                    [1]
17                -1  1     90880  models.common.C3                        [256, 128, 1, False]
18                -1  1    147712  models.common.Conv                      [128, 128, 3, 2]
19          [-1, 14]  1         0  models.common.Concat                    [1]
20                -1  1    296448  models.common.C3                        [256, 256, 1, False]
21                -1  1    590336  models.common.Conv                      [256, 256, 3, 2]
22          [-1, 10]  1         0  models.common.Concat                    [1]
23                -1  1   1182720  models.common.C3                        [512, 512, 1, False]
24      [17, 20, 23]  1     18879  models.yolo.Detect                      [2, [[10, 13, 16,
30, 33, 23], [30, 61, 62, 45, 59, 119], [116, 90, 156, 198, 373, 326]], [128, 256, 512]]
Model Summary: 270 layers, 7025023 parameters, 7025023 gradients, 16.0 GFLOPs

Transferred 342/349 items from models/yolov5s.pt
Scaled weight_decay = 0.00046875
optimizer: SGD with parameter groups 57 weight (no decay), 60 weight, 60 bias
train: Scanning '/home/zonesion/aiedge-exp/object_detection_yolov5-6.1/object_train.cache'
            images and labels... 378 found, 0 missing, 0 empty, 0 corrup
val: Scanning '/home/zonesion/aiedge-exp/object_detection_yolov5-6.1/object_val.cache'
            images and labels... 42 found, 0 missing, 0 empty, 0 corrupt: 10
Plotting labels to runs/train/yolov5s-mask3/labels.jpg...

AutoAnchor: 6.01 anchors/target, 1.000 Best Possible Recall (BPR). Current anchors are a
good fit to dataset
Image sizes 640 train, 640 val
Using 2 dataloader workers
Logging results to runs/train/yolov5s-mask3
Starting training for 30 epochs...
```

YOLOv5 模型會自動載入 GPU 進行訓練，訓練 30 個迴圈大約需要 20 min。

（3）訓練完成後系統會顯示口罩檢測模型的平均準確率等指標，如下所示：

```
Epoch    gpu_mem    box      obj       cls        total     labels        img_size
28/29    4.04G    0.0203   0.01142   0.002378   0.0341    48
640: 100%|██████████████████████████████████████████████| 122/122 [00:13<00:00,  9.38it/s]
Class     Images     Labels          P          R       mAP@.5   mAP@.5:.95:
    100%|██████████████████████████████████████████████| 7/7 [00:00<00:00, 10.68it/s]
all      270        360       0.97      0.888      0.94      0.647

Epoch    gpu_mem    box      obj       cls        total     labels        img_size
29/29    4.04G    0.02022   0.01136   0.002838   0.03442    67
640: 100%|██████████████████████████████████████████████| 122/122 [00:13<00:00,  9.36it/s]
Class     Images     Labels          P          R       mAP@.5   mAP@.5:.95:
    100%|██████████████████████████████████████████████| 7/7 [00:00<00:00,  7.07it/s]
all      270        360       0.97      0.884      0.945      0.638
mask     270        103       0.965     0.816      0.908      0.567
30 epochs completed in 0.129 hours.

Optimizer stripped from runs/train/yolov5s-mask/weights/last.pt, 25.1MB
Optimizer stripped from runs/train/yolov5s-mask/weights/best.pt, 25.1MB
```

（4）使用 tree 命令查看 yolov5s-mask 模型的專案目錄，其中的 best.pt 就是最終的模型檔案，如下所示：

```
$ tree runs/train/yolov5s-mask
runs/train/yolov5s-mask
├── confusion_matrix.png
├── events.out.tfevents.1678938893.zonesion.22408.0
├── F1_curve.png
├── hyp.yaml
├── labels_correlogram.jpg
├── labels.jpg
├── opt.yaml
├── P_curve.png
├── PR_curve.png
├── R_curve.png
├── results.csv
├── results.png
├── train_batch0.jpg
```

```
├── train_batch1.jpg
├── train_batch2.jpg
├── val_batch0_labels.jpg
├── val_batch0_pred.jpg
├── val_batch1_labels.jpg
├── val_batch1_pred.jpg
├── val_batch2_labels.jpg
├── val_batch2_pred.jpg
└── weights
    ├── best.pt
    └── last.pt
```

3.3.2.5 模型驗證

（1）若沒有訓練出最終模型，則可以解壓本專案提供的最終模型檔案 mask_model/yolov5s-mask.pt，並將其複製到 object_detection_yolov5-6.1 目錄後再進行步驟（3）所示的操作。

```
$ cd ~/aiedge-exp
$ unzip mask_model.zip
$ cd ~/aiedge-exp/object_detection_yolov5-6.1
$ cp -a ../mask_model/yolov5s-mask.pt .
```

（2）若透過前面的步驟完成口罩檢測模型的訓練，則將最終模型檔案命名為 yolov5s- mask.pt，並複製到 object_detection_yolov5-6.1 目錄下。

```
$ cd ~/aiedge-exp/object_detection_yolov5-6.1/runs/train/yolov5s-mask/weights
$ mv best.pt yolov5s-mask.pt
$ cd ~/aiedge-exp/object_detection_yolov5-6.1
$ cp -a ../object_detection_yolov5-6.1/runs/train/yolov5s-mask/weights/yolov5s-mask.pt .
```

（3）在資料集中選擇一張測試樣圖，如 dataset/images/test_00000100.jpg。

（4）在 SSH 終端輸入以下命令對測試樣圖進行測試，執行命令與結果如下：

```
$ cd ~/aiedge-exp/object_detection_yolov5-6.1
$ conda activate py36_tf25_torch110_cuda113_cv345
$ python3 detect.py --source ./dataset/images/test_00000100.jpg --weights ./yolov5s-mask.pt
            --img-size 640 --view-img --name yolov5s-mask
```

```
detect: weights=['./yolov5s-mask.pt'], source=./dataset/images/test_00000100.jpg,
            data=data/coco128.yaml, imgsz=[640, 640], conf_thres=0.25, iou_thres=0.45,
            max_det=1000, device=, view_img=True, save_txt=False, save_conf=False,
            save_crop=False, nosave=False, classes=None, agnostic_nms=False,
            augment=False, visualize=False, update=False, project=runs/detect,
            name=yolov5s-mask, exist_ok=False, line_thickness=3, hide_labels=False,
            hide_conf=False, half=False, dnn=False
requirements: /home/zonesion/hou/object_detection_yolov5-6.1/requirements.txt not found,
check failed.
YOLOv5 2022-2-22 torch 1.10.1 CUDA:0 (NVIDIA GeForce RTX 3080, 10010MiB)

Fusing layers...
Model Summary: 213 layers, 7015519 parameters, 0 gradients, 15.8 GFLOPs
qt.qpa.xcb: QXcbConnection: XCB error: 145 (Unknown), sequence: 179, resource id: 0,
         major code: 139 (Unknown), minor code: 20
image 1/1 /home/zonesion/hou/object_detection_yolov5-6.1/dataset/images/test_00000100.jpg:
         576x640 1 face_mask, Done. (0.010s)
Speed: 0.3ms pre-process, 9.6ms inference, 1.2ms NMS per image at shape (1, 3, 640, 640)
Results saved to runs/detect/yolov5s-mask
```

執行完成後會在專案目錄 runs/detect/yolov5s-mask 生成辨識結果檔案 test_00000100.jpg。

（5）開啟 test_00000100.jpg 後可以看到影像中的辨識準確度，如圖 3.39 所示。

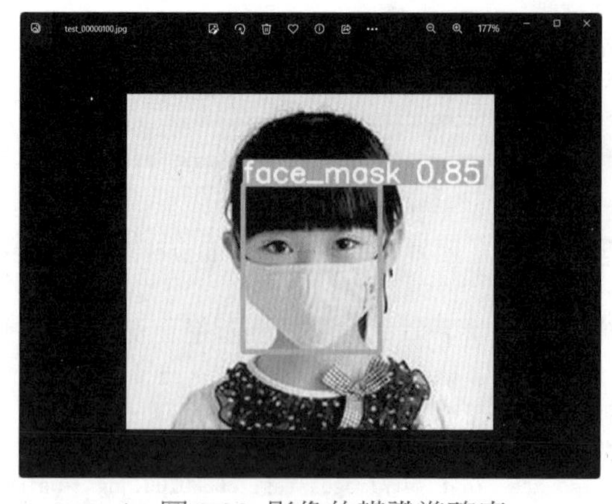

▲ 圖 3.39 影像的辨識準確度

3.3.3 本節小結

本節主要介紹了 PyTorch 框架和 YOLOv5 模型，結合案例學習了基於 YOLOv5 的物件辨識演算法模型部署、訓練及驗證的開發步驟。

3.3.4 思考與擴充

（1）在 YOLOv5 模型的專案中，如何實現資料格式的轉換？

（2）在 YOLOv5 模型的專案中，如何建立口罩檢測模型的超參數設定檔？如何透過修改模型的超參數配置，避免在訓練過程中因影像翻轉影響導致的模型精度變差？

3.4 YOLOv3 模型的推理與驗證

模型的推理是深度學習模型在實際應用中進行驗證的重要步驟，推理是指將訓練好的模型應用於新的、未見過的資料，從而生成預測或分類。模型的推理與驗證是確保深度學習模型在實際應用中能夠有效發揮作用的關鍵環節，透過持續監控和評估模型在真實場景中的表現，可以確保模型在不同情況下都能夠可靠工作。推理與驗證的作用如下：

（1）實際性能評估：模型的推理可用來評估模型在真實世界情境中的實際性能，這是與模型在訓練集和驗證集上的性能進行對比的重要步驟。

（2）泛化能力檢測：模型在推理時需要展現其泛化能力，即對未見過的資料的良好適應能力。透過在真實環境中驗證模型，可以評估其對新資料的預測能力。

（3）性能監控：模型在實際應用中可能會受到各種影響，如資料分佈的變化、雜訊的增加等。透過定期對模型進行推理與驗證，可以監控模型的性能並及時檢測到性能下降或其他問題。

（4）使用者體驗：在實際應用中，使用者可以對模型的性能和可靠性有更直接的感受。推理與驗證可以確保模型能夠提供令人滿意的使用者體驗。

（5）模型部署決策：模型的推理與驗證結果可能影響到模型是否能夠成功部署到生產環境中，在推理階段發現的問題可能需要進一步最佳化或調整模型框架。

（6）安全性和堅固性檢測：在實際應用中，模型可能面臨不同類型的攻擊或異常情況，推理與驗證有助評估模型在這些情況下的安全性和堅固性。

（7）決策支援：模型的推理與驗證結果可以為業務決策提供支援，模型在實際應用中的表現直接關係到其對業務目標的貢獻。

本節的基礎知識如下：

- 掌握模型從開發框架到推理框架的轉換過程。
- 了解常用的行動端邊緣推理框架。
- 結合案例掌握基於 YOLOv3 的 NCNN 推理框架部署、模型轉換及推理與驗證的過程。

3.4.1 原理分析與開發設計

3.4.1.1 推理框架概述

1）從開發框架到推理框架

深度學習演算法從研發到應用有兩個環節：第一個環節是設計並訓練模型；第二個環節是把模型部署到產品上。推理通常被認為是部署到產品後框架所需實現的運算。為一個模型輸入一個給定的資料後得出一個資料，類似於推理獲得了一個結果。不過上述提到的第一個環節，其實也涉及推理，只不過在訓練時的推理只是其中的一小部分而已。訓練通常包括推理 + 資料集 + 損失函式。能夠執行這個推理過程的軟體可以認為是一個推理框架。

將深度學習模型安置在恰當「工作職位」上的過程稱為模型部署。模型部署除了要將模型的能力植入「需要它的地方」，往往還伴隨著對模型的最佳化，如運算元融合與重排、權重量化、知識蒸餾等。

模型的設計和訓練往往比較複雜，這一過程依賴於靈活的模型開發框架（如 PyTorch、TensorFlow 等）。模型的使用則追求在特定場合下實現模型的「瘦身」及其加速推理，這需要專用推理框架（如 TensorRT、ONNX Runtime 等）的支援。從這個角度理解，模型部署過程也可以看成模型從開發框架到特定推理框架的轉換過程。

開發框架和推理框架的多樣性有利於模型的設計和使用，但也給模型轉換帶來的挑戰。為了簡化模型的部署過程，Facebook 和微軟在 2017 年共同發佈了一種深度學習模型的中間表示形式，即 ONNX（Open Neural Network eXchange）。這樣一來，許多的模型訓練框架和推理框架只需要與 ONNX 建立聯繫，就可以實現模型格式的相互轉換。ONNX 的出現使模型部署逐漸形成了「模型開發框架→模型中間表示→模型推理框架」的範式，如圖 3.40 所示。

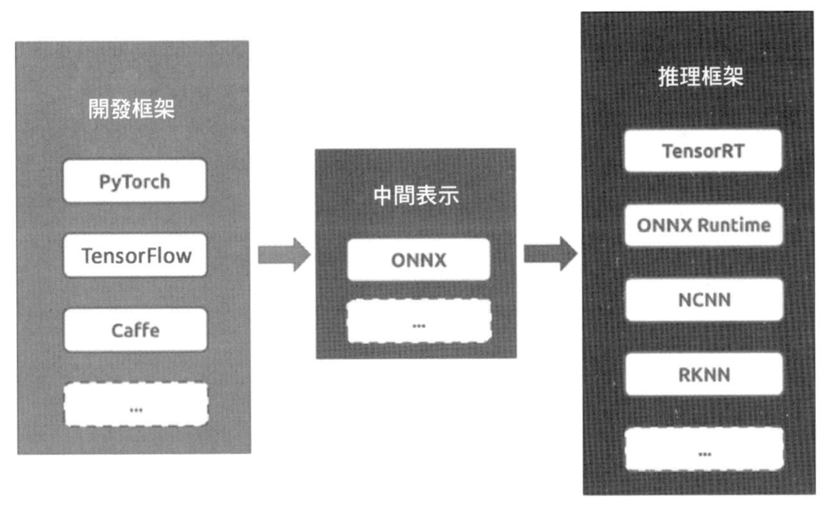

▲ 圖 3.40 「模型開發框架→模型中間表示→模型推理框架」的範式

2）行動端推理框架

自深度學習推出以來，龐大的計算量，以及對部署裝置較高的算力需求，使得深度學習模型通常部署在雲端，這種方式的效率和即時性都存在嚴重的問題。一般來說部署深度學習模型需要特定框架的支援，如 TensorFlow、PyTorch、Caffe 等，但這些框架都是 GPU 算力導向的，對於行動端尤其是僅有 CPU 的嵌入式平臺裝置並未做出最佳化。輕量化的模型結構設計雖然可以有效提升模型推理速度，但要在價格低廉、算力較弱的嵌入式裝置達到即時性的要求，需要解決以下問題：

（1）算力問題：目前行動端能夠堆積的算力不如伺服器端那麼靈活，算力較小。

（2）功耗問題：大量的算力必然導致較高的功耗。

（3）模型太大：假設一個應用場景需要一個模型，內建上百百萬位元組的模型是不可行的，下載上百百萬位元組的模型也不方便。這使得模型部署通常選取價格較高的、帶有 GPU 的裝置。近年來，業界推出了一些行動端或 CPU 最佳化導向的深度學習框架，如 TensorFlow Lite、NCNN 等前向推理框架，這類框架均對模型推理及部署進行了最佳化，可支援模型的 Int8 量化、卷積運算加速等。

3.4.1.2 常用框架

（1）TensorRT。TensorRT 是 NVIDIA 公司推出的 GPU 算力導向的推理框架，在伺服器端和嵌入式裝置上都有非常好的效果，但底層不開放原始碼。TensorRT 的合作方非常多，主流的框架都支持 TensorRT。TensorRT 的主要呼叫套件語言有 Python、C/C++，支援的模型包括 TensorFlow 1.x、TensorFlow 2.x、PyTorch、ONNX、PaddlePaddle、MXNet、Caffe、Theano、Torch、Lasagne、Blocks。

（2）TensorFlow Lite（TF-Lite）。TF-Lite 是 Google 針對行動端推出的推理框架，功能非常強大，原因是在 Keras、TensorFlow 等模型中都能使

用 TF-Lite，而且有專門的 TPU 和 Android 平臺，這種「一行龍」式的服務讓 TensorFlow 在部署方面非常有利。Keras、TensorFlow 等模型可以透過指令稿很快進行 TF-Lite 的轉換。TF-Lite 的主要呼叫套件語言有 Python、C/C++、Java，支援的模型包括 Keras、TensorFlow、ONNX。

（3）OpenVINO。OpenVINO 是 Intel 的推理框架，是一個功能超級強大的推理部署工具，提供了很多便利的工具，如提供了深度學習推理套件（DLDT），該套件可以將各種開放原始碼框架訓練好的模型進行線上部署，除此之外，還包含了影像處理工具套件 OpenCV、視訊處理工具套件 Media SDK。OpenVINO 主要部署在 Intel 的加速棒或工控機上，主要呼叫套件語言有 C/C++、Python，支援的模型包括 TensorFlow、PyTorch、ONNX、MXNet、PaddlePaddle。

（4）NCNN。NCNN（NVIDIA CUDA Convolutional Neural Network）是騰訊推出的推理框架，其推理速度超過了 TF-Lite，使用人數很多，運算元很多，社區做得也非常棒，是一個使用非常廣的推理框架。NCNN 可支援 x86、GPU，在嵌入式裝置、手機上的表現非常好。NCNN 的主要呼叫套件語言有 C/C++、Python，支援的模型包括 TensorFlow、ONNX、PyTorch、MXNet、Darknet、Caffe。

（5）Tenigne Lite。Tenigne Lite 是 OpenAILab 推出的邊緣端推理框架，OpenCV 在嵌入式裝置上首推 Tenigne Lite，該框架對 RISC-V、CUDA、TensorRT、NPU 的支持非常不錯。Tenigne Lite 的主要呼叫套件語言有 C/C++、Python，支援的模型包括 TensorFlow、ONNX、Darknet、MXNet、NCNN、Caffe、TF-Lite、NCNN。

（6）NNIE。NNIE（Neural Network Inference Engine）是海思 SVP 開發框架中的處理單元之一，主要用於深度學習卷積神經網路的加速處理，可用於影像分類、物件辨識等 AI 應用場景。NNIE 支援大部分公開的卷積神經網路模型，如 AlexNet、VGG16、ResNet18、ResNet50、GoogleNet 等分類網路，Faster R-CNN、YOLO、SSD、RFCN 等檢測目標網路，以及 FCN、SegNet 等分割場景網路。目前 NNIE 及工具鏈僅支援 Caffe 框架，使用其他框架的模型需要轉化為 Caffe 框架的模型。

（7）RKNN。RKNN（Rockchip Neural Network）是 Rockchip 推出的用於嵌入式裝置的神經網路框架。Rockchip 提供的 RKNN-Toolkit 開發套件可進行模型轉換、推理執行和性能評估，支援將 Caffe、TensorFlow、TensorFlow Lite、ONNX、Darknet 框架的模型轉換與 RKNN 框架的模型。

3.4.1.3 NCNN 框架

NCNN 是由騰訊優圖實驗室於 2017 年公佈的開放原始碼專案，旨在提供一個行動端、無第三方依賴、跨平臺導向的高性能神經網路前向推理框架。與 TF-lite 不同的是，NCNN 主要針對 CPU 進行最佳化，使用 ARM NEON 指令集實現 CNN 中的卷積層、全連接層、池化層等。NCNN 有自己的模型檔案格式，支援大部分主流深度學習框架訓練的模型轉換，也支援 Int8 量化和推理，在最新的版本中實現了 Int8 Winograd-f43 的核心最佳化，極大提升了量化模型在 ARM 上的推理速度。NCNN 框架的模型轉換與部署流程如圖 3.41 所示，將訓練後的模型透過 NCNN 框架模型轉換工具，生成對應格式的檔案，再透過模型載入完成部署，對比 TF-Lite，NCNN 對模型的訓練框架和部署平臺沒有太多要求，極大滿足了使用者在各類型裝置上部署模型的需求。

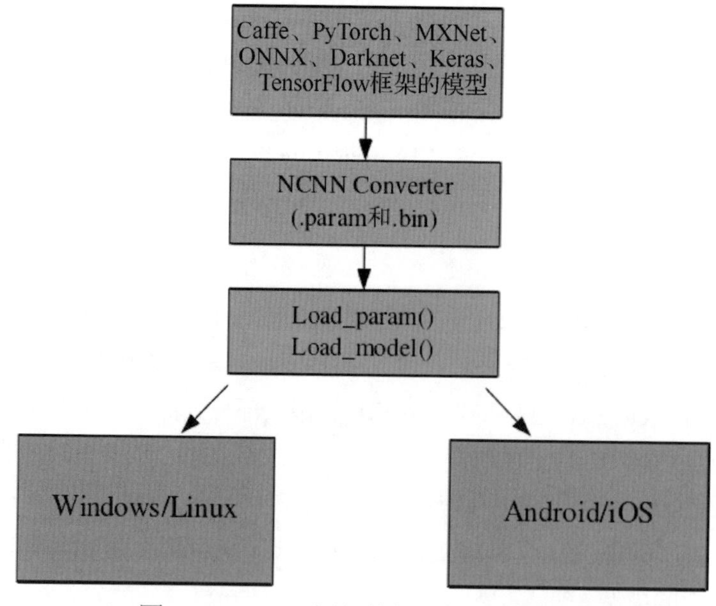

▲ 圖 3.41 NCNN 框架的模型轉換與部署流程

NCNN 框架主要在神經網路推理方面進行了最佳化，圍繞輕量級設計最佳化了神經網路推理的性能，在多種硬體平臺上進行了性能最佳化、硬體加速，使得嵌入式裝置和行動裝置能夠高效率地進行深度學習推理。NCNN 框架的原理和特點如下：

（1）輕量級設計：NCNN 是一個輕量級的深度學習框架，專注於神經網路推理任務，輕量級設計可以使其在資源受限的裝置（如行動裝置和嵌入式裝置）上執行。

（2）支援多個硬體平臺：NCNN 支援多種硬體平臺，包括 CPU、GPU（CUDA）、Vulkan 和 OpenCL，這使不同的硬體裝置可以充分發揮自身的優勢，具有較大的靈活性。

（3）性能最佳化：NCNN 採用多種性能最佳化技術，提高了神經網路推理的速度，包括：

① 記憶體最佳化：透過對記憶體分配和管理進行最佳化，NCNN 減小了模型推理過程中的記憶體佔用量。

② 指令級最佳化：透過對模型計算過程進行指令級最佳化，NCNN 盡可能地減小了計算銷耗，提高了推理速度。

（4）平行計算：NCNN 充分利用了硬體的平行計算能力，如 CUDA 和 Vulkan，加快了神經網路的計算過程。

（5）硬體加速：NCNN 可以透過硬體加速推理，如利用 NVIDIA 的 CUDA 技術和 Vulkan 圖形 API，能夠充分發揮硬體的性能優勢。

（6）支援常見深度學習模型：NCNN 支援常見的深度學習模型，包括卷積神經網路（CNN）和循環神經網路（RNN），可以高效實現這些模型，以及支援這些模型的載入和推理。

（7）非同步推理：NCNN 支援非同步推理，可在模型推理的同時執行其他任務，有助提高系統的整體效率，尤其是在需要處理多個任務的場景下。

3.4.2 開發步驟與驗證

3.4.2.1 專案部署

1）硬體部署

詳見 2.1.2.1 節。

2）專案部署

（1）執行 MobaXterm 工具，透過 SSH 登入到深度學習伺服器。

（2）在 SSH 終端執行以下命令，建立開發專案目錄。

```
$ mkdir -p ~/aiedge-exp
```

（3）透過 SSH 將本開發專案程式上傳到 ~/aicam-exp 目錄下，並採用 unzip 命令按照 3.2.2.1 節的方法進行解壓縮。

（4）在 SSH 終端輸入以下命令解壓縮開發專案和訓練好的交通辨識模型檔案。

```
$ cd ~/aiedge-exp
$ unzip darknet2ncnn.zip
$ unzip traffic_model.zip
```

（5）將 traffic_model 資料夾內的模型設定檔 yolov3-tiny-traffic.cfg 和模型檔案 yolov3-tiny-traffic_final.weight 複製到 darknet2ncnn/data 目錄下（上述檔案也可以透過 3.2 節的專案獲得）。

```
$ cd ~/aiedge-exp/darknet2ncnn
$ cp -a ../traffic_model/* ./data
```

3.4.2.2 模型轉換

透過 voc2yolo 將 voc 格式的資料集轉為 yolo 格式的資料集，並切分訓練集和驗證集。

（1）在 SSH 終端輸入以下命令進行模型轉換。

```
$ cd ~/aiedge-exp/darknet2ncnn
$ chmod 755 tools/*
$ ./tools/darknet2ncnn data/yolov3-tiny-traffic.cfg data/yolov3-tiny-traffic_final.weights
                        yolov3-tiny-traffic.param yolov3-tiny-traffic.bin 1
Loading cfg...
WARNING: The ignore_thresh=0.700000 of yolo0 is too high. An alternative value 0.25 is
written instead.
WARNING: The ignore_thresh=0.700000 of yolo1 is too high. An alternative value 0.25 is
written instead.
Loading weights...
Converting model...
47 layers, 49 blobs generated.
NOTE: The input of darknet uses: mean_vals=0 and norm_vals=1/255.f.
NOTE: Remember to use ncnnoptimize for better performance.
```

（2）在 SSH 終端輸入以下命令進行模型最佳化。

```
$ cd ~/aiedge-exp/darknet2ncnn
$ chmod 755 tools/*
$ ./tools/ncnnoptimize yolov3-tiny-traffic.param yolov3-tiny-traffic.bin yolov3-tiny-traffic-opt.
                                        param yolov3-tiny-traffic-opt.bin 0
fuse_convolution_batchnorm 0_26 0_26_bn
fuse_convolution_batchnorm 2_38 2_38_bn
fuse_convolution_batchnorm 4_50 4_50_bn
fuse_convolution_batchnorm 6_62 6_62_bn
fuse_convolution_batchnorm 8_74 8_74_bn
fuse_convolution_batchnorm 10_86 10_86_bn
fuse_convolution_batchnorm 12_98 12_98_bn
fuse_convolution_batchnorm 13_108 13_108_bn
fuse_convolution_batchnorm 14_116 14_116_bn
fuse_convolution_batchnorm 18_146 18_146_bn
fuse_convolution_batchnorm 21_160 21_160_bn
fuse_convolution_activation 0_26 0_26_bn_leaky
fuse_convolution_activation 2_38 2_38_bn_leaky
fuse_convolution_activation 4_50 4_50_bn_leaky
fuse_convolution_activation 6_62 6_62_bn_leaky
fuse_convolution_activation 8_74 8_74_bn_leaky
```

```
fuse_convolution_activation 10_86 10_86_bn_leaky
fuse_convolution_activation 12_98 12_98_bn_leaky
fuse_convolution_activation 13_108 13_108_bn_leaky
fuse_convolution_activation 14_116 14_116_bn_leaky
fuse_convolution_activation 18_146 18_146_bn_leaky
fuse_convolution_activation 21_160 21_160_bn_leaky
shape_inference
input = data
extract = output
estimated memory footprint = 26180.77 KB = 25.57 MB
mac = 2724074496 = 2724.07 M
```

（3）在專案目錄下生成模型檔案 yolov3-tiny-traffic-opt.bin/yolov3-tiny-traffic-opt.param。

3.4.2.3 模型測試

（1）在 SSH 終端輸入以下命令編譯模型測試程式，編譯成功後會在 build 目錄下生成測試執行檔案 objectdet。

```
$ cd ~/aiedge-exp/darknet2ncnn
$ mkdir -p build
$ cd build
$ cmake ..
-- The C compiler identification is GNU 7.5.0
-- The CXX compiler identification is GNU 7.5.0
-- Check for working C compiler: /usr/bin/cc
-- Check for working C compiler: /usr/bin/cc -- works
-- Detecting C compiler ABI info
-- Detecting C compiler ABI info - done
-- Detecting C compile features
-- Detecting C compile features - done
-- Check for working CXX compiler: /usr/bin/c++
-- Check for working CXX compiler: /usr/bin/c++ -- works
-- Detecting CXX compiler ABI info
-- Detecting CXX compiler ABI info - done
-- Detecting CXX compile features
-- Detecting CXX compile features - done
```

```
-- Found OpenCV: /usr/local (found version "4.5.3")
-- Found OpenMP_C: -fopenmp (found version "4.5")
-- Found OpenMP_CXX: -fopenmp (found version "4.5")
-- Found OpenMP: TRUE (found version "4.5")
OPENMP FOUND
-- Configuring done
-- Generating done
-- Build files have been written to: /home/zonesion/lusi/darknet2ncnn/build
$ make
Scanning dependencies of target objectdet
[ 50%] Building CXX object CMakeFiles/objectdet.dir/yolov3.cpp.o
[100%] Linking CXX executable objectdet
[100%] Built target objectdet
```

（2）執行測試程式，對測試影像進行推理與驗證。

```
$ cd ~/aiedge-exp/darknet2ncnn/build
$ ./objectdet ../data/test.jpg
5 = 0.90109 at 424.53 158.23 108.40 x 93.63
3 = 0.89344 at 147.85 142.42 111.84 x 116.85
4 = 0.51185 at 272.18 143.81 126.40 x 119.39
```

推理與驗證結果如圖 3.42 所示。

▲ 圖 3.42 推理與驗證結果

3.4.3 本節小結

本節首先介紹了推理框架的相關內容，包括開發框架到推理框架的轉換過程、常用的推理框架，然後透過具體的案例介紹了基於 YOLOv3 的 NCNN 框架部署、模型轉換、推理與驗證。

3.4.4 思考與擴充

（1）TensorFlow、PyTorch 與 NCNN 有何不同？

（2）如何實現從開發框架到推理框架的轉換？

3.5 YOLOv5 模型的推理與驗證

本節的基礎知識如下：

- 掌握模型推理與驗證的作用。
- 掌握 RKNN 框架的原理與應用。
- 結合口罩檢測案例，掌握基於 YOLOv5 的 RKNN 框架部署、模型轉換、推理與驗證的過程。

3.5.1 原理分析與開發設計

3.5.1.1 RKNN 框架簡介

RKNN 框架旨在支援深度學習模型的開發，包括模型訓練、模型最佳化、模型轉換和模型載入等。RKNN 框架可以將 TensorFlow、PyTorch、ONNX 等框架的模型轉換成 RKNN 框架的模型，並載入到 NPU（神經網路處理器）上進行計算，同時還可以對模型進行剪枝、量化等最佳化操作，提升模型的執行效率。RKNN 框架可以在 Android、Linux 等多個平臺上執行，通常在邊緣裝置（如手機、嵌入式裝置等）上執行的深度學習任務應用，如人臉辨識、物體檢測、語音辨識等。

RKNN 框架為深度學習模型在 NPU 上的推理提供了便利。為了在 RKNN 框架中使用其他框架訓練的模型，RKNN 官方發佈了 RKNN-Toolkit 開發套件，該開發套件提供了一系列 Python API，用於支援模型轉換、模型量化、模型推理以及模型的狀態檢測等。

RKNN 框架模型的典型部署過程是：模型配置→模型載入→模型建構→模型匯出。

（1）模型配置：用於設定模型轉換參數，包括輸入資料平均值、量化類型、量化演算法和模型部署平臺等。

（2）模型載入：將轉換前的模型載入到程式中，目前 RKNN 框架支援 ONNX、PyTorch、TensorFlow、Caffe 等框架的模型載入轉換。值得一提的是，模型載入是整個轉換過程中的關鍵步驟，這一步允許工程人員自行指定模型載入的輸出層及其名稱，可以決定原始模型的哪些部分參與模型轉換過程。

（3）模型建構：指定模型是否進行量化，以及用於量化校正的資料集。

（4）模型匯出：儲存轉換後的模型。

3.5.1.2 開發設計

1）RKNN 框架的口罩檢測模型

mask_test.py 用於在 RKNN 框架下進行口罩檢測，程式如下：

```
import cv2
import time
import random
import numpy as np
from rknnlite.api import RKNNLite

"""
```

```
RK3588 yolov5s 口罩檢測模型
"""


def get_max_scale(img, max_w, max_h):
    h, w = img.shape[:2]
    scale = min(max_w / w, max_h / h, 1)
    return scale


def get_new_size(img, scale):
    return tuple(map(int, np.array(img.shape[:2][::-1]) * scale))


class AutoScale:
    def __init__(self, img, max_w, max_h):
        self._src_img = img
        self.scale = get_max_scale(img, max_w, max_h)
        self._new_size = get_new_size(img, self.scale)
        self.__new_img = None

    @property
    def size(self):
        return self._new_size

    @property
    def new_img(self):
        if self.__new_img is None:
            self.__new_img = cv2.resize(self._src_img, self._new_size)
        return self.__new_img


def sigmoid(x):
    return 1 / (1 + np.exp(-x))


def filter_boxes(boxes, box_confidences, box_class_probs, conf_thres):
    # 條件機率，在該區域存在物體的機率是某個類別的機率
    box_scores = box_confidences * box_class_probs
    box_classes = np.argmax(box_scores, axis=-1)            # 找出機率最大的類別索引
    box_class_scores = np.max(box_scores, axis=-1)          # 最大類別對應的機率值
    pos = np.where(box_class_scores >= conf_thres)          # 找出機率大於設定值的類別
    #pos = box_class_scores >= OBJ_THRESH
    boxes = boxes[pos]
```

```
    classes = box_classes[pos]
    scores = box_class_scores[pos]
    return boxes, classes, scores

def nms_boxes(boxes, scores, iou_thres):
    x = boxes[:, 0]
    y = boxes[:, 1]
    w = boxes[:, 2]
    h = boxes[:, 3]

    areas = w * h
    order = scores.argsort()[::-1]

    keep = []
    while order.size > 0:
        i = order[0]
        keep.append(i)

        xx1 = np.maximum(x[i], x[order[1:]])
        yy1 = np.maximum(y[i], y[order[1:]])
        xx2 = np.minimum(x[i] + w[i], x[order[1:]] + w[order[1:]])
        yy2 = np.minimum(y[i] + h[i], y[order[1:]] + h[order[1:]])

        w1 = np.maximum(0.0, xx2 - xx1 + 0.00001)
        h1 = np.maximum(0.0, yy2 - yy1 + 0.00001)
        inter = w1 * h1

        ovr = inter / (areas[i] + areas[order[1:]] - inter)
        inds = np.where(ovr <= iou_thres)[0]
        order = order[inds + 1]
    keep = np.array(keep)
    return keep

def plot_one_box(x, img, color=None, label=None, line_thickness=None):
    tl = line_thickness or round(0.002 * (img.shape[0] + img.shape[1]) / 2) + 1
    color = color or [random.randint(0, 255) for _ in range(3)]
    c1, c2 = (int(x[0]), int(x[1])), (int(x[2]), int(x[3]))
    cv2.rectangle(img, c1, c2, color, thickness=tl, lineType=cv2.LINE_AA)
    if label:
```

```
        tf = max(tl - 1, 1)
        t_size = cv2.getTextSize(label, 0, fontScale=tl / 3, thickness=tf)[0]
        c2 = c1[0] + t_size[0], c1[1] - t_size[1] - 3
        cv2.rectangle(img, c1, c2, color, -1, cv2.LINE_AA)
        cv2.putText(img, label, (c1[0], c1[1] - 2), 0, tl / 3, [225, 255, 255],
                    thickness=tf, lineType=cv2.LINE_AA)
    return img

def letterbox(img, new_wh=(416, 416), color=(114, 114, 114)):
    a = AutoScale(img, *new_wh)
    new_img = a.new_img
    h, w = new_img.shape[:2]
    new_img = cv2.copyMakeBorder(new_img, 0, new_wh[1] - h, 0, new_wh[0] -
                                 w, cv2.BORDER_CONSTANT, value=color)
    return new_img, (new_wh[0] / a.scale, new_wh[1] / a.scale)

def load_model_npu(PATH, npu_id):
    rknn = RKNNLite()
    devs = rknn.list_devices()
    device_id_dict = {}
    for index, dev_id in enumerate(devs[-1]):
        if dev_id[:2] != 'TS':
            device_id_dict[0] = dev_id
        if dev_id[:2] == 'TS':
            device_id_dict[1] = dev_id
    print('-->loading model : ' + PATH)
    rknn.load_rknn(PATH)
    print('--> Init runtime environment on: ' + device_id_dict[npu_id])
    ret = rknn.init_runtime(device_id=device_id_dict[npu_id])
    if ret != 0:
        print('Init runtime environment failed')
        exit(ret)
    print('done')
    return rknn

def load_rknn_model(PATH):
    rknn = RKNNLite()
    ret = rknn.load_rknn(PATH)
    if ret != 0:
```

```
        print('load rknn model failed')
        exit(ret)
    print('done')
    ret = rknn.init_runtime()
    if ret != 0:
        print('Init runtime environment failed')
        exit(ret)
    print('done')
    return rknn

class RKNNDetector:
    def __init__(self, model, wh, masks, anchors, names):
        self.wh = wh
        self._masks = masks
        self._anchors = anchors
        self.names = names
        if isinstance(model, str):
            model = load_rknn_model(model)
        self._rknn = model
        self.draw_box = True

    def _predict(self, img_src, _img, gain, conf_thres=0.4, iou_thres=0.45):
        src_h, src_w = img_src.shape[:2]
        _img = cv2.cvtColor(_img, cv2.COLOR_BGR2RGB)
        t0 = time.time()
        # 呼叫 NPU 進行推理
        pred_onx = self._rknn.inference(inputs=[_img])
        print("inference time:\t", time.time() - t0)
        # 處理推理結果
        boxes, classes, scores = [], [], []
        for t in range(3):
            input0_data = sigmoid(pred_onx[t][0])
            input0_data = np.transpose(input0_data, (1, 2, 0, 3))
            grid_h, grid_w, channel_n, predict_n = input0_data.shape
            anchors = [self._anchors[i] for i in self._masks[t]]
            box_confidence = input0_data[..., 4]
            box_confidence = np.expand_dims(box_confidence, axis=-1)
            box_class_probs = input0_data[..., 5:]
            box_xy = input0_data[..., :2]
```

```
            box_wh = input0_data[..., 2:4]
            col = np.tile(np.arange(0, grid_w), grid_h).reshape(-1, grid_w)
            row = np.tile(np.arange(0, grid_h).reshape(-1, 1), grid_w)
            col = col.reshape((grid_h, grid_w, 1, 1)).repeat(3, axis=-2)
            row = row.reshape((grid_h, grid_w, 1, 1)).repeat(3, axis=-2)
            grid = np.concatenate((col, row), axis=-1)
            box_xy = box_xy * 2 - 0.5 + grid
            box_wh = (box_wh * 2) ** 2 * anchors
            box_xy /= (grid_w, grid_h)      #計算原尺寸的中心
            box_wh /= self.wh               #計算原尺寸的長寬
            box_xy -= (box_wh / 2)
            box = np.concatenate((box_xy, box_wh), axis=-1)
            res = filter_boxes(box, box_confidence, box_class_probs, conf_thres)
            boxes.append(res[0])
            classes.append(res[1])
            scores.append(res[2])
        boxes, classes, scores = np.concatenate(boxes), np.concatenate(classes),
np.concatenate(scores)
        nboxes, nclasses, nscores = [], [], []
        for c in set(classes):
            inds = np.where(classes == c)
            b = boxes[inds]
            c = classes[inds]
            s = scores[inds]
            keep = nms_boxes(b, s, iou_thres)
            nboxes.append(b[keep])
            nclasses.append(c[keep])
            nscores.append(s[keep])
        if len(nboxes) < 1:
            return [], []
        boxes = np.concatenate(nboxes)
        classes = np.concatenate(nclasses)
        scores = np.concatenate(nscores)
        label_list = []
        box_list = []
        for (x, y, w, h), score, cl in zip(boxes, scores, classes):
            x *= gain[0]
            y *= gain[1]
            w *= gain[0]
```

```
            h *= gain[1]
            x1 = max(0, np.floor(x).astype(int))
            y1 = max(0, np.floor(y).astype(int))
            x2 = min(src_w, np.floor(x + w + 0.5).astype(int))
            y2 = min(src_h, np.floor(y + h + 0.5).astype(int))
            label_list.append(self.names[cl])
            box_list.append((x1, y1, x2, y2))
            if self.draw_box:
                new_img = plot_one_box((x1, y1, x2, y2), img_src, label=self.names[cl])
        return new_img, label_list, box_list

    def predict_resize(self, img_src, conf_thres=0.4, iou_thres=0.45):
        """
        預測一幅影像，前置處理使用 resize
        return: labels,boxes
        """
        _img = cv2.resize(img_src, self.wh)
        gain = img_src.shape[:2][::-1]
        return self._predict(img_src, _img, gain, conf_thres, iou_thres, )

    def predict(self, img_src, conf_thres=0.4, iou_thres=0.45):
        """
        預測一幅影像，前置處理保持長寬比
        return: labels,boxes
        """
        _img, gain = letterbox(img_src, self.wh)
        return self._predict(img_src, _img, gain, conf_thres, iou_thres)
    def close(self):
        self._rknn.release()
    def __enter__(self):
        return self
    def __exit__(self, exc_type, exc_val, exc_tb):
        self.close()
    def __del__(self):
        self.close()

if __name__ == '__main__':
    #RKNN_MODEL_PATH = r"./yolov5s-traffic_light.rknn"
    RKNN_MODEL_PATH = r"./yolov5s-mask.rknn"
```

```
    SIZE = (640, 640)
    CLASSES = ('face','face_mask')
    MASKS = [[0, 1, 2], [3, 4, 5], [6, 7, 8]]
    ANCHORS = [[10, 13], [16, 30], [33, 23], [30, 61], [62, 45], [59, 119], [116, 90], [156,
198], [373, 326]]
    model = load_rknn_model(RKNN_MODEL_PATH)
    detector = RKNNDetector(model, SIZE, MASKS, ANCHORS, CLASSES)
    img = cv2.imread("./mask.jpg")
    new_img, labels, boxes = detector.predict(img)
    print('labels:', labels)
    print('boxes:', boxes)
    if len(new_img) > 0:
        cv2.imshow('result', new_img)
        while True:
            if cv2.waitKey(100)== ord('q'):
                break
        cv2.destroyAllWindows()
```

2）NCNN 框架的推理演算法

透過檢測指令稿程式 yolov5.cpp 載入 yolov5s-mask.ncnn.bin、yolov5s-mask.ncnn.param 模型檔案，並進行演算法推理，程式如下：

```
static const char* class_names[] = {"face","face_mask"};

struct Object
{
    cv::Rect_<float> rect;
    int label;
    float prob;
};

static inline float intersection_area(const Object& a, const Object& b)
{
    cv::Rect_<float> inter = a.rect & b.rect;
    return inter.area();
}
```

```
static void qsort_descent_inplace(std::vector<Object>& faceobjects, int left, int right)
{
    int i = left;
    int j = right;
    float p = faceobjects[(left + right) / 2].prob;

    while (i <= j)
    {
        while (faceobjects[i].prob > p)
            i++;
        while (faceobjects[j].prob < p)
            j--;
        if (i <= j)
        {
            //swap
            std::swap(faceobjects[i], faceobjects[j]);
            i++;
            j--;
        }
    }
    #pragma omp parallel sections
    {
        #pragma omp section
        {
            if (left < j) qsort_descent_inplace(faceobjects, left, j);
        }
        #pragma omp section
        {
            if (i < right) qsort_descent_inplace(faceobjects, i, right);
        }
    }
}

static void qsort_descent_inplace(std::vector<Object>& faceobjects)
{
    if (faceobjects.empty())
        return;

    qsort_descent_inplace(faceobjects, 0, faceobjects.size() - 1);
```

```
}

static void nms_sorted_bboxes(const std::vector<Object>& faceobjects, std::vector<int>&
picked, float nms_threshold, bool agnostic = false)
{
    picked.clear();

    const int n = faceobjects.size();

    std::vector<float> areas(n);
    for (int i = 0; i < n; i++)
    {
        areas[i] = faceobjects[i].rect.area();
    }

    for (int i = 0; i < n; i++)
    {
        const Object& a = faceobjects[i];

        int keep = 1;
        for (int j = 0; j < (int)picked.size(); j++)
        {
            const Object& b = faceobjects[picked[j]];

            if (!agnostic && a.label != b.label)
                continue;

            //intersection over union
            float inter_area = intersection_area(a, b);
            float union_area = areas[i] + areas[picked[j]] - inter_area;
            //float IoU = inter_area / union_area
            if (inter_area / union_area > nms_threshold)
                keep = 0;
        }

        if (keep)
            picked.push_back(i);
    }
}
```

```
static inline float sigmoid(float x)
{
    return static_cast<float>(1.f / (1.f + exp(-x)));
}

static void generate_proposals(const ncnn::Mat& anchors, int stride, const ncnn::Mat& in_
pad, const ncnn::Mat& feat_blob, float prob_threshold, std::vector<Object>& objects)
{
    const int num_grid_x = feat_blob.w;
    const int num_grid_y = feat_blob.h;
    const int num_anchors = anchors.w / 2;
    const int num_class = feat_blob.c / num_anchors - 5;
    const int feat_offset = num_class + 5;

    for (int q = 0; q < num_anchors; q++)
    {
        const float anchor_w = anchors[q * 2];
        const float anchor_h = anchors[q * 2 + 1];

        for (int i = 0; i < num_grid_y; i++)
        {
            for (int j = 0; j < num_grid_x; j++)
            {
                //find class index with max class score
                int class_index = 0;
                float class_score = -FLT_MAX;
                for (int k = 0; k < num_class; k++)
                {
                    float score = feat_blob.channel(q * feat_offset + 5 + k).row(i)[j];
                    if (score > class_score)
                    {
                        class_index = k;
                        class_score = score;
                    }
                }

                float box_score = feat_blob.channel(q * feat_offset + 4).row(i)[j];
                float confidence = sigmoid(box_score) * sigmoid(class_score);
```

```
                if (confidence >= prob_threshold)
                {
                    //yolov5/models/yolo.py Detect forward
                    //y = x[i].sigmoid()
                    //y[..., 0:2] = (y[..., 0:2] * 2. - 0.5 + self.grid[i].to(x[i].device))
* self.stride[i]
                    //y[..., 2:4] = (y[..., 2:4] * 2) ** 2 * self.anchor_grid[i]  #wh

                    float dx = sigmoid(feat_blob.channel(q * feat_offset + 0).row(i)[j]);
                    float dy = sigmoid(feat_blob.channel(q * feat_offset + 1).row(i)[j]);
                    float dw = sigmoid(feat_blob.channel(q * feat_offset + 2).row(i)[j]);
                    float dh = sigmoid(feat_blob.channel(q * feat_offset + 3).row(i)[j]);

                    float pb_cx = (dx * 2.f - 0.5f + j) * stride;
                    float pb_cy = (dy * 2.f - 0.5f + i) * stride;

                    float pb_w = pow(dw * 2.f, 2) * anchor_w;
                    float pb_h = pow(dh * 2.f, 2) * anchor_h;

                    float x0 = pb_cx - pb_w * 0.5f;
                    float y0 = pb_cy - pb_h * 0.5f;
                    float x1 = pb_cx + pb_w * 0.5f;
                    float y1 = pb_cy + pb_h * 0.5f;

                    Object obj;
                    obj.rect.x = x0;
                    obj.rect.y = y0;
                    obj.rect.width = x1 - x0;
                    obj.rect.height = y1 - y0;
                    obj.label = class_index;
                    obj.prob = confidence;

                    objects.push_back(obj);
                }
            }
        }
    }
}
```

```
/*
cv::Mat numpy_uint8_3c_to_cv_mat(py::array_t<unsigned char>& input) {

    if (input.ndim() != 3)
        throw std::runtime_error("3-channel image must be 3 dims ");
    py::buffer_info buf = input.request();
    cv::Mat mat(buf.shape[0], buf.shape[1], CV_8UC3, (unsigned char*)buf.ptr);
    return mat;
}

py::array_t<unsigned char> cv_mat_uint8_3c_to_numpy(cv::Mat& input) {

    py::array_t<unsigned char> dst = py::array_t<unsigned char>({ input.rows,input.cols,3},
input.data);
    return dst;
}
*/

static int detect_yolov5(const cv::Mat& bgr, std::vector<Object>& objects)
{
    ncnn::Net yolov5;

    //yolov5.opt.use_vulkan_compute = true;
    yolov5.load_param("./yolov5s-mask.ncnn.param");
    yolov5.load_model("./yolov5s-mask.ncnn.bin");

    const int target_size = 640;
    const float prob_threshold = 0.25f;
    const float nms_threshold = 0.45f;

    //yolov5/models/common.py DetectMultiBackend
    const int max_stride = 64;

    int img_w = bgr.cols;
    int img_h = bgr.rows;

    //letterbox pad to multiple of max_stride
    int w = img_w;
```

```
int h = img_h;

float scale = 1.f;

if (w > h)
{
    scale = (float)target_size / w;
    w = target_size;
    h = h * scale;
}
else
{
    scale = (float)target_size / h;
    h = target_size;
    w = w * scale;
}
printf("scale=%.4f\n", scale);

ncnn::Mat in = ncnn::Mat::from_pixels_resize(bgr.data, ncnn::Mat::PIXEL_BGR2RGB,
                                              bgr.cols, bgr.rows, target_size, target_size);

//pad to target_size rectangle
//yolov5/utils/datasets.py letterbox
int wpad = (w + max_stride - 1) / max_stride * max_stride - w;
int hpad = (h + max_stride - 1) / max_stride * max_stride - h;
printf("wpad=%d, hpad=%d\n", wpad, hpad);
ncnn::Mat in_pad;
ncnn::copy_make_border(in, in_pad, hpad / 2, hpad - hpad / 2, wpad / 2,
                       wpad - wpad / 2, ncnn::BORDER_CONSTANT, 114.f);

const float norm_vals[3] = {1 / 255.f, 1 / 255.f, 1 / 255.f};
in_pad.substract_mean_normalize(0, norm_vals);
ncnn::Extractor ex = yolov5.create_extractor();
ex.set_light_mode(true);
ex.set_num_threads(2);
ex.input("in0", in_pad);
std::vector<Object> proposals;
//anchor setting from yolov5/models/yolov5s.yaml
```

```
//stride 8
{
    ncnn::Mat out;
    ex.extract("out0", out);

    ncnn::Mat anchors(6);
    anchors[0] = 10.f;
    anchors[1] = 13.f;
    anchors[2] = 16.f;
    anchors[3] = 30.f;
    anchors[4] = 33.f;
    anchors[5] = 23.f;

    std::vector<Object> objects8;
    generate_proposals(anchors, 8, in_pad, out, prob_threshold, objects8);

    proposals.insert(proposals.end(), objects8.begin(), objects8.end());
}

//stride 16
{
    ncnn::Mat out;
    ex.extract("out1", out);

    ncnn::Mat anchors(6);
    anchors[0] = 30.f;
    anchors[1] = 61.f;
    anchors[2] = 62.f;
    anchors[3] = 45.f;
    anchors[4] = 59.f;
    anchors[5] = 119.f;

    std::vector<Object> objects16;
    generate_proposals(anchors, 16, in_pad, out, prob_threshold, objects16);

    proposals.insert(proposals.end(), objects16.begin(), objects16.end());
}

//stride 32
```

```
{
    ncnn::Mat out;
    ex.extract("out2", out);

    ncnn::Mat anchors(6);
    anchors[0] = 116.f;
    anchors[1] = 90.f;
    anchors[2] = 156.f;
    anchors[3] = 198.f;
    anchors[4] = 373.f;
    anchors[5] = 326.f;

    std::vector<Object> objects32;
    generate_proposals(anchors, 32, in_pad, out, prob_threshold, objects32);

    proposals.insert(proposals.end(), objects32.begin(), objects32.end());
}

//sort all proposals by score from highest to lowest
qsort_descent_inplace(proposals);

//apply nms with nms_threshold
std::vector<int> picked;
nms_sorted_bboxes(proposals, picked, nms_threshold);

int count = picked.size();

objects.clear();
objects.resize(count);
for (int i = 0; i < count; i++)
{
    objects[i] = proposals[picked[i]];

    //adjust offset to original unpadded
    float x0 = (objects[i].rect.x - (wpad / 2)) / scale;
    float y0 = (objects[i].rect.y - (hpad / 2)) / scale;
    float x1 = (objects[i].rect.x + objects[i].rect.width - (wpad / 2)) / scale;
    float y1 = (objects[i].rect.y + objects[i].rect.height - (hpad / 2)) / scale;
    /*
```

```
        float x0 = (objects[i].rect.x) / scale;
        float y0 = (objects[i].rect.y) / scale;
        float x1 = (objects[i].rect.x + objects[i].rect.width) / scale;
        float y1 = (objects[i].rect.y + objects[i].rect.height) / scale;
        */
        //clip
        x0 = std::max(std::min(x0, (float)(img_w - 1)), 0.f);
        y0 = std::max(std::min(y0, (float)(img_h - 1)), 0.f);
        x1 = std::max(std::min(x1, (float)(img_w - 1)), 0.f);
        y1 = std::max(std::min(y1, (float)(img_h - 1)), 0.f);

        objects[i].rect.x = x0;
        objects[i].rect.y = y0;
        objects[i].rect.width = x1 - x0;
        objects[i].rect.height = y1 - y0;

    }

    return 0;
}

static void draw_objects(const cv::Mat& bgr, const std::vector<Object>& objects)
{
    cv::Mat image = bgr.clone();
    for (size_t i = 0; i < objects.size(); i++)
    {
        const Object& obj = objects[i];
        fprintf(stderr, "%d = %.5f at %.2f %.2f %.2f x %.2f\n", obj.label, obj.prob,
                obj.rect.x, obj.rect.y, obj.rect.width, obj.rect.height);
        cv::rectangle(image, obj.rect, cv::Scalar(0, 0, 255));
        char text[256];
        sprintf(text, "%s %.1f%%", class_names[obj.label], obj.prob * 100);

        int baseLine = 0;
        cv::Size label_size = cv::getTextSize(text, cv::FONT_HERSHEY_SIMPLEX,
                                              0.5, 1, &baseLine);

        int x = obj.rect.x;
        int y = obj.rect.y - label_size.height - baseLine;
```

```
        if (y < 0)
            y = 0;
        if (x + label_size.width > image.cols)
            x = image.cols - label_size.width;

        cv::rectangle(image, cv::Rect(cv::Point(x, y), cv::Size(label_size.width,
                label_size.height + baseLine)), cv::Scalar(255, 255, 255), -1);

        cv::putText(image, text, cv::Point(x, y + label_size.height),
                cv::FONT_HERSHEY_SIMPLEX, 0.5, cv::Scalar(0, 0, 0));
    }
    cv::imwrite("result.jpg", image);
    cv::imshow("result", image);
    cv::waitKey(0);
}

int main(int argc, char** argv)
{
    if (argc != 2)
    {
        fprintf(stderr, "Usage: %s [imagepath]\n", argv[0]);
        return -1;
    }

    const char* imagepath = argv[1];

    cv::Mat m = cv::imread(imagepath, 1);
    cv::Mat m2 = m.clone();
    if (m.empty())
    {
        fprintf(stderr, "cv::imread %s failed\n", imagepath);
        return -1;
    }
    std::vector<Object> objects;
    detect_yolov5(m, objects);
    draw_objects(m2, objects);
    return 0;
```

3.5.2 開發步驟與驗證

3.5.2.1 硬體部署

詳見 2.1.2.1 節。

3.5.2.2 NCNN 框架模型開發驗證

1）專案部署

（1）執行 MobaXterm 工具，透過 SSH 登入深度學習伺服器。

（2）在 SSH 終端執行以下命令，建立開發專案目錄。

```
$ mkdir -p ~/aiedge-exp/yolov5-ncnn
```

（3）透過 SSH 將 object_detection_yolov5-6.1.zip（使用基於 PyTorch 的 YOLOv5 演算法訓練口罩檢測模型）、mask_model.zip（訓練好的模型檔案）和 yolov5_ncnn_cpp.zip（用來將訓練好的 YOLOv5 口罩檢測模型轉為 NCNN 框架使用的 bin 模型）上傳到 ~/aiedge-exp/yolov5-ncnn 目錄下，並採用 unzip 命令進行解壓縮。

2）模型轉換

將訓練好的口罩檢測模型轉為 NCNN 框架使用的 bin 模型。

（1）在 SSH 終端輸入以下命令，啟動伺服器上的 py36_tf25_torch110_cuda113_cv345 環境，並跳躍到 yolov5-6.1 目錄，將訓練好的模型檔案複製到專案中。

```
$ conda activate py36_tf25_torch110_cuda113_cv345
$ cd ~/aiedge-exp/yolov5-ncnn
$ cd object_detection_yolov5-6.1
$ cp ../mask_model/yolov5s-mask.pt ./
```

（2）使用 object_detection_yolov5-6.1 中的 export.py 檔案將訓練好的 yolov5s-mask.pt 模型轉為 TorchScript 模型。當生成 TorchScript 模型後，在下次轉換前必須先刪除當前的 TorchScript 模型再進行轉換，否則會顯示出錯。程式如下：

```
$ python3 export.py --weights ./yolov5s-mask.pt --include torchscript --train
export: data=data/coco128.yaml, weights=['./yolov5s-mask.pt'], imgsz=[640, 640], batch_
size=1,
                                  device=cpu, half=False, inplace=False, train=True,
                                  optimize=False, int8=False, dynamic=False, simplify=False,
                                  opset=12, verbose=False, workspace=4, nms=False,
                                  agnostic_nms=False, topk_per_class=100, topk_all=100,
                                  iou_thres=0.45, conf_thres=0.25, include=['torchscript']
YOLOv5 2022-2-22 torch 1.10.1 CPU

Fusing layers...
Model Summary: 213 layers, 7015519 parameters, 0 gradients, 15.8 GFLOPs

PyTorch: starting from yolov5s-mask.pt with output shape (1, 3, 80, 80, 7) (14.4 MB)

TorchScript: starting export with torch 1.10.1...
TorchScript: export success, saved as yolov5s-mask.torchscript (28.3 MB)

Export complete (1.34s)
Results saved to /home/zonesion/hou/object_detection_yolov5-6.1
Detect:         python detect.py --weights yolov5s-mask.torchscript
PyTorch Hub:  model = torch.hub.load('ultralytics/yolov5', 'custom', 'yolov5s-mask.
torchscript')
Validate:       python val.py --weights yolov5s-mask.torchscript
Visualize:      https://netron.app
```

（3）進入 yolov5_ncnn 目錄，使用 pnnx 命令將上一步生成的 TorchScript 模型轉為 NCNN 框架使用的 bin 模型和 param 模型。

```
$ cd ~/aiedge-exp/yolov5-ncnn/yolov5_ncnn_cpp
$ cp ../object_detection_yolov5-6.1/yolov5s-mask.torchscript ./
$ conda activate py36_tf25_torch110_cuda113_cv345
$ chmod +x pnnx/*
```

```
$ pnnx/pnnx ./yolov5s-mask.torchscript inputshape=[1,3,640,640] inputshape2=[1,3,320,320]
pnnxparam = ./yolov5s-mask.pnnx.param
pnnxbin = ./yolov5s-mask.pnnx.bin
pnnxpy = ./yolov5s-mask_pnnx.py
ncnnparam = ./yolov5s-mask.ncnn.param
ncnnbin = ./yolov5s-mask.ncnn.bin
ncnnpy = ./yolov5s-mask_ncnn.py
optlevel = 2
device = cpu
inputshape = [1,3,640,640]f32
inputshape2 = [1,3,320,320]f32
customop =
moduleop =
############pass_level0
inline module = models.common.Bottleneck
inline module = models.common.C3
inline module = models.common.Concat
inline module = models.common.Conv
inline module = models.common.SPPF
inline module = models.yolo.Detect
inline module = utils.activations.SiLU
inline module = models.common.Bottleneck
inline module = models.common.C3
inline module = models.common.Concat
inline module = models.common.Conv
inline module = models.common.SPPF
inline module = models.yolo.Detect
inline module = utils.activations.SiLU
190  191  input.10  195  196  input.24  205  206  input.8  214  215
input.12  219  220  221  222  225  226  227  input.14  232  233  input.6  237  238
input.16  247  248  input.20  257  258  input.22  262  263  264  265  271  272
input.26  276  277  278  279  282  283  284  input.28  289  290  input.18  294  295
input.30  304  305  input.34  315  316  input.2  320  321  322  323  329  330
input.4  334  335  336  337  343  344  input.36  348  349  350  351  354  355  356
input.38  361  362  input.32  366  367  input.40  376  377  input.42  385  386
input.54  390  391  392  393  396  397  398  input.44  403  404  input.46  412  413
input.52  415  416  417  input.56  422  423  input.48  427  428  input.50  132
input.58  440  441  input.64  448  449  input.66  453  454  455  458  459  460
input.68  465  466  input.60  470  471  input.62  143  input.70  483  484
```

```
input.74  491  492  input.76  496  497  498  501  502  503  input.78  508  509
input.72  513  514  515  input.80  526  527  input.84  534  535
input.86  539  540  541  544  545  546  input.88  551  552  input.82  556  557  558
input.90  569  570  input.5  577  578  input.7  582  583  584  587  588  589
input.3  594  595  input.1  609  bs.1  ny.1  nx.1  620  622  623  624  bs0.1  ny0.1
nx0.1  635  637  638  639  bs1.1  ny1.1  nx1.1  650  652  653  172  173  174
----------------

assign dynamic shape info
#############pass_level1
no attribute value
no attribute value
#############pass_level2
#############pass_level3
assign unique operator name pnnx_unique_0 to model.9.m
assign unique operator name pnnx_unique_1 to model.9.m
#############pass_level4
#############pass_level5
#############pass_ncnn
```

（4）查看生成的 yolov5s-mask.ncnn.bin 和 yolov5s-mask.ncnn.param 檔案。

```
$ ll yolov5s-mask.ncnn.*
-rw-rw-r-- 1 zonesion zonesion   14050412 3月  16 15:01 yolov5s-mask.ncnn.bin
-rw-rw-r-- 1 zonesion zonesion   13826 3月     16 15:01 yolov5s-mask.ncnn.param
```

3）模型測試

（1）在 SSH 終端輸入以下命令編譯模型測試程式，編譯成功後會在 build 目錄下生成測試執行檔案 yolov5.bin。

```
$ cd ~/aiedge-exp/yolov5-ncnn/yolov5_ncnn_cpp
$ mkdir build
$ cd build
$ cmake ..
-- The C compiler identification is GNU 7.5.0
-- The CXX compiler identification is GNU 7.5.0
-- Check for working C compiler: /usr/bin/cc
```

```
-- Check for working C compiler: /usr/bin/cc -- works
-- Detecting C compiler ABI info
-- Detecting C compiler ABI info - done
-- Detecting C compile features
-- Detecting C compile features - done
-- Check for working CXX compiler: /usr/bin/c++
-- Check for working CXX compiler: /usr/bin/c++ -- works
-- Detecting CXX compiler ABI info
-- Detecting CXX compiler ABI info - done
-- Detecting CXX compile features
-- Detecting CXX compile features - done
-- Found OpenCV: /usr/local (found version "3.4.5")
-- Found OpenMP_C: -fopenmp (found version "4.5")
-- Found OpenMP_CXX: -fopenmp (found version "4.5")
-- Found OpenMP: TRUE (found version "4.5")
OPENMP FOUND
-- Configuring done
-- Generating done
-- Build files have been written to: /home/zonesion/aiedge-exp/yolov5_ncnn_cpp/build
$ make -j4
Scanning dependencies of target yolov5.bin
[ 50%] Building CXX object CMakeFiles/yolov5.bin.dir/yolov5.cpp.o
[100%] Linking CXX executable yolov5.bin
[100%] Built target yolov5.bin
```

（2）執行測試程式對測試影像進行推理與驗證。

```
$ cd ~/aiedge-exp/yolov5-ncnn/yolov5_ncnn_cpp
$ cp build/yolov5.bin ./
$ chmod +x yolov5.bin
$ ./yolov5.bin data/test.jpg
```

從推理與驗證結果可以看出，檢測指令稿程式 yolov5.cpp 載入了 yolov5s-mask.ncnn.bin 和 yolov5s-mask.ncnn.param，並對本地的測試影像 test.jpg 進行了辨識。口罩檢測結果如圖 3.43 所示。

▲ 圖 3.43 口罩檢測結果（採用 NCNN 框架模型）

3.5.2.3 RKNN 框架模型開發驗證

1）專案部署

（1）透過 SSH 登入到電腦或虛擬機器。

（2）在 SSH 終端執行以下命令，建立開發專案目錄。

```
$ mkdir -p ~/aiedge-exp/yolov5-rknn
```

（3）透過 SSH 將 mask_model_rknn.zip（訓練好的模型檔案）和本開發專案程式中的 rknn_convert_tools.zip 上傳到 ~/aiedge-exp/yolov5-rknn 目錄下，並採用 unzip 命令進行解壓縮。

（4）將 mask_model_rknn 資料夾中的模型檔案複製到剛剛解壓出來的 rknn_convert_tools 目錄中。

```
$ cd ~/aiedge-exp/yolov5-rknn/rknn_convert_tools
$ cp -a ../mask_model_rknn/yolov5s-mask.pt ./
```

2）模型轉換

首先使用 ONNX 框架將訓練好的口罩檢測模型 yolov5s-mask.pt 轉為 onnx 格式的模型，然後使用 RKNN 框架將 onnx 格式的模型轉為 rknn 格式的模型。

（1）在 SSH 終端輸入以下命令，將 yolov5s-mask.pt 模型轉為 yolov5s-mask.onnx 模型，程式如下：

```
$ cd ~/aiedge-exp/yolov5-rknn/rknn_convert_tools
$ conda activate py38_tf25_torch110_cuda113_cv412
$ python3 models/export.py --weights ./yolov5s-mask.pt --img 640 --batch 1
Namespace(add_image_preprocess_layer=False, batch_size=1, device='cpu', dynamic=False,
          grid=False, ignore_output_permute=False, img_size=[640, 640],
          rknn_mode=False, weights='./yolov5s-mask.pt')
YOLOv5 torch 1.10.1 CPU

Fusing layers...
/home/zonesion/miniconda3/envs/py38_tf25_torch110_cuda113_cv412/lib/python3.8/site-
packages/
             torch/functional.py:445: UserWarning: torch.meshgrid: in an upcoming release,
it will
             be required to pass the indexing argument. (Triggered internally at
             /opt/conda/conda-bld/pytorch_1639180588308/work/aten/src/ATen/native/
             TensorShape.cpp:2157.)
return _VF.meshgrid(tensors, **kwargs)  #type: ignore[attr-defined]
Model Summary: 213 layers, 7015519 parameters, 0 gradients

Starting ONNX export with onnx 1.9.0...
/home/zonesion/miniconda3/envs/py38_tf25_torch110_cuda113_cv412/lib/python3.8/site-
packages/
          torch/onnx/symbolic_helper.py:381: UserWarning: You are trying to export the model
          with onnx:Resize for ONNX opset version 10. This operator might cause results to
          not match the expected results by PyTorch.
ONNX's Upsample/Resize operator did not match Pytorch's Interpolation until opset 11.
Attributes
          to determine how to transform the input were added in onnx:Resize in opset 11 to
          support Pytorch's behavior (like coordinate_transformation_mode and nearest_mode).
We recommend using opset 11 and above for models using this operator.
```

```
    warnings.warn("You are trying to export the model with " + onnx_op + " for ONNX opset version "
ONNX export success, saved as ./yolov5s-mask.onnx
```

（2）修改 rknn_convert_tools/config.yaml 檔案，設定 RKNN 框架模型轉換參數，程式如下：

```
running:
    model_type: onnx
    export: True
    inference: True
    eval_perf: False

parameters:
    onnx:
        model: './yolov5s-mask.onnx'                    # 設定輸入的 ONNX 框架模型路徑
    rknn:
        path: './yolov5s-mask.rknn'                     # 設定輸出的 RKNN 框架模型路徑
config:
    mean_values: [0, 0, 0] #123.675 116.28 103.53 58.395 #0 0 0 255
    std_values: [255, 255, 255]
    quant_img_RGB2BGR: True #'2 1 0' #'0 1 2' '2 1 0'
    target_platform: 'rk3588'                           # 設定 RKNN 裝置為 RK3588
    quantized_dtype:'asymmetric_quantized-8'#asymmetric_quantized-u8,dynamic_fixed_point-8,
                                                    dynamic_fixed_point-16
    optimization_level: 1

build:
    do_quantization: True
    dataset: './single_dataset.txt'

export_rknn:
    export_path: './yolov5s-mask.rknn'

init_runtime:
    target: null
    device_id: null
    perf_debug: False
```

```
    eval_mem: False
    async_mode: False

img: &img
    path: './mask.jpg'                              #設定測試影像的路徑

inference:
    inputs: *img
    data_type: 'uint8'
    data_format: 'nhwc' #'nchw', 'nhwc'
    inputs_pass_through: None

eval_perf:
    inputs: *img
    data_type: 'uint8'
    data_format: 'nhwc'
    is_print: True
```

（3）在 SSH 終端輸入以下命令，將 yolov5s-mask.onnx 轉為 RKNN 框架模型，程式如下：

```
$ cd ~/aiedge-exp/yolov5-rknn/rknn_convert_tools
$ chmod +x rknn_convert.sh rknn_convert
$ conda activate py38_tf25_torch110_cuda113_cv412
$ ./rknn_convert.sh
[935320] WARNING: file already exists but should not:
                  /tmp/_MEItJyZ3G/rknn/api/fuse_rules.cpython-38-x86_64-linux-gnu.so
[935320] WARNING: file already exists but should not:
                  /tmp/_MEItJyZ3G/rknn/api/graph_optimizer.cpython-38-x86_64-linux-gnu.so
[935320] WARNING: file already exists but should not:
                  /tmp/_MEItJyZ3G/rknn/api/hybrid_proposal.cpython-38-x86_64-linux-gnu.so
[935320] WARNING: file already exists but should not:
                  /tmp/_MEItJyZ3G/rknn/api/ir_graph.cpython-38-x86_64-linux-gnu.so
[935320] WARNING: file already exists but should not:
                  /tmp/_MEItJyZ3G/rknn/api/ir_utils.cpython-38-x86_64-linux-gnu.so
[935320] WARNING: file already exists but should not:
                  /tmp/_MEItJyZ3G/rknn/api/load_checker.cpython-38-x86_64-linux-gnu.so
[935320] WARNING: file already exists but should not:
```

```
                    /tmp/_MEItJyZ3G/rknn/api/mmse_quant.cpython-38-x86_64-linux-gnu.so
[935320] WARNING: file already exists but should not:
                    /tmp/_MEItJyZ3G/rknn/api/quant_optimizer.cpython-38-x86_64-linux-gnu.so
[935320] WARNING: file already exists but should not:
                    /tmp/_MEItJyZ3G/rknn/api/quant_utils.cpython-38-x86_64-linux-gnu.so
[935320] WARNING: file already exists but should not:
                    /tmp/_MEItJyZ3G/rknn/api/quantizer.cpython-38-x86_64-linux-gnu.so
[935320] WARNING: file already exists but should not:
                    /tmp/_MEItJyZ3G/rknn/api/rknn_base.cpython-38-x86_64-linux-gnu.so
[935320] WARNING: file already exists but should not:
                    /tmp/_MEItJyZ3G/rknn/api/rknn_log.cpython-38-x86_64-linux-gnu.so
[935320] WARNING: file already exists but should not:
                    /tmp/_MEItJyZ3G/rknn/api/rknn_platform.cpython-38-x86_64-linux-gnu.so
[935320] WARNING: file already exists but should not:
                    /tmp/_MEItJyZ3G/rknn/api/rknn_runtime.cpython-38-x86_64-linux-gnu.so
[935320] WARNING: file already exists but should not:
                    /tmp/_MEItJyZ3G/rknn/api/rknn_utils.cpython-38-x86_64-linux-gnu.so
[935320] WARNING: file already exists but should not:
                    /tmp/_MEItJyZ3G/rknn/api/session.cpython-38-x86_64-linux-gnu.so
[935320] WARNING: file already exists but should not:
                    /tmp/_MEItJyZ3G/rknn/api/simulator.cpython-38-x86_64-linux-gnu.so
[935320] WARNING: file already exists but should not:
                    /tmp/_MEItJyZ3G/rknn/api/sparse_weight.cpython-38-x86_64-linux-gnu.so
model_type is onnx
W __init__: rknn-toolkit2 version: 1.4.0-22dcfef4
--> config model
W config: The quant_img_RGB2BGR of input 0 is set to True, which means that the RGB2BGR
conversion will be done first when the quantized image is loaded (only valid for jpg/jpeg/
png/bmp, npy will ignore this flag).
Special note here, if quant_img_RGB2BGR is True and the quantized image is jpg/jpeg/png/
bmp, the mean_values / std_values in the config corresponds the order of BGR.
done
--> Loading model
W load_onnx: It is recommended onnx opset 12, but your onnx model opset is 10!
W load_onnx: Model converted from pytorch, 'opset_version' should be set 12 in torch.onnx.
export for successful convert!
More details can be found in examples/pytorch/resnet18_export_onnx
done
--> Building model
```

```
I base_optimize ...
I base_optimize done.
I
I fold_constant ...
I fold_constant done.
I
I correct_ops ...
I correct_ops done.
I
I fuse_ops ...
I fuse_ops results:
I squeeze_to_4d_transpose: remove node = [], add node = ['Transpose_204_squeeze0',
                           'Transpose_204_squeeze1']
I squeeze_to_4d_transpose: remove node = [], add node = ['Transpose_201_squeeze0',
                           'Transpose_201_squeeze1']
I squeeze_to_4d_transpose: remove node = [], add node = ['Transpose_198_squeeze0',
                           'Transpose_198_squeeze1']
I fuse_two_reshape: remove node = ['Reshape_203', 'Reshape_200', 'Reshape_197']
I fold_constant ...
I fold_constant done.
I fuse_ops done.
I
I sparse_weight ...
I sparse_weight done.
I
Analysing : 100%|██████████████████████████████| 151/151 [00:00<00:00, 1635.27it/s]
Quantizating :100%|██████████████████████████████| 151/151 [00:00<00:00, 431.85it/
s]
（......省略部分過程）
--> skip eval_perf
```

（4）最終會在 ~/aiedge-exp/yolov5-rknn/rknn_convert_tools 目錄下生成轉換好的 RKNN 框架模型 yolov5s-mask.rknn。

```
$ tree -L 1
├── config.yaml
├── mask.jpg
├── models
├── readme.md
```

```
├── rknn_convert
├── rknn_convert.sh
├── single_dataset.txt
├── utils
├── yolov5s-mask.onnx
├── yolov5s-mask.pt
└── yolov5s-mask.rknn
```

3）模型測試

以下步驟是在 GW3588 邊緣計算閘道中進行的。

（1）執行 MobaXterm 工具，透過 SSH 登入到 GW3588 邊緣計算閘道。

（2）在 SSH 終端執行以下命令，建立開發專案目錄。

```
$ mkdir -p ~/aiedge-exp
```

（3）透過 SSH 將本開發專案程式中的 yolov5_rknn_demo.zip 上傳到 ~/aiedge-exp 目錄下。

（4）在 SSH 終端輸入以下命令，解壓縮開發專案和訓練好的口罩檢測模型檔案。

```
$ cd ~/aiedge-exp
$ unzip yolov5_rknn_demo.zip
```

將轉換完成的 yolov5s-mask.rknn 檔案透過 SSH 上傳到 yolov5_rknn_demo 資料夾。

（5）在 SSH 終端輸入以下命令，測試 RKNN 口罩檢測模型，程式如下：

```
$ cd ~/aiedge-exp/yolov5_rknn_demo
$ python3 mask_test.py
done
done
I RKNN: [15:09:04.146] RKNN Runtime Information: librknnrt version:
                    1.4.0 (a10f100eb@2022-09-09T09:07:14)
```

```
I RKNN: [15:09:04.146] RKNN Driver Information: version: 0.7.2
I RKNN: [15:09:04.147] RKNN Model Information: version: 1, toolkit version:
                    1.4.0-22dcfef4(compiler version: 1.4.0 (3b4520e4f@2022-09-05T20:52:35)),
                    target: RKNPU v2, target platform: rk3588, framework name: ONNX,
                    framework layout: NCHW
done
inference time:  0.041899681091308594
labels: ['face_mask']
boxes: [(386, 113, 601, 425)]
```

從推理與驗證結果可以看出，推理程式透過 NPU 載入了 RKNN 框架模型，並對本地的測試影像 mask.jpg 進行了辨識，口罩檢測結果如圖 3.44 所示，按「Q」鍵可關閉檢測結果視窗。

▲ 圖 3.44 口罩檢測結果（採用 RKNN 框架模型）

3.5.3 本節小結

本節主要介紹了嵌入式推理框架 RKNN，結合案例學習了基於 YOLOv5 的 RKNN 框架部署、模型轉換、推理與驗證的過程。

3.5.4 思考與擴充

（1）在 SSH 終端輸入什麼命令可以將 yolov5s-mask.onnx 轉為 RKNN 框架模型？

（2）透過 SSH 登入到什麼裝置才能實現模型測試？

3.6 YOLOv3 模型的介面應用

模型介面通常是指用於與深度學習模型進行互動的程式設計介面，包括模型的載入、推理、設定參數等功能。模型介面的設計與所用的深度學習框架、函式庫或硬體平臺有關，以下是一些常見的模型介面：

（1）TensorFlow Serving API：TensorFlow Serving 提供了一個用於部署和服務 TensorFlow 模型的 API，允許透過 RESTful API 或 gRPC 介面請求進行模型推理。

（2）TensorFlow Lite Interpreter API：TensorFlow Lite 提供了一個用於在行動端和嵌入式裝置上進行推理的 API，包括 TensorFlow Lite Interpreter，允許載入並在裝置上執行 TensorFlow Lite 模型。

（3）PyTorch TorchScript API：PyTorch 不僅提供了 TorchScript 格式，這是一種用於序列化和儲存 PyTorch 模型的格式，還提供了用於在 Python 中載入和執行 TorchScript 格式模型的 API。

（4）ONNX Runtime API：ONNX Runtime 是一個跨平臺的推理引擎，支援 ONNX 框架模型，它提供了用於載入和執行 ONNX 框架模型的 API，可以在不同的深度學習框架之間共用模型。

（5）Keras Model API：Keras 是一個高層次的深度學習框架，其 Model API 提供了載入、訓練和推理模型的方法。Keras 框架模型可以在 TensorFlow、Theano 等框架上執行。

（6）Scikit-learn Estimator API：Scikit-learn 是一個用於深度學習的 Python 函式庫，其 Estimator API 提供了一致的介面，用於載入、訓練和評估模型。

（7）Caffe2 API：Caffe2 是一個輕量級的深度學習框架，它有自己的 C++ 和 Python 介面。Caffe2 API 允許使用者載入和執行 Caffe2 模型。

（8）RKNN-Toolkit API：對於瑞芯微晶片（Rockchip）上的模型推理，RKNN-Toolkit 提供了用於載入和執行 RKNN 框架模型的 API。

模型介面提供了一種與深度學習模型進行互動的方式，使得模型能夠在不同的應用場景和硬體裝置上得以部署和執行。在選擇介面時，需要考慮模型的訓練框架、推理引擎和目標硬體平臺的相容性。

本節的基礎知識如下：

- 了解常見模型介面。
- 掌握 NCNN 框架的模型介面設計。
- 結合交通號誌辨識案例，掌握基於 YOLOv3 的 NCNN 框架模型介面設計過程。

3.6.1 原理分析與開發設計

3.6.1.1 NCNN 框架模型的推理過程

NCNN 是一個為行動端進行了極致最佳化的高性能神經網路前向計算框架。從設計之初，NCNN 框架就全面考慮了行動端的部署和使用。NCNN 框架不依賴於第三方，具有跨平臺特性，其在行動端的速度快於其他的開放原始碼框架。基於 NCNN 框架，開發者能夠將深度學習演算法輕鬆地移植到行動端，開發人工智慧 App，將人工智慧帶到使用者的指尖。NCNN 框架目前已在騰訊的多款應用中使用，如 QQ、Qzone、微信、天天 P 圖等。

3.2 節的專案完成了交通號誌辨識的模型介面函式庫，透過 Python 程式可以呼叫交通號誌辨識模型進行推理。NCNN 框架模型的推理過程如圖 3.45 所示。

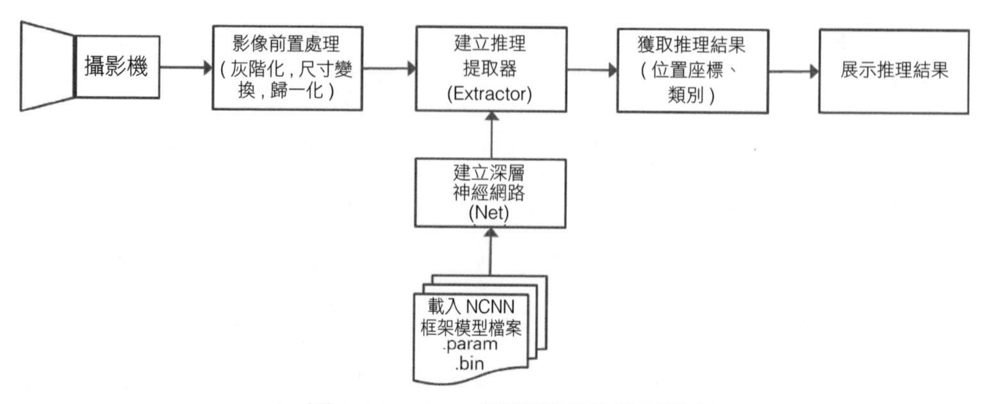

▲ 圖 3.45 NCNN 框架模型的推理過程

推理流程的描述如下：

（1）獲取攝影機的影像資料。

（2）影像前置處理：對影像資料進行灰度化處理，並採用雙線插值等演算法進行影像尺寸變換，最後進行影像資料的歸一化處理。

（3）載入 NCNN 框架模型檔案：使用建立的 Net 神經網路物件，載入 .param 檔案和 .bin 檔案，其中，.param 檔案是神經網路結構說明，.bin 檔案是神經網路權重參數檔案。

（4）建立深層神經網路：呼叫 NCNN 框架介面，建立 Net 神經網路物件。

（5）建立推理提取器（Extractor）：呼叫 NCNN 框架介面建立推理提取器，設定提取器執行緒數量和輕量的模式開關等。

（6）獲取推理結果：為提取器輸入經過步驟（2）處理後的影像資料，輸出推理結果，推理結果包括目標物體座標、類別等資料。

（7）展示推理結果：將推理結果繪製在原始影像上向客戶展示，並傳回推理結果資料。

3.6.1.2 交通號誌辨識模型的介面設計

NCNN 框架是基於 C/C++ 實現的，而 AiCam 平臺的演算法層採用 Python 語言，需要採用 pybind11 將 NCNN 框架模型的介面封裝成 so 函式庫，供 Python 演算法呼叫。

AiCam 平臺對 NCNN 框架模型介面的呼叫與傳回的資料做了標準化處理，傳回的資料採用 JSON 格式，方便 AiCam 平臺的演算法層進行解析。NCNN 框架模型介面函式如下（以交通號誌辨識模型為例）：

```
PYBIND11_MODULE(trafficdet, m) {
    py::class_<TrafficDet>(m, "TrafficDet")
        .def(py::init<>())
        .def("init", &TrafficDet::init)
        .def("detect", &TrafficDet::detect_yolov3);
}
```

模型介面的傳回如下：

```
// 函式：ModelDetector.detect(image)
// 參數：image 表示影像資料
// 結果：JSON 格式的字串
//code：200 表示執行成功，301 表示執行失敗
{
    "code" : 200,                           // 傳回碼
    "msg" : "SUCCESS",                      // 傳回訊息
    "result" : {                            // 傳回結果
        "obj_list" : [                      // 傳回內容
            {
                "location" : {              // 目標座標
                    "height" : 58,
                    "left" : 215,
                    "top" : 137,
                    "width" : 45
                },
                "mark" : [                  // 目標關鍵點
                {
                    "x" : 227,
```

```
                "y" : 160
            },
            {
                "x" : 247,
                "y" : 157
            },
            {
                "x" : 239,
                "y" : 169
            },
            {
                "x" : 231,
                "y" : 180
            },
            {
                "x" : 249,
                "y" : 177
            }
        ],
        "score" : 0.99148327112197876   //置信度
    }],
    "name" : "one",                         //目標名稱
    "obj_num" : 1                           //目標數量
  },
  "time" : 17.5849609375                    //推理時間（ms）
}
```

3.6.1.3 交通號誌辨識模型的演算法設計

1）模型介面演算法

交通號誌辨識模型介面對模型的呼叫和推理相關方法進行了封裝，並編譯成 so 函式庫，供 AiCam 平臺的演算法層呼叫。演算法檔案（traffic_detection_interface\cpp\traffic_detection.cpp）的相關程式如下：

```
#########################################################################################
#檔案：traffic_detection.cpp
#說明：交通號誌辨識模型介面
```

```
##########################################################################################
#include "traffic_detection.h"
// 資料資訊清單
std::vector<std::string> data_info{};

// 將 Numpy 三通道 UINT8 型態資料轉為 OpenCV 矩陣資料
cv::Mat numpy_uint8_3c_to_cv_mat(py::array_t<unsigned char>& input) {

    if (input.ndim() != 3)
        throw std::runtime_error("3-channel image must be 3 dims ");
    py::buffer_info buf = input.request();
    cv::Mat mat(buf.shape[0], buf.shape[1], CV_8UC3, (unsigned char*)buf.ptr);

    return mat;
}

// 將 OpenCV 矩陣資料轉為 Numpy 三通道 UINT8 型態資料
py::array_t<unsigned char> cv_mat_uint8_3c_to_numpy(cv::Mat& input) {
    py::array_t<unsigned char> dst = py::array_t<unsigned char>({ input.rows,input.cols,3},
input.data);
    return dst;
}

// 建立交通號誌辨識物件
TrafficDet::~TrafficDet(){
    yolov3.clear();
}
// 初始化交通號誌辨識物件
int TrafficDet::init(std::string model_path)
{
    // 模型參數檔案
    std::string mask_det_param = model_path + "/yolov3-tiny-traffic-opt.param";
    // 模型權重檔案
    std::string mask_det_bin= model_path + "/yolov3-tiny-traffic-opt.bin";
    // 載入模型參數
    yolov3.load_param(mask_det_param.data());
    // 載入模型權重
    yolov3.load_model(mask_det_bin.data());
}
```

```
std::string TrafficDet::detect_yolov3(py::array_t<unsigned char>& pyimage)
{
    Json::Value root;
    double dStart = ncnn::get_current_time();

    Json::Value obj_list, result;
    // 資料格式轉換
    cv::Mat input_image = numpy_uint8_3c_to_cv_mat(pyimage);
    const int target_size = 416;

    int img_w = input_image.cols;
    int img_h = input_image.rows;
    // 影像資料尺寸裁剪
    ncnn::Mat in = ncnn::Mat::from_pixels_resize(input_image.data, ncnn::Mat::PIXEL_BGR2RGB,
                                                 img_w, img_h, target_size, target_size);
    // 影像資料歸一化處理
    const float mean_vals[3] = {127.5f, 127.5f, 127.5f};
    const float norm_vals[3] = {0.007843f, 0.007843f, 0.007843f};
    in.substract_mean_normalize(mean_vals, norm_vals);
    // 建立推理提取器
    ncnn::Extractor ex = yolov3.create_extractor();
    // 輸入影像資料
    ex.input("data", in);
    // 獲取推理結果
    ncnn::Mat out;
    //ex.extract("detection_out", out);
    ex.extract("output", out);

    static const char* class_names[] = {"background","red","green","left","straight","right"};
    // 遍歷推理結果，組裝 JSON 物件
    for (int i = 0; i < out.h; i++)
    {
        const float* values = out.row(i);

        Json::Value obj;
        int idx = (int)values[0];
        // 設定目標名稱
        if (idx >= 0 && idx < sizeof class_names / sizeof class_names[0]) {
```

```
            obj["name"] = class_names[idx];
        } else obj["name"] = "unknow";
        // 設定置信度的分值
        obj["score"] = values[1];
        // 設定目標位置座標
        int x = (int)(values[2] * img_w);
        int y = (int)(values[3] * img_h);
        int w = (int)(values[4] * img_w - x);
        int h = (int)(values[5] * img_h - y);
        Json::Value location;
        location["left"] = x;
        location["top"] = y;
        location["width"] = w;
        location["height"] = h;

        obj["location"] = location;
        obj_list.append(obj);

    }
    result["obj_num"] = obj_list.size();
    result["obj_list"] = obj_list;

    root["result"] = result;
    root["code"] = 200;
    root["msg"] = "SUCCESS";
    root["time"] = ncnn::get_current_time()-dStart;
    std::string r =  root.toStyledString();

    return r;
}
// 使用 pybind11 載入交通號誌辨識模型
PYBIND11_MODULE(trafficdet, m) {
    py::class_<TrafficDet>(m, "TrafficDet")
        .def(py::init<>())
        .def("init", &TrafficDet::init)
        .def("detect", &TrafficDet::detect_yolov3);
}
```

2）模型介面測試程式

交通號誌辨識模型介面測試程式（traffic_detection_interface\traffic_detection.py）的程式如下：

```
###########################################################################################
# 檔案：traffic_detection.py
# 說明：交通號誌辨識模型介面測試程式
###########################################################################################
import numpy as np
import cv2 as cv
import os
import json
c_dir = os.path.split(os.path.realpath(__file__))[0]
import trafficdet

# 單元測試，如果處理類別中引用了檔案，則在單元測試中要修改檔案路徑
if __name__=='__main__':
    # 讀取測試影像
    img = cv.imread("./test.jpg")
    # 建立影像處理物件
    img_object=trafficdet.TrafficDet()
    img_object.init(c_dir+'/models/traffic_detection')
    result=img_object.detect(img)
    print(result)
```

3.6.2 開發步驟與驗證

3.6.2.1 專案部署

1）硬體部署

詳見 2.1.2.1 節。

2）專案部署

（1）執行 MobaXterm 工具，透過 SSH 登入到邊緣計算閘道。

（2）在 SSH 終端執行以下命令，建立開發專案目錄。

```
$ mkdir -p ~/aiedge-exp
```

（3）透過 SSH 將本開發專案程式上傳到 ~/aicam-exp 目錄下，並採用 unzip 命令進行解壓縮。

3.6.2.2 模型介面編譯

在 SSH 終端輸入以下命令，編譯模型介面。

```
$ cd ~/aiedge-exp/traffic_detection_interface
$ mkdir -p build
$ cd build
$ cmake ..
  ……
  ……
-- Build files have been written to: /home/zonesion/aiedge-exp/traffic_detection_interface/
build
$ make
Scanning dependencies of target trafficdet
[ 50%] Building CXX object CMakeFiles/trafficdet.dir/cpp/traffic_detection.cpp.o
[100%] Linking CXX shared module trafficdet.cpython-35m-x86_64-linux-gnu.so
[100%] Built target trafficdet
```

編譯完成後會在 build 目錄中生成 trafficdet.cpython-35m-x86_64-linux-gnu.so（模型介面函式庫檔案）。注意：在 ARM 平臺中的檔案為 trafficdet.cpython-35m-aarch64-linux-gnu.so。

3.6.2.3 模型介面測試

在 SSH 終端輸入以下命令，執行模型介面。

```
$ cd ~/aiedge-exp/traffic_detection_interface
$ cp build/*.so ./
$ conda activate py36_tf114_torch15_cpu_cv345    //Ubuntu 20.04 作業系統下需要切換環境
$ python3 traffic_detection.py
{
```

```
    "code" : 200,
    "msg" : "SUCCESS",
    "result" : {
        "obj_list" : [
        {
            "location" : {
                "height" : 93,
                "left" : 425,
                "top" : 158,
                "width" : 107
            },
            "name" : "right",
            "score" : 0.90108931064605713
        },
        {
            "location" : {
                "height" : 115,
                "left" : 149,
                "top" : 143,
                "width" : 109
            },
            "name" : "left",
            "score" : 0.89343971014022827
        },
        {
            "location" : {
                "height" : 116,
                "left" : 273,
                "top" : 145,
                "width" : 124
            },
            "name" : "straight",
            "score" : 0.51184618473052979
        }],
        "obj_num" : 3
    },
    "time" : 61.909912109375
}
```

執行成功後，程式將呼叫模型對測試影像進行推理。

3.6.3 本節小結

本節基於交通號誌辨識模型，介紹了 NCNN 框架模型介面演算法的設計，以及基於 YOLOv3 的 NCNN 框架模型介面設計。

3.6.4 思考與擴充

NCNN 框架模型是基於 C/C++ 實現的，而 AiCam 平臺的演算法層採用的是 Python 語言，如何封裝介面才能被 Python 程式呼叫？

3.7 YOLOv5 模型的介面應用

本節的基礎知識如下：

- 掌握模型介面的設計作用。
- 掌握 RKNN 框架模型介面演算法的設計。
- 結合口罩檢測模型，掌握基於 YOLOv5 的 RKNN 框架模型介面的設計過程。

3.7.1 原理分析與開發設計

3.7.1.1 口罩檢測模型的介面設計與演算法設計

1）口罩檢測模型的介面設計

NCNN 框架模型介面函式如下（以口罩檢測模型為例）：

```
PYBIND11_MODULE(maskdet, m) {
    py::class_<MaskDet>(m, "MaskDet")
        .def(py::init<>())
```

```
        .def("init", &MaskDet::init)
        .def("detect", &MaskDet::detect_yolov5);
}
```

模型介面的傳回如下：

```
// 函式：img_object.detect (image)
// 參數：image 表示影像資料
// 結果：JSON 格式的字串
//code：200 表示執行成功，301 表示執行失敗
{
    'code': 200,
    'msg': 'SUCCESS',
    'result': {
        "code" : 200,
        "msg" : "SUCCESS",
        "result" : {
            "obj_list" : [
            {
                "location" : {
                    "height" : 341.3070068359375,
                    "left" : 149.92092895507812,
                    "top" : 0,
                    "width" : 252.85385131835938
                },
                "name" : "face_mask",
                "score" : 0.9254075288772583
            }],
            "obj_num" : 1
        },
        "time" : 88.807861328125
    }
}
```

2）口罩檢測模型的演算法設計

口罩檢測模型介面實現了模型的呼叫和推理相關方法的封裝，編譯成 so 函式庫，AiCam 平臺的演算法層呼叫。演算法檔案（mask_detection_interface_ncnn\cpp\mask_detection.cpp）的相關程式如下：

```
##########################################################################################
# 檔案：mask_detection.cpp
# 說明：口罩檢測模型介面
##########################################################################################
#include "mask_detection.h"

std::vector<std::string> data_info{};
int MaskDet::init(std::string model_path)
{
    std::string mask_det_param = model_path + "/yolov5s-mask.ncnn.param";
    std::string mask_det_bin= model_path + "/yolov5s-mask.ncnn.bin";
    yolov5.opt.use_vulkan_compute = true;
    //yolov5.opt.use_bf16_storage = true;
    //yolov5.register_custom_layer("Yolov5Focus", Yolov5Focus_layer_creator);
    yolov5.load_param(mask_det_param.data());
    yolov5.load_model(mask_det_bin.data());
}

std::string MaskDet::detect_yolov5(py::array_t<unsigned char>& pyimage)
{
    Json::Value root;
    Json::Value obj_list, result;
    static const char* class_names[] = {"face","face_mask"};
    double dStart = ncnn::get_current_time();

    const int target_size = 640;
    const float prob_threshold = 0.25f;
    const float nms_threshold = 0.45f;

    //yolov5/models/common.py DetectMultiBackend
    const int max_stride = 64;
    cv::Mat input_image = numpy_uint8_3c_to_cv_mat(pyimage);

    int img_w = input_image.cols;
    int img_h = input_image.rows;

    //letterbox pad to multiple of max_stride
    int w = img_w;
    int h = img_h;
```

```
    float scale = 1.f;
    if (w > h)
    {
        scale = (float)target_size / w;
        w = target_size;
        h = h * scale;
    }
    else
    {
        scale = (float)target_size / h;
        h = target_size;
        w = w * scale;
    }

    ncnn::Mat in = ncnn::Mat::from_pixels_resize(input_image.data, ncnn::Mat::PIXEL_BGR2RGB,
                img_w, img_h, w, h);

    //pad to target_size rectangle
    //yolov5/utils/datasets.py letterbox
    int wpad = (w + max_stride - 1) / max_stride * max_stride - w;
    int hpad = (h + max_stride - 1) / max_stride * max_stride - h;
    ncnn::Mat in_pad;
    ncnn::copy_make_border(in, in_pad, hpad / 2, hpad - hpad / 2, wpad / 2, wpad - wpad / 2,
                        ncnn::BORDER_CONSTANT, 114.f);

    const float norm_vals[3] = {1 / 255.f, 1 / 255.f, 1 / 255.f};
    in_pad.substract_mean_normalize(0, norm_vals);

    ncnn::Extractor ex = yolov5.create_extractor();

    ex.set_light_mode(true);
    ex.set_num_threads(2);
    ex.input("in0", in_pad);

std::vector<Object> proposals;
//stride 8
    {
        ncnn::Mat out;
        ex.extract("out0", out);
```

```
        ncnn::Mat anchors(6);
        anchors[0] = 10.f;
        anchors[1] = 13.f;
        anchors[2] = 16.f;
        anchors[3] = 30.f;
        anchors[4] = 33.f;
        anchors[5] = 23.f;
        std::vector<Object> objects8;
        generate_proposals(anchors, 8, in_pad, out, prob_threshold, objects8);
        proposals.insert(proposals.end(), objects8.begin(), objects8.end());
    }
    result["obj_num"] = obj_list.size();
    result["obj_list"] = obj_list;

    root["result"] = result;
    root["code"] = 200;
    root["msg"] = "SUCCESS";
    root["time"] = ncnn::get_current_time()-dStart;
    std::string r =  root.toStyledString();

    return r;
}
PYBIND11_MODULE(maskdet, m) {
    py::class_<MaskDet>(m, "MaskDet")
        .def(py::init<>())
        .def("init", &MaskDet::init)
        .def("detect", &MaskDet::detect_yolov5);
}
```

3）測試程式

口罩檢測模型介面的測試程式（mask_detection_interface_ncnn\mask_detection.py）的程式如下：

```
##########################################################################################
# 檔案：mask_detection.py
# 說明：口罩檢測
##########################################################################################
```

```
import numpy as np
import cv2 as cv
import os
import json
c_dir = os.path.split(os.path.realpath(__file__))[0]
import maskdet

# 單元測試，如果處理類別中引用了檔案，則在單元測試中要修改檔案路徑
if __name__=='__main__':

    # 讀取測試影像
    img = cv.imread("./test.jpg")
    # 建立影像處理物件
    img_object=maskdet.MaskDet()
    img_object.init(c_dir+'/models/mask')
    result=img_object.detect(img)
    # 轉化為 JSON 格式的字串
    result=json.loads(result)
    # 列印結果
    print(result)
    # 獲取結果字典中需要標註的邊框的位置
    location=result["result"]["obj_list"][0]["location"]
    pos1=(int(location["left"]),int(location["top"]))
    pos2=(int(location["left"]+location["width"]),int(location["top"]+location["height"]))
    name=result["result"]["obj_list"][0]["name"]
    # 畫矩形框
    cv.rectangle(img,pos1,pos2,(255,0,0),thickness=1)
    # 增加文字
    cv.putText(img,name,pos2, cv.FONT_HERSHEY_SIMPLEX,0.5,(255,0,0),thickness=1)
    cv.imshow('result', img)
    key = cv.waitKey(0)
```

3.7.1.2 RKNN 框架模型推理

1）推理框架

RKNN 是 Rockchip 的 NPU 平臺使用的模型，模型檔案的副檔名為 .rknn。Rockchip 提供了完整的用於模型轉換的 Python 工具，方便使用者將自主研發的

演算法模型轉換成 RKNN 框架模型，同時 Rockchip 也提供了 C/C++ 和 Python API 介面。

RKNN-Toolkit 是一種用於在 PC、Rockchip NPU 平臺上進行模型轉換、推理和性能評估的開發套件，使用者透過該開發套件提供的 Python 介面可以便捷地完成以下功能：

（1）模型轉換：可以將 Caffe、TensorFlow、TensorFlow Lite、ONNX、Darknet、PyTorch、MXNet 框架模型轉成 RKNN 框架模型，支援 RKNN 框架模型的匯入匯出，後續能夠在 Rockchip NPU 平臺上載入使用。RKNN-Toolkit 從 1.2.0 版本開始支援多輸入模型，從 1.3.0 版本開始支援 PyTorch 和 MXNet 框架模型。

（2）量化功能：支援將浮點模型轉成量化模型，目前支援的量化方法有非對稱量化（asymmetric_quantized-u8）、動態定點量化（dynamic_fixed_point-8 和 dynamic_fixed_point-16）。RKNN-Toolkit 從 1.0.0 版本開始支援混合量化功能。

（3）模型推理：能夠在電腦上模擬 Rockchip NPU 平臺，執行 RKNN 框架模型並獲取推理結果；也可以將 RKNN 框架模型分發到指定的 NPU 裝置上進行推理。

（4）性能評估：能夠在電腦上模擬 Rockchip NPU 平臺，執行 RKNN 框架模型並評估模型性能（包括總耗時和每一層的耗時）；也可以將 RKNN 框架模型分發到指定 NPU 裝置上執行，以評估模型在實際裝置上執行時期的性能。

（5）記憶體評估：能夠對系統和 NPU 記憶體的消耗情況進行評估，使用該功能時，必須將 RKNN 框架模型分發到 Rockchip NPU 平臺上執行，並呼叫相關介面獲取記憶體使用資訊。從 0.9.9 版本開始，Rockchip 支援該功能。

（6）模型預先編譯：透過預先編譯技術生成的 RKNN 框架模型可以減少模型在硬體平臺上的載入時間。對於部分模型，還可以減小模型尺寸。但預先編譯後的 RKNN 框架模型只能在 Rockchip NPU 裝置上執行。目前只有 x86_64

Ubuntu 作業系統能夠直接從原始模型生成 RKNN 框架的預先編譯模型。RKNN-Toolkit 從 0.9.5 版本開始支援模型預先編譯功能，並在 1.0.0 版本中對預先編譯方法進行了升級，升級後的預先編譯模型無法與舊版本的驅動相容。從 1.4.0 版本開始，RKNN-Toolkit 也可以透過 Rockchip NPU 平臺將普通的 RKNN 框架模型轉成 RKNN 框架的預先編譯模型。

（7）模型分段：在多個模型同時執行時期，模型分段可以將單一模型分成多段在 Rockchip NPU 平臺上執行，透過調節多個模型佔用 Rockchip NPU 平臺的執行時間，避免因為一個模型佔用太多的執行時間使其他模型得不到及時執行。RKNN-Toolkit 從 1.2.0 版本開始支援該功能，該功能必須在 Rockchip NPU 平臺上使用，且 Rockchip NPU 的驅動版本要高於 0.9.8 版本。

2）推理流程

前面的專案完成了口罩檢測模型介面函式庫，透過 Python 程式可以呼叫口罩檢測模型進行推理。RKNN 框架模型的推理過程如圖 3.46 所示。

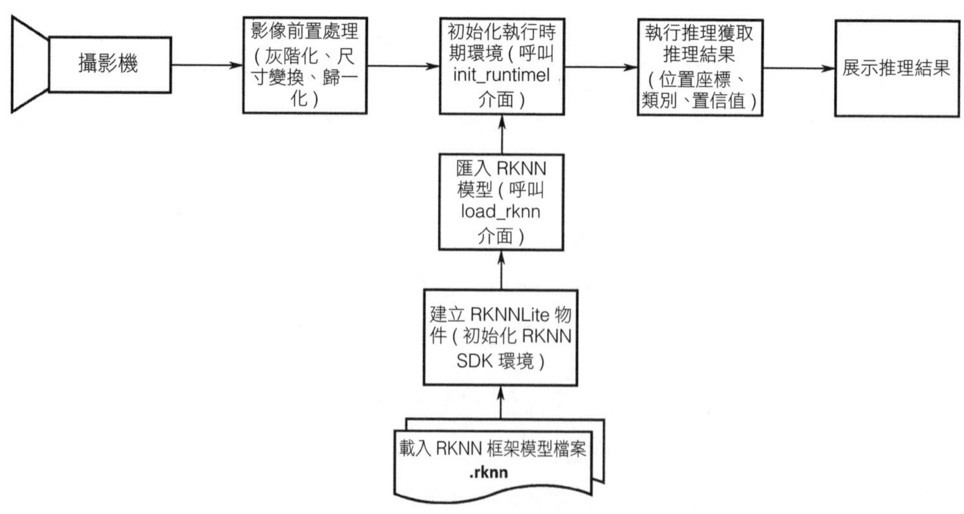

▲ 圖 3.46 RKNN 框架模型的推理過程

推理流程的描述如下：

（1）獲取攝影機的影像資料。

（2）影像前置處理：對影像資料進行灰度化處理，並採用雙線插值等演算法進行影像尺寸變換，最後進行影像資料的歸一化處理。

（3）載入 RKNN 框架模型檔案：載入 .rknn 模型檔案。

（4）建立 RKNNLite 物件：呼叫 RKNN 框架介面，建立 RKNNLite 物件。

（5）匯入 RKNN 模型：使用 RKNNLite 物件載入模型。

（6）初始化執行時期環境：呼叫 init_runtime 介面，初始化執行時期環境。

（7）執行推理獲取推理結果：呼叫 RKNNLite 物件相應的推理介面函式 inference 進行模型推理，輸出推理預測結果。推理結果包括目標物體的位置座標、類別和置信度等。

（8）展示推理結果：將推理結果繪製在原始影像上向客戶展示，並傳回推理結果資料。

3）口罩檢測模型的介面設計

AiCam 平臺對 RKNN 框架模型介面的呼叫與傳回的資料做了標準化處理，傳回的資料採用 JSON 格式。為了在應用層能夠更進一步地分析資料，模型傳回的設計如下（以口罩檢測模型為例）：

```
// 函式：mask_model.predict_resize(image)
// 參數：image 表示影像資料
// 結果：JSON 格式的字串
// 類別：{"face","face_mask"} 表示 {" 沒戴口罩 "," 戴口罩 "}
//code：200 表示執行成功
{
    "code" : 200,                                    // 傳回碼
    "msg" : "SUCCESS",                               // 傳回訊息
    "result" : {                                     // 傳回結果
        "obj_list" : [                               // 傳回內容
```

```
        {
            "location" : {                          // 目標座標
                "height" : 400,
                "left" : 1215,
                "top" : 1052,
                "width" : 570
            },
            "name" : "face_mask",                   // 目標名稱
            "score" : 0.9994969367980957            // 置信度
        }],
        "obj_num" : 1                               // 目標數量
    },
    "time" : 33.180908203125                        // 推理時間（ms）
}
```

4）口罩檢測模型的演算法設計

在 yolov5_rknn 模型推理程式中，同樣使用 _predict 函式進行模型的推理與驗證。為了使推理結果的結構更加清晰，在本專案程式中，對這個介面進行了進一步的封裝，具體以下程式所示：

```
######################################################################################
# 檔案：yolov5_rknn_detection.py
# 說明：口罩檢測模型
######################################################################################
import cv2
import time
import random
import numpy as np
import json
from copy import deepcopy
import base64
from rknnlite.api import RKNNLite

"""
RK3588 yolov5s 口罩檢測模型
"""
class AutoScale:
```

```
    def __init__(self, img, max_w, max_h):
        self._src_img = img
        self.scale = self.get_max_scale(img, max_w, max_h)
        self._new_size = self.get_new_size(img, self.scale)
        self.__new_img = None

    def get_max_scale(self,img, max_w, max_h):
        h, w = img.shape[:2]
        scale = min(max_w / w, max_h / h, 1)
        return scale

    def get_new_size(self,img, scale):
        return tuple(map(int, np.array(img.shape[:2][::-1]) * scale))

    @property
    def size(self):
        return self._new_size

    @property
    def new_img(self):
        if self.__new_img is None:
            self.__new_img = cv2.resize(self._src_img, self._new_size)
        return self.__new_img

class Yolov5_Rknn_Detection():
    def __init__(self, path, wh, masks, anchors, names):
        model = self.load_rknn_model(path)
        self.wh = wh
        self._masks = masks
        self._anchors = anchors
        self.names = names
        if isinstance(model, str):
            model = self.load_rknn_model(model)
        self._rknn = model
        self.draw_box = False

    def sigmoid(self,x):
        return 1 / (1 + np.exp(-x))
```

```
def filter_boxes(self,boxes, box_confidences, box_class_probs, conf_thres):
    # 條件機率，在區域存在物體的機率是某個類別的機率
    box_scores = box_confidences * box_class_probs
    box_classes = np.argmax(box_scores, axis=-1)            # 找出機率最大的類別索引
    box_class_scores = np.max(box_scores, axis=-1)          # 最大類別對應的機率值
    pos = np.where(box_class_scores >= conf_thres)          # 找出機率大於設定值的類別
    #pos = box_class_scores >= OBJ_THRESH
    boxes = boxes[pos]
    classes = box_classes[pos]
    scores = box_class_scores[pos]
    return boxes, classes, scores

def nms_boxes(self,boxes, scores, iou_thres):
    x = boxes[:, 0]
    y = boxes[:, 1]
    w = boxes[:, 2]
    h = boxes[:, 3]

    areas = w * h
    order = scores.argsort()[::-1]

    keep = []
    while order.size > 0:
        i = order[0]
        keep.append(i)

        xx1 = np.maximum(x[i], x[order[1:]])
        yy1 = np.maximum(y[i], y[order[1:]])
        xx2 = np.minimum(x[i] + w[i], x[order[1:]] + w[order[1:]])
        yy2 = np.minimum(y[i] + h[i], y[order[1:]] + h[order[1:]])

        w1 = np.maximum(0.0, xx2 - xx1 + 0.00001)
        h1 = np.maximum(0.0, yy2 - yy1 + 0.00001)
        inter = w1 * h1

        ovr = inter / (areas[i] + areas[order[1:]] - inter)
        inds = np.where(ovr <= iou_thres)[0]
        order = order[inds + 1]
    keep = np.array(keep)
```

```
    return keep

def plot_one_box(self,x, img, color=None, label=None, line_thickness=None, score=None):
    tl = line_thickness or round(0.002 * (img.shape[0] + img.shape[1]) / 2) + 1
    color = color or [random.randint(0, 255) for _ in range(3)]
    c1, c2 = (int(x[0]), int(x[1])), (int(x[2]), int(x[3]))
    cv2.rectangle(img, c1, c2, color, thickness=tl, lineType=cv2.LINE_AA)
    if label:
        tf = max(tl - 1, 1)
        text=label+' score:%.2f'% score     #輸出文字
        t_size = cv2.getTextSize(text, 0, fontScale=tl / 3, thickness=tf)[0]
        c2 = c1[0] + t_size[0], c1[1] - t_size[1] - 3
        cv2.rectangle(img, c1, c2, color, -1, cv2.LINE_AA)
        cv2.putText(img, text, (c1[0], c1[1] - 2), 0, tl / 3, [225, 255, 255],
                  thickness=tf, lineType=cv2.LINE_AA)
    return img

def letterbox(self,img, new_wh=(416, 416), color=(114, 114, 114)):
    a = AutoScale(img, *new_wh)
    new_img = a.new_img
    h, w = new_img.shape[:2]
    new_img = cv2.copyMakeBorder(new_img, 0, new_wh[1] - h, 0, new_wh[0] -
                              w, cv2.BORDER_CONSTANT, value=color)
    return new_img, (new_wh[0] / a.scale, new_wh[1] / a.scale)

def load_model_npu(self,PATH, npu_id):
    rknn = RKNNLite()
    devs = rknn.list_devices()
    device_id_dict = {}
    for index, dev_id in enumerate(devs[-1]):
        if dev_id[:2] != 'TS':
            device_id_dict[0] = dev_id
        if dev_id[:2] == 'TS':
            device_id_dict[1] = dev_id
    print('-->loading model : ' + PATH)
    rknn.load_rknn(PATH)
    print('--> Init runtime environment on: ' + device_id_dict[npu_id])
    ret = rknn.init_runtime(device_id=device_id_dict[npu_id])
    if ret != 0:
```

```
            print('Init runtime environment failed')
            exit(ret)
        print('done')
        return rknn

    def load_rknn_model(self,PATH):
        rknn = RKNNLite()
        ret = rknn.load_rknn(PATH)
        if ret != 0:
            print('load rknn model failed')
            exit(ret)
        print('done')
        ret = rknn.init_runtime()
        if ret != 0:
            print('Init runtime environment failed')
            exit(ret)
        print('done')
        return rknn

    def _predict(self, img_src, _img, gain, conf_thres=0.4, iou_thres=0.45):
        # 推理傳回的結構
        respond={
        "code" : None,              # 傳回碼
        "msg" : None,               # 傳回訊息
        "result" :                  # 傳回結果
            {
                "obj_list" :
                [
                    {"location" : {"left": None,"top": None,"width": None, "height": None},
                    "name": None,"score": None }   # 目標對應的位置、名稱、置信值
                ],
                "obj_num" : 1,
                "time" : None           # 推理時間
            },
        }
        # 準備推理
        src_h, src_w = img_src.shape[:2]
        _img = cv2.cvtColor(_img, cv2.COLOR_BGR2RGB)
        t0 = time.time()
```

```
        #在 Rockchip NPU 平臺上進行推理
        pred_onx = self._rknn.inference(inputs=[_img])
        respond["result"]["time"]=round((time.time() - t0)*1000,2)

        #處理推理結果
        boxes, classes, scores = [], [], []
        for t in range(3):
            input0_data = self.sigmoid(pred_onx[t][0])
            input0_data = np.transpose(input0_data, (1, 2, 0, 3))
            grid_h, grid_w, channel_n, predict_n = input0_data.shape
            anchors = [self._anchors[i] for i in self._masks[t]]
            box_confidence = input0_data[..., 4]
            box_confidence = np.expand_dims(box_confidence, axis=-1)
            box_class_probs = input0_data[..., 5:]
            box_xy = input0_data[..., :2]
            box_wh = input0_data[..., 2:4]
            col = np.tile(np.arange(0, grid_w), grid_h).reshape(-1, grid_w)
            row = np.tile(np.arange(0, grid_h).reshape(-1, 1), grid_w)
            col = col.reshape((grid_h, grid_w, 1, 1)).repeat(3, axis=-2)
            row = row.reshape((grid_h, grid_w, 1, 1)).repeat(3, axis=-2)
            grid = np.concatenate((col, row), axis=-1)
            box_xy = box_xy * 2 - 0.5 + grid
            box_wh = (box_wh * 2) ** 2 * anchors
            box_xy /= (grid_w, grid_h)  #計算原尺寸的中心
            box_wh /= self.wh  #計算原尺寸的長寬
            box_xy -= (box_wh / 2)
            box = np.concatenate((box_xy, box_wh), axis=-1)
            res = self.filter_boxes(box, box_confidence, box_class_probs, conf_thres)
            boxes.append(res[0])
            classes.append(res[1])
            scores.append(res[2])
        boxes, classes, scores = np.concatenate(boxes), np.concatenate(classes),
np.concatenate(scores)
        nboxes, nclasses, nscores = [], [], []
        for c in set(classes):
            inds = np.where(classes == c)
            b = boxes[inds]
            c = classes[inds]
            s = scores[inds]
```

```
            keep = self.nms_boxes(b, s, iou_thres)
            nboxes.append(b[keep])
            nclasses.append(c[keep])
            nscores.append(s[keep])
        if len(nboxes) < 1:
            respond["code"]=201
            respond["msg"]="NO_OBJECT"
            respond["result"]["obj_list"]=[]
            respond["result"]["obj_num"]=0
            return respond
        boxes = np.concatenate(nboxes)
        classes = np.concatenate(nclasses)
        scores = np.concatenate(nscores)

        # 傳回的推理結果
        respond["code"]=200
        respond["msg"]="SUCCESS"
        respond["result"]["obj_list"]=[]
        respond["result"]["obj_num"]=len(nboxes)
        for (x, y, w, h), score, cl in zip(boxes, scores, classes):
            obj={}
            x *= gain[0]
            y *= gain[1]
            w *= gain[0]
            h *= gain[1]
            x1 = max(0, np.floor(x).astype(int))
            y1 = max(0, np.floor(y).astype(int))
            x2 = min(src_w, np.floor(x + w + 0.5).astype(int))
            y2 = min(src_h, np.floor(y + h + 0.5).astype(int))

            obj["location"]={"left":int(x1),"top":int(y1),"width":int(x2-x1),
"height":int(y2-y1)}
            obj["name"]=self.names[cl]
            obj["score"]=float(score)
            respond["result"]["obj_list"].append(obj)

            if self.draw_box:
                self.plot_one_box((x1, y1, x2, y2), img_src, label=self.names[cl],
score=score)
```

```
        return respond

    def predict_resize(self, img_src, conf_thres=0.4, iou_thres=0.45):
        """
        預測一幅影像，前置處理使用 resize
        return: labels,boxes,scores
        """
        _img = cv2.resize(img_src, self.wh)
        gain = img_src.shape[:2][::-1]
        return self._predict(img_src, _img, gain, conf_thres, iou_thres, )

    def predict(self, img_src, conf_thres=0.4, iou_thres=0.45):
        """
        預測一幅影像，前置處理保持長寬比
        return: labels,boxes,scores
        """
        _img, gain = self.letterbox(img_src, self.wh)
        return self._predict(img_src, _img, gain, conf_thres, iou_thres)

    def close(self):
        self._rknn.release()

    def __enter__(self):
        return self

    def __exit__(self, exc_type, exc_val, exc_tb):
        self.close()

    def __del__(self):
        self.close() 測試程式
```

口罩檢測模型介面的單元測試程式如下：

```
if __name__ == '__main__':
    RKNN_MODEL_PATH = r"./yolov5s-mask.rknn"
    SIZE = (640, 640)
    MASKS = [[0, 1, 2], [3, 4, 5], [6, 7, 8]]
    ANCHORS = [[10, 13], [16, 30], [33, 23], [30, 61], [62, 45], [59, 119], [116, 90],
               [156, 198], [373, 326]]
```

```
CLASSES = ('face','face_mask')
#初始化
detector = Yolov5_Rknn_Detection(RKNN_MODEL_PATH, SIZE, MASKS, ANCHORS,
                                  CLASSES)

#讀取測試影像
img = cv2.imread("./test.jpg")
respond= detector.predict_resize(img)
print(respond)
```

3.7.2 開發步驟與驗證

3.7.2.1 專案部署

1）硬體部署

詳見 2.1.2.1 節。

2）專案部署

（1）執行 MobaXterm 工具，透過 SSH 登入到邊緣計算閘道。

（2）在 SSH 終端中執行以下命令，建立開發專案目錄。

```
$ mkdir -p ~/aiedge-exp
```

（3）透過 SSH 將本開發專案程式上傳到 ~/aicam-exp 目錄下，並採用 unzip 命令進行解壓縮。

```
$ unzip mask_detection_interface_ncnn.zip
$ unzip mask_detection_interface_rknn.zip
```

3.7.2.2 NCNN 框架模型介面開發驗證

1）參數配置

（1）開啟 mask_detection_interface_ncnn/cpp/mask_detection.cpp 檔案，修改 198、199 行的程式，將需要載入的 bin 和 param 檔案名稱修改為轉換好的 NCNN 框架模型檔案名稱。

```
# 修改的內容
std::string mask_det_param = model_path + "/yolov5s-mask.ncnn.param";
std::string mask_det_bin= model_path + "/yolov5s-mask.ncnn.bin";
```

（2）若模型類別不同，則需要修改 mask_detection_interface_ncnn/cpp/mask_detection. cpp 檔案中的其他類別名稱（對應於程式碼的 211 行），即：

```
static const char* class_names[] = {"face","face_mask"};
```

2）編譯 SO 函式庫檔案

透過 cmake 工具直接進行編譯，如果已經存在 build 目錄，則需要刪除該目錄後重新建立 build 目錄，否則會無法生成 so 檔案（這是因為無法替換已經存在的 so 檔案）。編譯完成後生成的 so 檔案為 build/maskdet.cpython-38-aarch64-linux-gnu.so，將其複製到 build 的上一級目錄。

```
$ cd ~/aiedge-exp/mask_detection_interface_ncnn
$ mkdir build && cd build
$ cmake ..
$ make -j4
Scanning dependencies of target maskdet
[ 50%] Building CXX object CMakeFiles/maskdet.dir/cpp/mask_detection.cpp.o
[100%] Linking CXX shared module maskdet.cpython-38-aarch64-linux-gnu.so
[100%] Built target maskdet
$ cp maskdet.cpython-38-aarch64-linux-gnu.so ../
```

3）介面測試

（1）執行 MobaXterm 工具，透過 SSH 登入到 GW3588 邊緣計算閘道。

（2）在 SSH 終端輸入以下命令，測試口罩檢測模型。

```
$ cd ~/aiedge-exp/mask_detection_interface_ncnn
$ python3 mask_detection.py
{
    "code" : 200,
    "msg" : "SUCCESS",
    "result" : {
        "obj_list" : [
        {
            "location" : {
                "height" : 317.11355590820312,
                "left" : 390.5355224609375,
                "top" : 112.06796264648438,
                "width" : 209.1123046875
            },
            "name" : "face_mask",
            "score" : 0.90283530950546265
        }],
        "obj_num" : 1
    },
    "time" : 88.098876953125
}
```

從測試結果可以看出，mask_detection.py 載入了 maskdet.cpython-38-aarch64-linux-gnu.so 檔案，並對本地影像 test.jpg 進行了檢測。口罩檢測結果如圖 3.47 所示。

▲ 圖 3.47 口罩檢測結果

3.7.2.3 RKNN 框架模型測試

在 GW3588 邊緣計算閘道的 SSH 終端輸入以下命令，測試口罩檢測模型介面，程式將開啟測試影像進行口罩檢測並將傳回檢測結果。

```
$ cd ~/aiedge-exp/mask_detection_interface_rknn/
$ python3 yolov5_rknn_detection.py
{'code': 200, 'msg': 'SUCCESS', 'result': {'obj_list': [{'location': {'left': 449, 'top': 0,
    'width': 142, 'height': 84}, 'name': 'face', 'score': 0.8385043144226074},
    {'location': {'left': 804, 'top': 181, 'width': 130, 'height': 175},
    'name': 'face', 'score': 0.8091291785240173}, {'location': {'left': 348, 'top': 76,
    'width': 148, 'height': 168}, 'name': 'face_mask', 'score': 0.7066819071769714}],
    'obj_num': 2, 'time': 46.52}}
```

3.7.3 本節小結

本節以口罩檢測模型為例，介紹了 RKNN 框架模型的介面設計與演算法設計，以及基於 YOLOv5 的 RKNN 框架模型的介面設計過程。

3.7.4 思考與擴充

如何在 SSH 終端進行 yolov5_rknn 模型的推理？

3.8 YOLOv3 模型的演算法設計

在電腦和深度學習領域中，模型通常用來解決特定的問題，一般是數學方程式、規則和計算步驟的組合。模型可以使得電腦執行某些任務或預測某些結果。以下是一些常見的模型演算法類型：

（1）線性回歸（Linear Regression）：用於建構輸入特徵與輸出標籤之間的線性關係，如預測房價等。

（2）邏輯回歸（Logistic Regression）：用於解決二分類問題，透過將輸入映射到一個機率範圍內，從而預測某個樣本屬於哪一類。

（3）決策樹（Decision Trees）：使用樹狀結構進行決策，每個節點代表一個特徵，每個分支代表一個可能的決策，用於分類和回歸。

（4）支援向量機（Support Vector Machines，SVM）：用於分類，透過在特徵空間中找到一個最佳超平面來進行分類。

（5）神經網路（Neural Networks）：由神經元和層組成的模型，用於處理複雜的非線性關係，廣泛用於深度學習領域。

（6）K 近鄰演算法（K-Nearest Neighbors，KNN）：透過查詢最接近輸入樣本的 k 個鄰居來進行分類或回歸。

（7）聚類演算法（Clustering Algorithms）：用於將資料分為不同的簇，如 K 平均值聚類（K-Means Clustering）。

（8）單純貝氏（Naive Bayes）：在貝氏定理的基礎上處理分類問題。

（9）隨機森林（Random Forest）：由多個決策樹組成的整合模型，用於提高預測的準確性和堅固性。

（10）強化學習演算法（Reinforcement Learning Algorithms）：用於訓練智慧體進行決策，以最大化累積獎勵，如 Q 學習、深度強化學習演算法。

本節的基礎知識如下：

- 掌握基於 AiCam 平臺的機器視覺應用框架。
- 基於 YOLOv3 和 AiCam 平臺，掌握交通號誌辨識模型的演算法設計過程。

3.8.1 原理分析與開發設計

3.8.1.1 基於 AiCam 平臺的機器視覺應用框架

AiCam 平臺採用 RESTful 介面，以 Flask 服務的方式為應用層提供演算法服務。根據實際的 AI 應用邏輯，AiCam 平臺的演算法提供兩種互動介面，分別用於處理即時推理和單次推理的應用場景。AiCam 平臺的開發框架見圖 2.4 所示。

3.8.1.2 基於 YOLOv3 和 AiCam 平臺的交通號誌辨識模型的演算法設計

1）模型推理

透過前面的專案我們完成了交通號誌辨識模型的介面函式庫，透過 Python 程式可以呼叫模型進行推理：

```
#即時視頻界面：@__app.route('/stream/<action>')
#image：攝影機即時傳遞過來的影像
#param_data：必須為 None
result = self.traffic_model.detect(image)
```

傳回的結果如下：

```
//函式：TrafficDet.detect(image)
//參數：image 表示影像資料
//結果：JSON 格式的字串
//code：200 表示執行成功，301 表示執行失敗
{
    "code" : 200,                                    //傳回碼
    "msg" : "SUCCESS",                               //傳回訊息
    "result" : {                                     //傳回結果
        "obj_list" : [                               //傳回內容
        {
            "location" : {                           //目標座標
                "height" : 400,
                "left" : 1215,
                "top" : 1052,
                "width" : 570
            },
            "name" : "right",                        //目標名稱
            "score" : 0.9994969367980957             //置信度
        }],
        "obj_num" : 1                                //目標數量
    },
    "time" : 33.180908203125                         //推理時間（ms）
}
```

2）介面設計

交通號誌辨識模型演算法透過 AiCam 平臺的即時推理介面對視訊流內的影像進行即時計算和辨識，攝影機擷取到的視訊影像透過交通號誌辨識模型演算法進行即時計算，傳回標註了辨識框和辨識內容的結果影像，並即時串流推送到應用端以視訊的方式顯示，同時將計算的結果資料（交通號誌座標、關鍵點、名稱、推理時間、置信度等）傳回到應用端，用於業務的處理。交通號誌辨識模型演算法的詳細邏輯如下：

（1）開啟邊緣計算閘道的攝影機，獲取即時的視訊影像。

（2）將即時視訊影像推送給演算法介面的 inference 方法。

（3）呼叫演算法介面的 inference 方法進行影像處理。

（4）演算法介面的 inference 方法傳回 base64 編碼的結果影像、結果資料。

（5）AiCam 平臺將傳回的結果影像和結果資料拼接為 text/event-stream 資料流程，供應用層呼叫。

（6）應用層透過 EventSource 介面獲取即時推送的演算法資料流程（結果影像和結果資料）。

（7）應用層解析資料流程，提取出結果影像和結果資料進行應用展示。

3）交通號誌辨識模型演算法設計

交通號誌辨識模型演算法透過 AiCam 平臺的即時推理介面對視訊流內的影像進行即時計算和辨識，演算法檔案（traffic_detection\traffic_detection.py）的相關程式如下：

```
##########################################################################################
# 檔案：traffic_detection.py
# 說明：交通號誌辨識模型
##########################################################################################
from PIL import Image,ImageDraw,ImageFont
import numpy as np
import cv2 as cv
import os
import json
import base64
c_dir = os.path.split(os.path.realpath(__file__))[0]

class TrafficDetection(object):
    def __init__(self, model_path="models/traffic_detection"):
        self.model_path = model_path
        self.traffic_model = TrafficDet()
        self.traffic_model.init(self.model_path)
```

```
def image_to_base64(self, img):
    image = cv.imencode('.jpg', img, [cv.IMWRITE_JPEG_QUALITY, 60])[1]
    image_encode = base64.b64encode(image).decode()
    return image_encode

def base64_to_image(self, b64):
    img = base64.b64decode(b64.encode('utf-8'))
    img = np.asarray(bytearray(img), dtype="uint8")
    img = cv.imdecode(img, cv.IMREAD_COLOR)
    return img

def draw_pos(self, img, objs):
    img_rgb = cv.cvtColor(img, cv.COLOR_BGR2RGB)
    pilimg = Image.fromarray(img_rgb)
    # 建立 ImageDraw 繪圖類別
    draw = ImageDraw.Draw(pilimg)
    # 設定字型
    font_size = 20
    font_path = c_dir+"/font/wqy-microhei.ttc"
    font_hei = ImageFont.truetype(font_path, font_size, encoding="utf-8")

    for obj in objs:
        loc = obj["location"]
        draw.rectangle((loc["left"], loc["top"], loc["left"]+loc["width"],
                    loc["top"]+loc["height"]), outline='green',width=2)
        msg =  obj["name"]+": %.2f"%obj["score"]
        draw.text((loc["left"], loc["top"]-font_size*1), msg, (0, 255, 0), font=font_hei)
    result = cv.cvtColor(np.array(pilimg), cv.COLOR_RGB2BGR)
    return result
def inference(self, image, param_data):
    #code：辨識成功傳回 200
    #msg：相關提示訊息
    #origin_image：原始影像
    #result_image：處理之後的影像
    #result_data：結果資料
    return_result = {'code': 200, 'msg': None, 'origin_image': None, 'result_image': None,
                'result_data': None}

    # 即時視頻界面：@__app.route('/stream/<action>')
```

```
        #image：攝影機即時傳遞過來的影像
        #param_data：必須為 None
        result = self.traffic_model.detect(image)
        result = json.loads(result)
        if result["code"] == 200 and result["result"]["obj_num"] > 0:
            r_image = self.draw_pos(image, result["result"]["obj_list"])
        else:
            r_image = image
        return_result["code"] = result["code"]
        return_result["msg"] = result["msg"]
        return_result["result_image"] = self.image_to_base64(r_image)
        return_result["result_data"] = result["result"]
        return return_result
```

4）單元測試

交通號誌辨識模型演算法檔案提供了單元測試程式，相關程式如下：

```
# 單元測試，如果處理類別中引用了檔案，則在單元測試中要修改檔案路徑
if __name__=='__main__':

    from trafficdet import TrafficDet
    # 建立視訊捕捉物件
    cap=cv.VideoCapture(0)
    if cap.isOpened()!=1:
        pass
    # 迴圈獲取影像、處理影像、顯示影像
    while True:
        ret,img=cap.read()
        if ret==False:
            break
        # 建立影像處理物件
        img_object=TrafficDetection(c_dir+'/models/traffic_detection')
        # 呼叫影像處理函式對影像進行加工處理
        result=img_object.inference(img,None)
        frame = img_object.base64_to_image(result["result_image"])
        # 影像顯示
        cv.imshow('frame',frame)
        key=cv.waitKey(1)
```

```
        if key==ord('q'):
            break
    cap.release()
    cv.destroyAllWindows()
else :
    from .trafficdet import TrafficDet
```

3.8.2 開發步驟與驗證

3.8.2.1 專案部署

1）硬體部署

詳見 2.1.2.1 節。

2）專案部署

（1）執行 MobaXterm 工具，透過 SSH 登入到邊緣計算閘道。

（2）在 SSH 終端執行以下命令，建立開發專案目錄。

```
$ mkdir -p ~/aiedge-exp
```

（3）透過 SSH 將本開發專案程式上傳到 ~/aicam-exp 目錄下，並採用 unzip 命令進行解壓縮。

3.8.2.2 演算法測試

在 SSH 終端輸入以下命令執行交通號誌辨識演算法，對交通號誌辨識演算法進行測試。本專案將讀取測試影像，提交給交通號誌辨識演算法介面進行辨識，完成後將結果影像在視窗顯示（見圖 3.48），並傳回對比結果資訊。對比結果資訊如下：

```
$ cd ~/aiedge-exp/traffic_detection
$ conda activate py36_tf114_torch15_cpu_cv345    //Ubuntu 20.04 作業系統下需要切換環境
$ python3 traffic_detection.py
```

```
{'obj_list': [{'location': {'height': 98, 'top': 151, 'left': 142, 'width': 121},
        'name': 'left', 'score': 0.9995859265327454}, {'location': {'height': 99,
        'top': 154, 'left': 424, 'width': 112}, 'name': 'right', 'score':
0.9991812109947205},
        {'location': {'height': 93, 'top': 155, 'left': 283, 'width': 110},
        'name': 'straight', 'score': 0.9880151152610779}, {'location': {'height': 65,
        'top': 154, 'left': -4, 'width': 123}, 'name': 'green', 'score':
0.9801796674728394}],
        'obj_num': 4}
```

▲ 圖 3.48 交通號誌辨識結果

3.8.3 本節小結

本節介紹了基於 AiCam 平臺的機器視覺應用框架，並基於 YOLOv3 和 AiCam 平臺詳細剖析了交通號誌辨識模型的演算法設計。

3.8.4 思考與擴充

請闡述透過 AiCam 平臺的即時推理介面對視訊流內的交通號誌進行即時計算和辨識的演算法邏輯。

3.9 YOLOv5 模型的演算法設計

本節的基礎知識如下：

- 掌握基於 AiCam 平臺的機器視覺應用框架。
- 結合 AiCam 平臺，掌握利用 NCNN 框架實現基於 YOLOv5 的口罩檢測模型演算法的設計過程。
- 結合 AiCam 平臺，掌握利用 RKNN 框架實現基於 YOLOv5 的口罩檢測模型演算法的設計過程。

3.9.1 原理分析與開發設計

3.9.1.1 基於 AiCam 平臺的機器視覺應用框架

詳見 3.8.1.1 節。

3.9.1.2 基於 YOLOv5 和 AiCam 平臺的口罩檢測模型的演算法設計

1）模型推理

本節以基於 NCNN 框架的口罩檢測模型為例介紹。透過前面的專案我們完成了口罩檢測模型的介面函式庫，透過 Python 程式可以呼叫 so 檔案進行推理，程式如下：

```
# 即時視頻界面：@__app.route('/stream/<action>')
#image：攝影機即時傳遞過來的影像
#param_data：必須為 None
result = self.mask_model.detect(image)
傳回的結果如下：
// 函式：MaskDet.detect(image)
// 參數：image 表示影像資料
// 結果：JSON 格式的字串
//code：200 表示執行成功，301 表示執行失敗
{
```

```
    "code" : 200,
    "msg" : "SUCCESS",
    "result" : {
        "obj_list" : [
        {
            "location" : {
                "height" : 341.3070068359375,
                "left" : 149.92092895507812,
                "top" : 0,
                "width" : 252.85385131835938
            },
            "name" : "face_mask",
            "score" : 0.9254075288772583
        }],
        "obj_num" : 1
    },
    "time" : 88.807861328125
}
```


2）介面設計

口罩檢測模型演算法透過 AiCam 平臺的即時推理介面對視訊流內的影像進行即時計算和辨識，攝影機擷取到的視訊影像透過口罩檢測模型演算法進行即時計算，傳回標註了辨識框和辨識內容的結果影像，並即時串流推送到應用端以視訊的方式顯示，同時將計算的結果資料（口罩座標、關鍵點、名稱、推理時間、置信度等）傳回到應用端，用於業務的處理。口罩檢測模型演算法的詳細邏輯和交通號誌辨識模型演算法相同，詳見 3.8.1.2 節。

3.9.1.3 基於 NCNN 框架與 RKNN 框架的口罩檢測模型演算法設計

1）基於 NCNN 框架的口罩檢測模型演算法設計

透過 AiCam 平臺的即時推理介面對視訊流內的影像進行即時計算和檢測，演算法檔案（mask_detection_yolov5_ncnn\algorithm\mask_detection_yolov5_ncnn\mask_detection_yolov5_ncnn.py）的相關程式如下：

```
##############################################################################
# 檔案：mask_detection.py
# 說明：口罩檢測模型
##############################################################################
from PIL import Image,ImageDraw,ImageFont
import numpy as np
import cv2 as cv
import os
import json
import base64
c_dir = os.path.split(os.path.realpath(__file__))[0]

class MaskDetection(object):
    def __init__(self, model_path="models/mask_detection"):
        self.model_path = model_path
        self.mask_model = MaskDet()
        self.mask_model.init(self.model_path)

    def image_to_base64(self, img):
        image = cv.imencode('.jpg', img, [cv.IMWRITE_JPEG_QUALITY, 60])[1]
        image_encode = base64.b64encode(image).decode()
        return image_encode

    def base64_to_image(self, b64):
        img = base64.b64decode(b64.encode('utf-8'))
        img = np.asarray(bytearray(img), dtype="uint8")
        img = cv.imdecode(img, cv.IMREAD_COLOR)
        return img

    def draw_pos(self, img, objs):
        img_rgb = cv.cvtColor(img, cv.COLOR_BGR2RGB)
        pilimg = Image.fromarray(img_rgb)
        # 建立 ImageDraw 繪圖類別
        draw = ImageDraw.Draw(pilimg)
        # 設定字型
        font_size = 20
        font_path = c_dir+"/../../font/wqy-microhei.ttc"
        font_hei = ImageFont.truetype(font_path, font_size, encoding="utf-8")
```

```
        for obj in objs:
            if obj["name"] == 'face':
                loc = obj["location"]
                draw.rectangle((loc["left"], loc["top"], loc["left"]+loc["width"],
                            loc["top"]+loc["height"]), outline='green',width=2)
                msg =  obj["name"]+": %.2f"%obj["score"]
                draw.text((loc["left"], loc["top"]-font_size*1), msg, (0, 255, 0),
font=font_hei)
            else:
                loc = obj["location"]
                draw.rectangle((loc["left"], loc["top"], loc["left"]+loc["width"],
                            loc["top"]+loc["height"]), outline='green',width=2)
                msg =  obj["name"]+": %.2f"%obj["score"]
                draw.text((loc["left"], loc["top"]-font_size*1), msg, (255, 0, 255),
font=font_hei)
        result = cv.cvtColor(np.array(pilimg), cv.COLOR_RGB2BGR)
        return result
    def inference(self, image, param_data):
        #code：檢測成功傳回 200
        #msg：相關提示訊息
        #origin_image：原始影像
        #result_image：處理之後的影像
        #result_data：結果資料
        return_result = {'code': 200, 'msg': None, 'origin_image': None,
                    'result_image': None, 'result_data': None}

        # 即時視頻界面：@__app.route('/stream/<action>')
        #image：攝影機即時傳遞過來的影像
        #param_data：必須為 None
        result = self.mask_model.detect(image)
        result = json.loads(result)
        if result["code"] == 200 and result["result"]["obj_num"] > 0:
                    r_image = self.draw_pos(image, result["result"]["obj_list"])
        else:
            r_image = image
        return_result["code"] = result["code"]
        return_result["msg"] = result["msg"]
        return_result["result_image"] = self.image_to_base64(r_image)
        return_result["result_data"] = result["result"]
```

```
        return return_result
# 單元測試，如果處理類別中引用了檔案，則在單元測試中要修改檔案路徑
if __name__=='__main__':
    import time
    from maskdet import MaskDet
    # 建立視訊捕捉物件
    cap=cv.VideoCapture("rtsp://admin:zonesion123@192.168.1.64/h265/ch1/sub/av_stream")
    cap.set(cv.CAP_PROP_FPS, 10)    # 每秒顯示畫面，單位為幀 / 秒
    if cap.isOpened()!=1:
        pass
    # 建立影像處理物件
    img_object=MaskDetection(c_dir+'/../../models/mask_detection')
    # 迴圈獲取影像、處理影像、顯示影像
    while True:
        ret,img=cap.read()
        # 判斷讀取到影像
        if ret is None or np.size(img)==0:
            continue

        img = cv.resize(img,  (640, 480))
        if ret==False:
            break
        # 呼叫影像處理函式對影像進行加工處理
        result=img_object.inference(img,None)
        frame = img_object.base64_to_image(result["result_image"])
        if len(frame) != 0:
            # 影像顯示
            cv.imshow('frame',frame)
        else:
            cv.imshow('frame',img)
        key=cv.waitKey(1)
        if key==ord('q'):
            break
    cap.release()
    cv.destroyAllWindows()
else :
    from .maskdet import MaskDet
```

2）基於 RKNN 框架的口罩檢測模型演算法設計

口罩檢測模型演算法透過 AiCam 框架的即時推理介面對視訊流內的影像進行即時計算和檢測，演算法檔案（mask_detection_yolov5_rknn\algorithm\mask_detection_rk3588\mask_detection_ rk3588.py）的相關程式如下：

```
###################################################################################
# 檔案：mask_detection_rk3588.py
# 說明：口罩檢測
###################################################################################
from PIL import Image,ImageDraw,ImageFont
import numpy as np
import cv2 as cv
import os
import json
import base64
c_dir = os.path.split(os.path.realpath(__file__))[0]

class MaskDetectionRk3588(object):
    def __init__(self,  model_path="models/mask_detection"):
        # 載入口罩檢測模型
        self.mask_model = Yolov5_Rknn_Detection(model_path+'/yolov5s-mask.rknn',
                        wh=(640, 640), masks=[[0, 1, 2], [3, 4, 5], [6, 7, 8]],
                        anchors=[[10, 13], [16, 30], [33, 23], [30, 61], [62, 45], [59, 119],
                        [116, 90], [156, 198], [373, 326]], names =('face','face_mask'))
    # 對影像進行 base64 編碼
    def image_to_base64(self, img):
        image = cv.imencode('.jpg', img, [cv.IMWRITE_JPEG_QUALITY, 100])[1]
        image_encode = base64.b64encode(image).decode()
        return image_encode

    def base64_to_image(self, b64):
        img = base64.b64decode(b64.encode('utf-8'))
        img = np.asarray(bytearray(img), dtype="uint8")
        img = cv.imdecode(img, cv.IMREAD_COLOR)
        return img

    def draw_pos(self, img, objs):
```

```
        img_rgb = cv.cvtColor(img, cv.COLOR_BGR2RGB)
        pilimg = Image.fromarray(img_rgb)
        # 建立 ImageDraw 繪圖類別
        draw = ImageDraw.Draw(pilimg)
        # 設定字型
        font_size = 20
        font_path = c_dir+"/../../font/wqy-microhei.ttc"
        font_hei = ImageFont.truetype(font_path, font_size, encoding="utf-8")
        # 遍歷物件
        for obj in objs:
            # 判斷檢測物件是否為 mask
            if obj["name"] == 'face_mask':
                # 獲取物件圖中座標
                loc = obj["location"]
                # 根據座標對影像進行繪矩形框
                draw.rectangle((loc["left"], loc["top"], loc["left"]+loc["width"],
                            loc["top"]+loc["height"]), outline='green',width=2)
                #msg 表示在圖中展示的文字內容
                msg = obj["name"]+": %.2f"%obj["score"]
                # 將展示的文字內容顯示在影像中
                draw.text((loc["left"], loc["top"]-font_size*1), msg, (0, 255, 0),
font=font_hei)
        # 對影像進行 RGB 至 BGR 的轉換
        result = cv.cvtColor(np.array(pilimg), cv.COLOR_RGB2BGR)
        # 傳回結果影像
        return result

    def inference(self, image, param_data):
        #code：辨識成功傳回 200
        #msg：相關提示訊息
        #origin_image：原始影像
        #result_image：處理之後的影像
        #result_data：結果資料
        return_result = {'code': 200, 'msg': None, 'origin_image': None,
                    'result_image': None, 'result_data': None}

        # 即時視頻界面：@__app.route('/stream/<action>')
        #image：攝影機即時傳遞過來的影像
        #param_data：必須為 None
```

```
        result = self.mask_model.predict_resize(image)
        #當推理結果不為空且推理成功時，對檢測目標畫矩形框
        if result["code"] == 200 and result["result"]["obj_num"] > 0:
            #呼叫繪圖函式對影像進行畫圖
            r_image = self.draw_pos(image, result["result"]["obj_list"])
        else:
            #傳回原影像
            r_image = image

        return_result["code"] = result["code"]
        return_result["msg"] = result["msg"]
        return_result["result_image"] = self.image_to_base64(r_image)
        return_result["result_data"] = result["result"]
        return return_result
#單元測試，如果處理類別中引用了檔案，則在單元測試中要修改檔案路徑
if __name__ == '__main__':
    from yolov5_rknn_detection import Yolov5_Rknn_Detection

    #建立視訊捕捉物件
    cap=cv.VideoCapture(1)
    if cap.isOpened()!=1:
        pass
    #迴圈獲取影像、處理影像、顯示影像
    while True:
        ret,img=cap.read()
        if ret==False:
            break
        #建立影像處理物件
        img_object=MaskDetectionRk3588(c_dir+'/../../models/mask_detection')
        #呼叫影像處理函式對影像進行加工處理
        result=img_object.inference(img,None)
        frame = img_object.base64_to_image(result["result_image"])

        #影像顯示
        cv.imshow('frame',frame)
        key=cv.waitKey(1)
        if key==ord('q'):
            break
    cap.release()
```

```
    cv.destroyAllWindows()

else :
    from .yolov5_rknn_detection import Yolov5_Rknn_Detection
```

3.9.2 開發步驟與驗證

3.9.2.1 專案部署

1）硬體部署

詳見 2.1.2.1 節。

2）專案部署

（1）執行 MobaXterm 工具，透過 SSH 登入到邊緣計算閘道。

（2）在 SSH 終端執行以下命令，建立開發專案目錄。

```
$ mkdir -p ~/aiedge-exp
```

（3）透過 SSH 將本開發專案程式上傳到 ~/aicam-exp 目錄下，並採用 unzip 命令進行解壓縮。

```
$ unzip mask_detection_yolov5_ncnn.zip
$ unzip mask_detection_yolov5_rknn.zip
```

3.9.2.2 演算法測試

1）基於 NCNN 框架的口罩檢測模型演算法測試

在 SSH 終端輸入以下命令，執行口罩檢測模型演算法進行 yolov5_ncnn 模型的推理，並對推理的結果進行統一的介面封裝，封裝後的資訊如下：

```
$ cd ~/aiedge-exp/mask_detection_yolov5_ncnn/algorithm/mask_detection_yolov5_ncnn
$ python3 mask_detection_yolov5_ncnn.py
```

基於 NCNN 框架的口罩檢測模型演算法測試畫面如圖 3.49 所示。

▲ 圖 3.49 基於 NCNN 框架的口罩檢測模型演算法測試畫面

2）基於 RKNN 框架的口罩檢測模型演算法測試

在 SSH 終端輸入以下命令，執行口罩檢測模型演算法進行 yolov5_rknn 模型的推理，並對推理的結果進行統一的介面封裝，封裝後的資訊如下：

```
$ cd ~/aiedge-exp/mask_detection_yolov5_rknn/algorithm/mask_detection_rk3588
$ python3 mask_detection_rk3588.py
```

基於 RKNN 框架的口罩檢測模型演算法測試畫面如圖 3.50 所示。

▲ 圖 3.50 基於 RKNN 框架的口罩檢測模型演算法測試畫面

3.9.3 本節小結

本節介紹了基於 AiCam 平臺的機器視覺應用框架，並基於 YOLOv5 和 AiCam 平臺詳細剖析了口罩檢測模型的演算法設計。

3.9.4 思考與擴充

請闡述透過 AiCam 平臺的即時推理介面對視訊流內的口罩進行即時計算和檢測的演算法邏輯。

MEMO

MEMO

4 邊緣計算與人工智慧基礎應用程式開發

本章學習邊緣計算演算法與人工智慧基礎應用程式開發，共 8 個開發案例：

（1）人臉門禁應用程式開發：掌握基於深度學習 MobileFaceNet 模型實現人臉辨識的基本原理，結合模型演算法和 AiCam 平臺進行人臉門禁系統的開發。

（2）人體入侵監測應用程式開發：掌握基於深度學習 YOLOv3 模型實現人體檢測的基本原理，結合模型演算法和 AiCam 平臺進行人體入侵監測系統的開發。

（3）手勢開關風扇應用程式開發：掌握基於深度學習 NanoDet 模型實現手勢辨識的基本原理，結合模型演算法和 AiCam 平臺進行手勢開關風扇系統的開發。

（4）視覺火情監測應用程式開發：掌握基於深度學習 YOLOv3 模型實現火焰辨識的基本原理，結合模型演算法和 AiCam 框平臺進行視覺火情監測系統的開發。

（5）視覺車牌辨識應用程式開發：掌握基於深度學習 LPRNet 模型實現車牌辨識的基本原理，結合模型演算法和 AiCam 平臺進行視覺車牌辨識系統的開發。

（6）視覺智慧抄表應用程式開發：掌握基於百度數字辨識演算法實現數字辨識的基本原理，結合演算法和 AiCam 平臺進行視覺智慧抄表系統的開發。

（7）語音窗簾控制應用程式開發：掌握基於百度語音辨識介面實現語音辨識的基本原理，結合介面演算法和 AiCam 平臺進行語音窗簾控制系統的開發。

（8）語音環境播報應用程式開發：掌握基於百度語言合成介面實現語音合成的基本原理，結合介面演算法和 AiCam 平臺進行語音環境播報系統的開發。

4.1 人臉門禁應用程式開發

人臉門禁是一種結合了人臉辨識技術和門禁系統的裝置或系統，通常用於安全門、入口控制或身份驗證，透過掃描個體的臉部特徵來驗證其身份，根據驗證結果控制門禁的開啟或關閉。

人臉門禁主要用於增強安全性、提高便利性和進行身份驗證。人臉門禁在多個領域中有著廣泛的用途，例如：

- 在建築物入口，如公司大樓、政府辦公樓、學校和其他建築物的入口，使用人臉門禁控制人員進出，可確保只有授權人員能夠進入；
- 在公共交通站點，如地鐵站、火車站和公車站等公共交通站點，使用人臉門禁來控制乘客的進出，可確保只有購票或授權的乘客可以使用交通服務；

- 在機場和港口，使用人臉門禁來管理旅客的進出，可提高安全性和便捷性，同時監測潛在的安全威脅；
- 在體育場館和娛樂場所，如體育場館、演唱會場地和娛樂場所，使用人臉門禁來控制觀眾的進出，可確保只有購票或授權的人員可以進入場館；
- 在酒店和度假勝地，使用人臉門禁來管理客人的入住和退房，可提供更高級別的安全和便利性；
- 在醫療機構，使用人臉門禁可確保只有授權的人員能夠進入敏感區域，如手術室或病房；
- 在教育機構，使用人臉門禁管理校園的入口，可保護學生和教職員工的安全；
- 在商業和零售業，使用人臉門禁來控制員工和訪客的進出，可防止未經授權的人員進出。

人臉門禁的主要作用是在需要對進入特定區域或建築物的人員進行身份驗證和存取控制的場合提高安全性和便捷性，有助減少未經授權人員的進出，提高了安全性，減少了依賴於傳統身份驗證方法的複雜性。

本節的基礎知識如下：

- 了解人臉辨識技術在門禁系統中的應用。
- 掌握深度學習 MobileFaceNet 模型實現人臉辨識的基本原理。
- 結合模型演算法和 AiCam 平臺進行人臉門禁應用程式開發。

4.1.1 原理分析與開發設計

4.1.1.1 整體框架

人臉辨識是一種生物辨識技術，它透過分析和辨識人臉上的特徵來確認個體的身份。該技術是基於人臉的生物特徵（如臉部輪廓、眼睛、鼻子、嘴巴、眉毛等），以及這些特徵之間的相對位置和比例來辨識人臉的。

人臉辨識技術的發展歷史可以追溯到幾十年前，但真正的進展主要發生在近年來，特別是在電腦視覺和深度學習領域取得重大突破之後。人臉辨識技術的發展歷史如下：

（1）20 世紀 60 年代到 70 年代：早期的人臉辨識技術主要依賴於手工製作的特徵範本匹配方法，需要人工選擇和標註臉部特徵點，並計算特徵之間的距離和比例。

（2）20 世紀 80 年代到 90 年代：隨著電腦技術的發展，基於電腦的自動人臉辨識研究開始興起，這些研究主要依賴於傳統的影像處理和模式辨識技術，如主成分分析（PCA）和線性判別分析（LDA）。

（3）21 世紀初：人臉辨識技術獲得了顯著進展，特別是在深度學習和電腦視覺領域。支援向量機（SVM）等深度學習演算法的引入，極大地提高了人臉辨識的性能；3D 人臉辨識和紅外人臉辨識等新技術也開始出現，提高了在不同光照和姿勢條件下的辨識準確性。

（4）2010 年後：深度學習的興起改變了人臉辨識技術的格局，深度卷積神經網路（CNN）被廣泛用於人臉檢測和辨識，大大提高了人臉辨識的準確性。2014 年，Facebook 的 DeepFace 系統首次實現了與人眼相媲美的人臉辨識性能；2015 年，Google 發佈的 FaceNet 系統引入了三元損失函式，進一步提高了人臉辨識的準確性。2018 年以來，人臉辨識技術被廣泛用於安全領域、手機解鎖、社交媒體標註照片等。

人臉辨識技術的發展經歷了從基礎的手工方法到深度學習的演進。隨著技術的不斷進步，人臉辨識在安全、身份驗證、存取控制、社交媒體等領域中得到廣泛應用。人臉辨識範例如圖 4.1 所示。

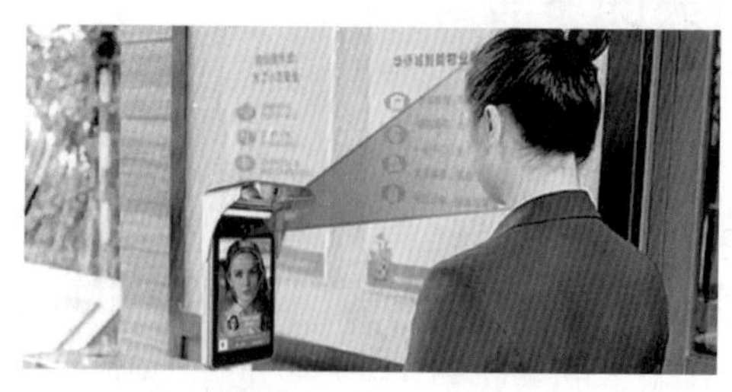

▲ 圖 4.1 人臉辨識範例

本專案採用兩種演算法實現人臉門禁。

（1）深度學習演算法：基於 MobileFaceNet 的人臉辨識模型實現人臉的即時、精準辨識。由於 MobileFaceNet 是開放原始碼框架、資料集受限等因素，辨識準確性不高，特別是非正臉狀態下辨識準確性較低。

（2）百度人臉演算法：基於百度雲端介面呼叫實現人臉辨識，準確性高，但只支持影像的辨識，且需要聯網雲端計算，有一定的延遲。

1）基於 MobileFaceNet 的人臉辨識模型

MobileFaceNet 是一種用於人臉辨識的輕量級深度神經網路框架，旨在行動端上實現高效的人臉辨識。MobileFaceNet 是專門針對資源受限的裝置（如智慧型手機、平板電腦和嵌入式裝置）設計的，用於在這些裝置上執行人臉辨識任務。MobileFaceNet 的目標是在保持高準確性的同時，減小網路的模型大小和計算複雜度，以適應這些資源受限的裝置。

MobileFaceNet 在 MobileNet V2 的基礎上，使用可分離卷積替代平均池化層，即使用一個 7×7×512（512 表示輸入特徵圖通道數目）的可分離卷積層代替了全域的平均池化層，這樣可以讓 MobileFaceNet 為不同點賦予不同的學習權重。MobileFaceNet 將一般人臉辨識模型中的全域平均池化層替換成全域深度卷積層，讓網路自動學習不同點的權重，以此提高模型的準確率。MobileFaceNet 的工作流程如圖 4.2 所示。

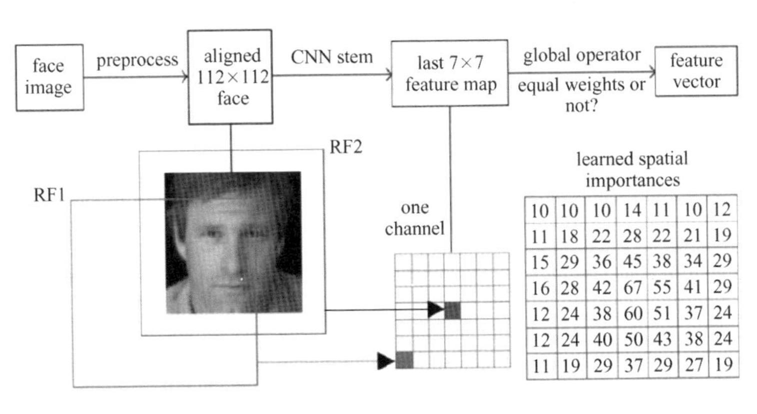

▲ 圖 4.2 MobileFaceNet 的工作流程

MobileFaceNet 的結構如圖 4.3 所示，採用了 MobileNet V2 中的 Bottleneck 作為建構模型的主要模組，但 MobileFaceNet 中 Bottlenecks 的擴充因數更小一點，並採用 PReLU 作為啟動函式，在開始階段使用快速下採樣策略，在後面幾層卷積層採用早期降維策略，在最後的線性全域深度卷積層後加入一個 1×1 的線性卷積層作為特徵輸出，損失函式採用的是 Insightface Loss。

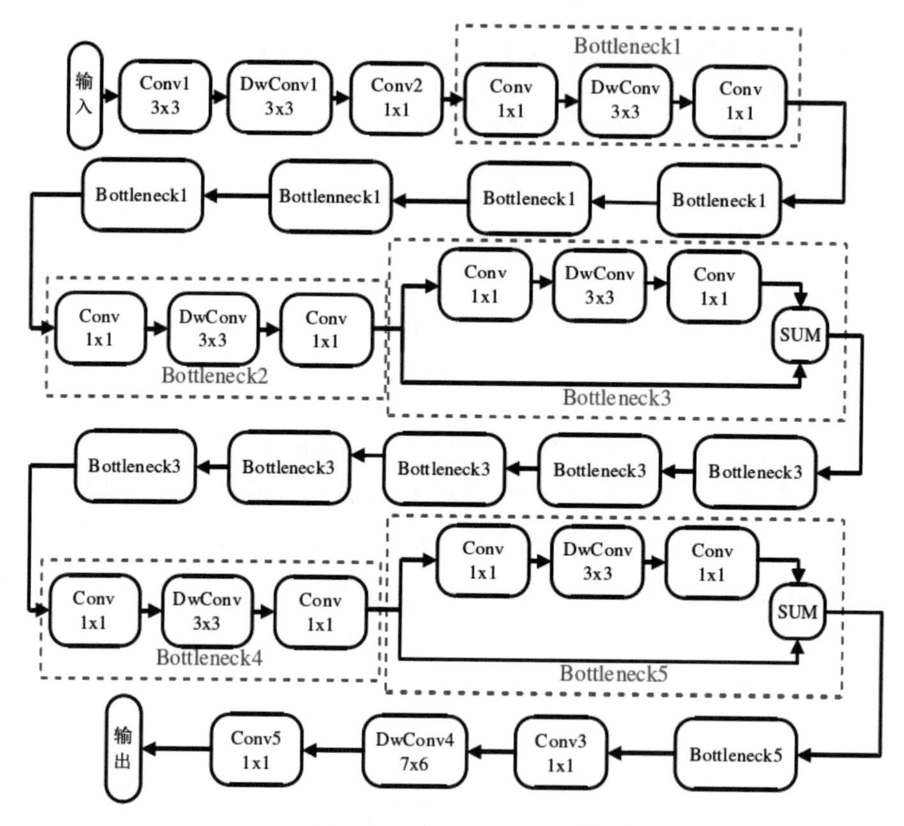

▲ 圖 4.3 MobileFaceNet 的結構

2）系統框架

從邊緣計算的角度看，人臉門禁系統可分為硬體層、邊緣層、應用層，如圖 4.4 所示。

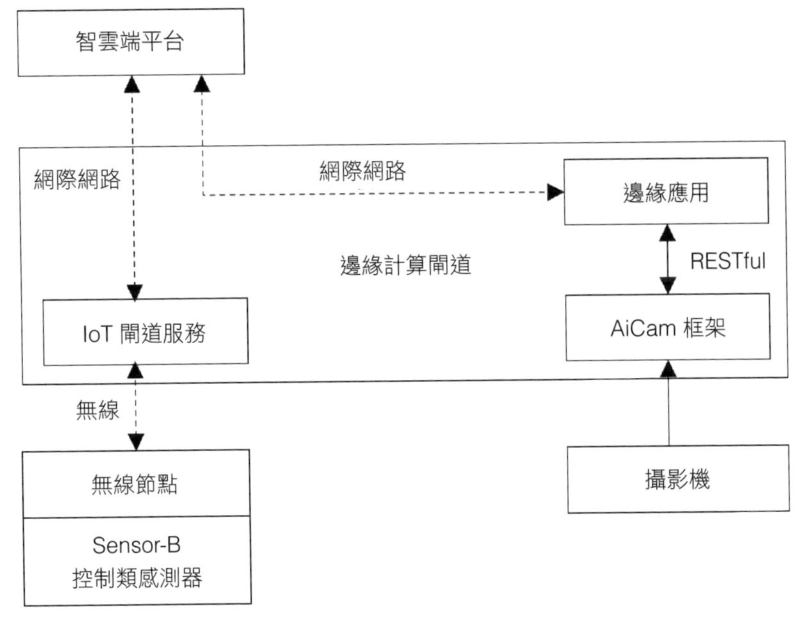

▲ 圖 4.4 人臉門禁系統的結構

（1）硬體層：無線節點和 Sensor-B 控制類感測器組成了人臉門禁系統的硬體層，透過 Sensor-B 控制類感測器的繼電器 K1 來控制門禁的開關。

（2）邊緣層：包括邊緣計算閘道內建 IoT 閘道服務和 AiCam 框架。IoT 閘道服務負責接收和下發無線節點的資料，發送給應用端或將資料發給雲端的智雲端平台。AiCam 框架內建了演算法、模型、視訊串流推送等服務，支援應用層的邊緣計算推理任務。

（3）應用層：透過智雲介面與 IoT 的硬體層互動（預設與雲端的智雲端平台的介面互動），透過 AiCam 的 RESTful 介面與演算法層互動。

4.1.1.2 硬體設計與通訊協定

1）系統硬體設計

本專案既可以採用 LiteB 無線節點、Sensor-B 控制類感測器來完成硬體的架設，也可以透過虛擬模擬軟體來建立一個虛擬的硬體裝置，如圖 4.5 所示。

▲ 圖 4.5 虛擬的硬體裝置

2）通訊協定設計

Sensor-B 控制類感測器的通訊協定如表 4.1 所示。

▼ 表 4.1 Sensor-B 控制類感測器的通訊協定

名稱	TYPE	參數	含義	許可權	說明
Sensor-B 控制類感測器	602	D1(OD1/CD1)	RGB	R/W	D1 的 Bit0 ～ Bit1 代表 RGB 三色燈的顏色狀態，00 表示關、01 表示紅色、10 表示綠色、11 表示藍色
		D1(OD1/CD1)	步進馬達	R/W	D1 的 Bit2 表示步進馬達的正反轉動狀態，0 表示正轉、1 表示反轉
		D1(OD1/CD1)	風扇 / 蜂鳴器	R/W	D1 的 Bit3 表示風扇 / 蜂鳴器的開關狀態，0 表示關閉，1 表示開啟
		D1(OD1/CD1)	LED	R/W	D1 的 Bit4 ～ Bit5 表示 LED1、LED2 的開關狀態，0 表示關閉，1 表示開啟
		D1(OD1/CD1)	繼電器	R/W	D1 的 Bit6 ～ Bit7 表示繼電器 K1、K2 的開關狀態，0 表示斷開，1 表示吸合
		V0	上報間隔	R/W	A0 ～ A7 和 D1 的循環上報時間間隔

繼電器 K1 用來模擬門禁的開關，相關命令如表 4.2 所示。

▼ 表 4.2 繼電器 K1 模擬門禁的開關命令

發送命令	接收結果	含義
{D1=?}	{D1=XX}	查詢門禁的當前開關狀態
{OD1=64,D1=?}	{D1=XX}	開啟門禁（Bit6 為 1 表示開啟門禁）
{CD1=64,D1=?}	{D1=XX}	關閉門禁（Bit6 為 0 標識關閉門禁）

4.1.1.3 功能設計與開發

1）系統框架設計

AiCam 平臺採用統一模型呼叫、統一硬體介面、統一演算法封裝和統一應用範本的設計模式，可以在嵌入式邊緣計算環境下進行快速的應用程式開發和專案實施。AiCam 平臺透過 RESTful 介面呼叫模型的演算法，即時傳回視訊影像的分析結果，透過物聯網雲端平台的應用介面與硬體連接和互動，最終實現各種應用。AiCam 平臺的開發框架請參考圖 2.4。

2）介面描述

本專案基於 AiCam 平臺開發，開發流程如下：

（1）專案配置。在 aicam 專案的設定檔（config\app.json）中增加攝影機。

```
{
    "max_load_algorithm_num":16,
    "cameras": {
        #攝影機 0：邊緣計算閘道附帶的 USB 攝影機 /dev/video0
        "0": "wc://0",
        #攝影機 1：海康威視錄影機通道 1 子碼流（從 33 開始）
        "1": "hk://admin:zonesion123@192.168.20.5/33/1",
        #攝影機 2：海康威視錄影機 RTSP 通道 1 子碼流（從 1 開始）
        "2": "rtsp://admin:zonesion123@192.168.20.5/Streaming/Channels/102"
        #攝影機 3：海康威視攝影機子碼流
        "3": "hk://admin:zonesion123@192.168.20.14/1/1",
```

```
        # 攝影機 4：海康威視攝影機 RTSP 子碼流
        "4": "rtsp://admin:zonesion123@192.168.20.14/h264/ch1/sub/av_stream"
    }
}
```

（2）增加模型。在 aicam 專案中增加模型檔案 models/face_recognition、人臉檢測模型檔案 face_det.bin/face_det.param、人臉辨識模型檔案 face_rec.bin/face_rec.param。

（3）增加演算法。在 aicam 專案中增加深度學習人臉辨識演算法 algorithm/face_recognition/face_recognition.py、百度人臉辨識演算法 algorithm/baidu_face_recognition/baidu_face_ recognition.py。

（4）增加應用。在 aicam 專案中增加前端應用 static/edge_door。

3）硬體通訊設計

前端應用中的硬體控制部分透過智雲 ZCloud API 連接到硬體系統，前端應用處理範例如下：

```
getConnect()
// 連接智雲端服務
function getConnect(){                                  // 建立連接服務的函式
    rtc = new WSNRTConnect(config.user.id, config.user.key)
    rtc.setServerAddr(config.user.addr);
    rtc.connect();
    rtc.onConnect = () => {                             // 連接成功回呼函式
        online = true
        setTimeout(() => {
            if(online){
                cocoMessage.success(` 資料服務連接成功！查詢資料中 ...`)
                // 發起資料查詢
                rtc.sendMessage(config.macList.mac_602,config.sensor.mac_602.query);
            }
        }, 200);
    }
    rtc.onConnectLost = () => {                         // 資料服務掉線回呼函式
```

```
        online = false
        cocoMessage.error(` 資料服務連接失敗！請檢查網路或 IDKEY...`)
    };

    rtc.onmessageArrive = (mac, dat) => {              // 訊息處理回呼函式
        if (dat[0] == '{' && dat[dat.length - 1] == '}') {
            // 截取背景傳回的 JSON 物件（去掉 {} 符號）後，以「,」分割為陣列
            let its = dat.slice(1,-1).split(',')
            for (let i = 0; i < its.length; i++) {    // 迴圈遍歷陣列的每一個值
                let t = its[i].split("=");             // 將每個值以「=」分割為陣列，
                if (t.length != 2) continue;
                //mac_602 控制類感測器
                if (mac == config.macList.mac_602) {
                    console.log(' 門鎖開關：',t);
                    if(t[0] == 'D1'){        // 開關控制
                        if (t[1] & 64) {
                            // 判斷接收到的命令是否跟上次的命令一樣
                            // 透過當前顯示的圖示可以得到上次命令的結果
                            if($('#icon').attr('src').indexOf('icon-off') > -1){
                                $('#icon').attr('src','./img/icon-on.gif')
                                setTimeout(() => {
                                    rtc.sendMessage(mac,config.sensor.mac_602.doorClose);
                                }, 5000);
                            }
                        }else{
                            // 判斷接收到的命令是否跟上次的命令一樣
                            // 透過當前顯示的圖示可以得到上次命令的結果
                            if($('#icon').attr('src').indexOf('icon-on') > -1){
                                $('#icon').attr('src','./img/icon-off.gif')
                            }
                        }
                    }
                }
            }
        }
    }
}
```

4）人臉辨識演算法的互動

前端應用的演算法採用 RESTful 介面獲取處理後的視訊流，傳回 base64 編碼的結果影像和結果資料。當檢測到註冊的人臉時將執行開啟門禁命令。存取 URL 位址的格式如下（IP 位址為邊緣計算閘道的位址）：

```
http://192.168.100.200:4001/stream/[algorithm_name]?camera_id=0
```

前端應用處理範例如下：

```
let linkData = [
    '/stream/index?camera_id=0',
    '/stream/face_recognition?camera_id=0'
]

// 請求影像流資源
let imgData = new EventSource(linkData[1])
// 對影像流傳回的資料進行處理
imgData.onmessage = function (res) {
    let {result_image} = JSON.parse(res.data)
    $('#img_box>img').attr('src', `data:image/jpeg;base64,${result_image}`)

    if (interactionThrottle) {
        interactionThrottle = false
        let {result_data} = JSON.parse(res.data)
        let html = `<div>${new Date().toLocaleTimeString()}————${JSON.stringify(result_
data)} </div>`
        $('#text-list').prepend(html);
        console.log(result_data);
        // 只有成功連接智雲端服務（online），且不處於節流（throttle）狀態，才能進行判斷
        if (result_data.obj_num > 0 && throttle && online) {
            throttle = false
            // 辨識到人臉且 name 不為 unknown，表示已註冊使用者，發送開啟閘門命令
            if(result_data.obj_list[0].name != 'unknow'){
                cocoMessage.success(` 使用者：${result_data.obj_list[0].name}!, 門鎖開啟中、請
稍等 ...`)
```

```
            rtc.sendMessage(config.macList.mac_602, config.sensor.mac_602.doorOpen);
        }else{
            // 辨識到人臉且 name 為 unknow，表示未註冊使用者，
            // 彈窗提示並截圖當前影像顯示到右側影像清單
            cocoMessage.error(` 非法使用者！請進行註冊 ...`)
            $('#result1').click()
        }
        setTimeout(() => {
            throttle = true
        }, 10000);
    }
    setTimeout(() => {
        interactionThrottle = true
    }, 1000);
  }
}
```

5）百度人臉辨識演算法互動

前端應用截取影像後，透過 Ajax 介面將影像以及包含百度帳號資訊的資料傳遞給演算法進行人臉辨識。百度人臉辨識演算法的互動參數如表 4.3 所示。

▼ 表 4.3 百度人臉辨識演算法的互動參數

參數	說明
url	"/file/baidu_face_recognition?camera_id=0"
method	'POST'
processData	false
contentType	false
dataType	'json'

（續表）

參數	說明

data	let img = $('.camera>img').attr('src')
	let blob = dataURItoBlob(img)
	var formData = new FormData();
	formData.append('file_name',blob,'image.png');
	//type=1 表示人臉辨識
	formData.append('param_data', JSON.stringify({"APP_ID":config.user.baidu_id,
	"API_KEY":config.user.baidu_apikey, "SECRET_KEY":config.user.baidu_secretkey,"type":1}));

success	function(res){} 內容： return_result = {'code': 200, 'msg': None, 'origin_image': None, 'result_image': None, 'result_data': None} 範例： code/msg：200 表示人臉註冊成功，404 表示沒有檢測到人臉，500 表示人臉辨識介面呼叫失敗。origin_image/result_image：表示原始影像和結果影像。result_data：演算法傳回的人臉資訊

前端應用拍照並上傳含有待辨識的人臉影像，演算法進行人臉辨識，當辨識到註冊的人臉後，將執行開啟閘門的命令：

```
$('#result2').click(function () {
    let img = $('#img_box>img').attr('src')
    let blob = dataURItoBlob(img)
    cocoMessage.info('正在辨識中，請稍等...')
    var formData = new FormData();
    formData.append('file_name',blob,'image.png');
    //type=1 表示人臉辨識
    formData.append('param_data', JSON.stringify({"APP_ID":config.user.baidu_id,
                "API_KEY":config.user.baidu_apikey,
                "SECRET_KEY":config.user.baidu_secretkey,"type":1}));
    $.ajax({
        url: '/file/baidu_face_recognition',
```

```
        method: 'POST',
        processData: false,                          // 必需的
        contentType: false,                          // 必需的
        dataType: 'json',
        data: formData,
        success: function(result) {
            console.log(result);
            if(result.code==200) {
                let img = 'data:image/jpeg;base64,' + result.origin_image;
                let html = `<div class="img-li">
                    <div  class="img-box">
                        <img src="${img}" alt=""  data-toggle="modal" data-
target="#myModal">
                    </div>
                    <div class="time"> 原始影像 <span></span><span>${new
                                        Date().toLocaleString()}</span></div>
                    </div>`
                    $('.list-box').prepend(html);

                    let img1 = 'data:image/jpeg;base64,' + result.result_image;
                    let html1 = `<div class="img-li">
                    <div  class="img-box">
                    <img src="${img1}" alt="" data-toggle="modal" data-target="#myModal">
                    </div>
                    <div class="time"> 辨識結果 <span></span><span>${new
                                  Date().toLocaleString()}</span></div>
                    </div>`
                    $('.list-box').prepend(html1);
                    // 將辨識到的人臉文字資訊著色到頁面上
                    let text = result.result_data.result.face_list[0].user_list[0].user_id
                    let html2 = `<div>${new Date().toLocaleTimeString()}——
                              辨識結果：${text}</div>`
                    $('#text-list').prepend(html2);
                    // 傳回成功的辨識結果則開啟閘門
                    if(result.result_image && online){
                    cocoMessage.success(` 使用者：${text}!, 門鎖開啟中、請稍等 ...`)
                    rtc.sendMessage(config.macList.mac_602, config.sensor.mac_602.doorOpen);
                }
            }else if(result.code==404){
```

```
                cocoMessage.error('使用者辨識失敗，請重新嘗試...')
                let img = 'data:image/jpeg;base64,' + result.origin_image;
                let html = `<div class="img-li">
                <div  class="img-box">
                    <img src="${img}" alt=""  data-toggle="modal" data-target="#myModal">
                </div>
                <div class="time">原始影像<span></span><span>${new
                                        Date().toLocaleString()}</span></div>
                </div>`
                $('.list-box').prepend(html);
            }else{
                cocoMessage.error('使用者辨識失敗，請重新嘗試...')
            }
            //請求影像流資源
            imgData.close()
            imgData = new EventSource(linkData[0])
            //對影像流傳回的資料進行處理
            imgData.onmessage = function (res) {
                let {result_image} = JSON.parse(res.data)
                $('#img_box>img').attr('src', `data:image/jpeg;base64,${result_image}`)
            }
        }, error: function(error){
            console.log(error);
            cocoMessage.error('介面呼叫失敗，請重新嘗試...')
            //請求影像流資源
            imgData.close()
            imgData = new EventSource(linkData[0])
            //對影像流傳回的資料進行處理
            imgData.onmessage = function (res) {
                let {result_image} = JSON.parse(res.data)
                $('#img_box>img').attr('src', `data:image/jpeg;base64,${result_image}`)
            }
        }
    });
})
```

6）人臉辨識演算法介面設計

人臉辨識演算法介面設計的程式如下：

```
#####################################################################################
# 檔案：face_recognition.py
# 說明：人臉辨識
#####################################################################################
import numpy as np
import cv2 as cv
import os
import json
import base64
import copy
import math
import time
from PIL import Image, ImageDraw, ImageFont

c_dir = os.path.split(os.path.realpath(__file__))[0]

def load_json_file(name):
    jo = {}
    if os.path.exists(name):
        with open(name,"r") as f:jo = json.loads(f.read())
    return jo

def save_json_file(name, jo):
    with open(name,"w") as f:f.write(json.dumps(jo))

def image_to_base64(img):
    image = cv.imencode('.jpg', img, [cv.IMWRITE_JPEG_QUALITY, 60])[1]
    image_encode = base64.b64encode(image).decode()
    return image_encode

def base64_to_image(b64):
    img = base64.b64decode(b64.encode('utf-8'))
    img = np.asarray(bytearray(img), dtype="uint8")
    img = cv.imdecode(img, cv.IMREAD_COLOR)
    return img
```

```
class FaceRecognition(object):
    def __init__(self, model_path="models/face_recognition"):
        self.facerec_model = Facerec()
        self.facerec_model.init(model_path)

        self.__features_file_name = c_dir+"/features.txt"
        self.name_feature =  load_json_file(self.__features_file_name)


    def __calculate_similarity(self, feature1, feature2):
        ''' 人臉相似度計算 '''
        inner_product = 0.0
        feature_norm1 = 0.0
        feature_norm2 = 0.0
        for i in range(len(feature1)):
            inner_product += feature1[i] * feature2[i]
            feature_norm1 += feature1[i] * feature1[i]
            feature_norm2 += feature2[i] * feature2[i]

        return inner_product / math.sqrt(feature_norm1) / math.sqrt(feature_norm2);

    def __find_name(self, feature):
        ''' 根據特徵碼匹配人名 '''
        mp = 0
        rname = "unknow"
        for name in self.name_feature.keys():
            f = self.name_feature[name]
            p = self.__calculate_similarity(f, feature)
            if p>0.5 and p > mp :
                rname = name
                mp = p
        return rname, mp

    def __draw_info(self, image, loc, msg):
        img_rgb = cv.cvtColor(image, cv.COLOR_BGR2RGB)
        pilimg = Image.fromarray(img_rgb)
        # 建立 ImageDraw 繪圖類別
        draw = ImageDraw.Draw(pilimg)
```

```
        # 設定字型
        font_size = 20
        font_path = c_dir+"/../../font/wqy-microhei.ttc"
        font_hei = ImageFont.truetype(font_path, font_size, encoding="utf-8")
        draw.rectangle((loc["left"], loc["top"], loc["left"]+loc["width"],
loc["top"]+loc["height"]),
                    outline='green',width=2)
        draw.text((loc["left"], loc["top"]-font_size*1), msg, (0, 255, 0), font=font_hei)
        result = cv.cvtColor(np.array(pilimg), cv.COLOR_RGB2BGR)
        return result

    def inference(self, image, param_data):
        #code：辨識成功傳回 200
        #msg：相關提示訊息
        #origin_image：原始影像
        #result_image：處理之後的影像
        #result_data：結果資料
        st = time.time()
        return_result = {'code': 200, 'msg': None, 'origin_image': None,
                    'result_image': None, 'result_data': None}
        # 應用請求介面：@__app.route('/file/<action>', methods=["POST"])
        #image：應用傳遞過來的資料（根據實際應用可能為影像、音訊、視訊、文字）
        #param_data：應用傳遞過來的參數，不能為空
        if param_data != None:
            # 人臉註冊
            if param_data["type"]==0 and "reg_name" in param_data:
                if param_data["reg_name"] in self.name_feature:
                    # 已經註冊
                    return_result["code"] = 202
                    return_result["msg"] = '%s 使用者已經註冊！ '%param_data["reg_name"]
                else:
                    image = np.asarray(bytearray(image), dtype="uint8")
                    image = cv.imdecode(image, cv.IMREAD_COLOR)
                    ret = self.facerec_model.feature(image)
                    jret = json.loads(ret)
                    if jret["code"] == 200:
                        if jret["result"]["obj_num"] > 0:
                            # 檢測到人臉
                            if jret["result"]["obj_num"] > 1:
```

```
                    #檢測到多個人臉
                    return_result["code"] = 204
                    return_result["msg"] = "找到多個人臉！"
                else:
                    feature = jret["result"]["obj_list"][0]["feature"] #特徵碼
                    self.name_feature[param_data["reg_name"]] = feature
                    save_json_file(self.__features_file_name, self.name_feature)
                    return_result["code"] = 200
                    return_result["msg"] = "註冊成功！"
                    #框出人臉位置
                    obj = jret["result"]["obj_list"][0]
                    result_img = self.__draw_info(image,obj["location"],
                                param_data["reg_name"])
                    return_result["result_image"] = image_to_base64(result_img)
                    return_result["origin_image"] = image_to_base64(image)
            else:
                #沒有檢測到人臉
                return_result["code"] = 404
                return_result["msg"] = "沒有找到人臉！"
        else:
            #c++介面呼叫錯誤
            return_result["code"] = jret["code"]
            return_result["msg"] = jret["msg"]
#人臉刪除
elif param_data["type"]==1 and "del_name" in param_data:
    if param_data["del_name"] in self.name_feature:
        del self.name_feature[param_data["del_name"]]
        return_result["code"] = 200
        return_result["msg"] = '刪除成功！'
        save_json_file(self.__features_file_name, self.name_feature)
    else:
        #刪除不存在的使用者，失敗
        return_result["code"] = 205
        return_result["msg"] = '未註冊，刪除失敗！'
#人臉查詢
elif param_data["type"]==2 and "find_name" in param_data:

    if param_data["find_name"] in self.name_feature:
        return_result["code"] = 200
```

```
                return_result["msg"] = '查詢'+param_data["find_name"]+'成功，已註冊'
            else:
                return_result["code"] = 205
                return_result["msg"] = '查詢'+param_data["find_name"]+'失敗，請先註冊'
        else:
            #參數錯誤
            return_result["code"] = 201
            return_result["msg"] = '參數錯誤！'

    #即時視頻界面：@__app.route('/stream/<action>')
    #image：攝影機即時傳遞過來的影像
    #param_data：必須為 None
    else:
        result = self.facerec_model.feature(image)
        jret = json.loads(result)
        result_img = image
        if jret['code'] == 200:
            face_list = [] #儲存辨識到已註冊的人臉資訊
            if jret["result"]["obj_num"] > 0:
                for obj in jret["result"]["obj_list"]:
                    name, pp = self.__find_name(obj["feature"])
                    face = {}
                    face["name"] = name
                    face["score"] = pp
                    face["location"] = obj["location"]
                    face_list.append(face)
                    show_text = name +":%.2f"%pp
                    result_img = self.__draw_info(result_img, obj["location"], show_text)
            r_data = {
                "obj_num":len(face_list),
                "obj_list":face_list
            }
            return_result["code"] = 200
            return_result["msg"] = "SUCCESS"
            return_result["result_image"] = image_to_base64(result_img)
            return_result["result_data"] = r_data
        else:
            #C++介面呼叫錯誤
            return_result["code"] = jret["code"]
```

```
                return_result["msg"] = jret["msg"]
            et = time.time()
            return_result["time"] = et - st
            return return_result

# 單元測試，如果處理類別中引用了檔案，則在單元測試中要修改檔案路徑
if __name__=='__main__':
    from facerec import Facerec

    # 讀取測試影像
    img = cv.imread("./test.jpg")
    with open("./test.jpg", "rb") as f:
        file_image = f.read()

    # 建立影像處理物件
    img_object=FaceRecognition(c_dir+'/../../models/face_recognition')
    # 呼叫影像處理函式對影像加工處理
    result=img_object.inference(img,None)
    print(" 辨識 1", result["code"], result["msg"], result["result_data"])
    cv.imshow('frame',base64_to_image(result["result_image"]))
    cv.waitKey(10000)

    result = img_object.inference(file_image, {"type":0, "reg_name":"abc"})
    print(" 註冊 ", result["code"], result["msg"])
    cv.imshow('frame',base64_to_image(result["result_image"]))
    cv.waitKey(10000)

    result=img_object.inference(img,None)
    print(" 辨識 2", result["code"], result["msg"], result["result_data"])
    cv.imshow('frame',base64_to_image(result["result_image"]))
    cv.waitKey(10000)

    result = img_object.inference(file_image, {"type":2, "find_name":"abc"})
    print(" 查詢 1", result["code"], result["msg"], result["result_data"])
    result = img_object.inference(file_image, {"type":1, "del_name":"abc"})
    print(" 刪除 ", result["code"], result["msg"], result["result_data"])
    result = img_object.inference(file_image, {"type":2, "find_name":"abc"})
    print(" 查詢 2", result["code"], result["msg"], result["result_data"])
    result = img_object.inference(img, None)
```

```
    print(" 辨識 3", result["code"], result["msg"], result["result_data"])
    # 影像顯示
    cv.imshow('frame',base64_to_image(result["result_image"]))
    cv.waitKey(10000)
    cv.destroyAllWindows()
else :
    from .facerec import Facerec
```

7）百度人臉辨識演算法介面設計

百度人臉辨識演算法介面設計的程式如下：

```
#######################################################################################
# 檔案：baidu_face_recognition.py
# 說明：人臉註冊與人臉對比
#######################################################################################
from PIL import Image, ImageDraw, ImageFont
import numpy as np
import cv2 as cv
import os,sys,time
import json
import base64
from aip import AipFace

class BaiduFaceRecognition(object):
    def __init__(self, font_path="font/wqy-microhei.ttc"):
        self.font_path = font_path

    def imencode(self, image_np):
        # 將 JPG 格式的影像編碼為資料流程
        data = cv.imencode('.jpg', image_np)[1]
        return data

    def image_to_base64(self, img):
        image = cv.imencode('.jpg', img, [cv.IMWRITE_JPEG_QUALITY, 60])[1]
        image_encode = base64.b64encode(image).decode()
        return image_encode

    def base64_to_image(self, b64):
```

```
        img = base64.b64decode(b64.encode('utf-8'))
        img = np.asarray(bytearray(img), dtype="uint8")
        img = cv.imdecode(img, cv.IMREAD_COLOR)
        return img

    def inference(self, image, param_data):
        #code：辨識成功傳回 200
        #msg：相關提示訊息
        #origin_image：原始影像
        #result_image：處理之後的影像
        #result_data：結果資料
        return_result = {'code': 200, 'msg': None, 'origin_image': None,
                         'result_image': None, 'result_data': None}

        #應用請求介面：@__app.route('/file/<action>', methods=["POST"])
        #image：應用傳遞過來的資料（根據實際應用可能為影像、音訊、視訊、文字）
        #param_data：應用傳遞過來的參數，不能為空
        if param_data != None:
            #讀取應用傳遞過來的影像
            image = np.asarray(bytearray(image), dtype="uint8")
            image = cv.imdecode(image, cv.IMREAD_COLOR)
            #對影像資料進行壓縮，方便傳輸
            img=self.image_to_base64(image)

            #type=0 表示註冊
            if param_data["type"] == 0:
                #呼叫百度人臉搜索與函式庫管理介面，透過以下使用者金鑰連接百度伺服器
                #APP_ID：百度應用 ID
                #API_KEY：百度 API_KEY
                #SECRET_KEY：百度使用者金鑰
                client=AipFace(param_data['APP_ID'],param_data['API_KEY'],
                               param_data['SECRET_KEY'])

                #配置可選參數
                options = {}
                options["user_info"] = param_data["userId"]    #使用者資訊
                options["quality_control"] = "NORMAL"          #影像品質正常
                options["liveness_control"] = "NONE"           #活體檢測
                options["action_type"] = "REPLACE"             #替換之前註冊使用者的資料
```

```
            imageType = "BASE64"

            # 群組名稱
            groupId = "zonesion"
            st = time.time()
            # 避免註冊多個使用者時 QPS（每秒查詢率）不足
            while True:
                # 帶有參數呼叫人臉註冊
                response = client.addUser(img, imageType, groupId, param_data["userId"],
options)

                if response['error_msg'] == 'SUCCESS':
                    return_result["code"] = 200
                    return_result["msg"] = "註冊成功，已成功增加至人臉庫！"
                    return_result["result_data"] = response
                    break
                time.sleep(1)
                if time.time() - st > 5:
                    return_result["code"] = 408
                    return_result["msg"] = "註冊逾時，請檢查網路！"
                    return_result["result_data"] = response
                    break
                if response['error_msg'] == 'pic not has face':
                    return_result["code"] = 404
                    return_result["msg"] = "未檢測到人臉！"
                    return_result["result_data"] = response
                    break
        else:
            # 呼叫百度人臉搜索與函式庫管理介面，透過以下使用者金鑰連接百度伺服器
            client = AipFace(param_data['APP_ID'], param_data['API_KEY'],
                             param_data['SECRET_KEY'])

            # 配置可選參數
            options = {}
            options["max_face_num"] = 10            # 檢測人臉的最大數量
            # 匹配設定值（設定設定值後，score 低於設定值的使用者將不會傳回資訊）
            options["match_threshold"] = 70
            # 影像品質控制，NONE 表示不要求影像品質，LOW 表示較低的影像品質要求，
            #NORMAL 表示一般的影像品質要求
```

```
        options["quality_control"] = "NORMAL"
        options["liveness_control"] = "NONE"    #活體檢測控制 NONE：不進行控制
        #傳回相似度最高的幾個使用者，預設為 1 個，最多傳回 50 個
        options["max_user_num"] = 3

        #搜索的群組清單，這裡只搜索 zonesion 群組中的使用者
        groupIdList = "zonesion"
        imageType = "BASE64"
        #帶有參數呼叫人臉搜索，M:N 辨識介面
        response = client.multiSearch(img, imageType, groupIdList, options)

        #應用部分
        if "error_msg" in response:
            if response['error_msg'] == 'pic not has face':
                return_result["code"] = 404
                return_result["msg"] = "未檢測到人臉！"
                return_result["result_data"] = response
                return_result["origin_image"] = self.image_to_base64(image)
                return return_result

            if response['error_msg'] == 'SUCCESS':
                #影像輸入
                img_rgb = cv.cvtColor(image, cv.COLOR_BGR2RGB)
                pilimg = Image.fromarray(img_rgb)
                #建立 ImageDraw 繪圖類別
                draw = ImageDraw.Draw(pilimg)
                #設定字型
                font_size = 20
                font_hei = ImageFont.truetype(self.font_path, font_size,
encoding="utf-8")

                #獲取資料
                #人臉資料清單
                face_list = response['result']['face_list']
                #人臉數量
                face_num = response['result']['face_num']
                for i in range(face_num):
                    loc = face_list[i]['location']
                    if len(face_list[i]['user_list'])>0:
```

```
                    #獲取辨識分數最高的人臉資料
                    user = face_list[i]['user_list'][0]
                    score = '%.2f'%user['score']
                    user_id = user['user_id']
                    group_id = user['group_id']
                    user_info = user['user_info']
                else:
                    score=user_id=group_id=user_info='none'
                #繪製矩形外框
                draw.rectangle((int(loc["left"]), int(loc["top"]),(int(loc
["left"]) +
                            int(loc["width"])), (int(loc["top"]) +
                            int(loc["height"]))), outline='green',width=2)
                #繪製字元
                draw.text((loc["left"], loc["top"]-font_size*4), '使用者
ID:'+user_id,
                        (0, 255, 0), font=font_hei)
                draw.text((loc["left"], loc["top"]-font_size*3), '使用者群組
ID:'+group_id,
                        (0, 255, 0), font=font_hei)
                draw.text((loc["left"], loc["top"]-font_size*2), '使用者
資訊：'+user_info,
                        (0, 255, 0), font=font_hei)
                draw.text((loc["left"], loc["top"]-font_size*1), '置信值：
'+score,
                        (0, 255, 0), font=font_hei)

            #輸出影像
            result = cv.cvtColor(np.array(pilimg), cv.COLOR_RGB2BGR)
            return_result["code"] = 200
            return_result["msg"] = "人臉辨識成功！"
            return_result["origin_image"] = self.image_to_base64(image)
            return_result["result_image"] = self.image_to_base64(result)
            return_result["result_data"] = response
        else:
            return_result["code"] = 500
            return_result["msg"] = "人臉介面呼叫失敗！"
            return_result["result_data"] = response
    else:
```

```
                return_result["code"] = 500
                return_result["msg"] = "百度介面呼叫失敗！"
                return_result["result_data"] = response

        #即時視頻界面：@__app.route('/stream/<action>')
        #image：攝影機即時傳遞過來的影像
        #param_data：必須為None
        else:
            return_result["result_image"] = self.image_to_base64(image)

        return return_result

#單元測試，如果處理類別中引用了檔案，則在單元測試中要修改檔案路徑
if __name__=='__main__':
    #建立影像處理物件
    img_object = BaiduFaceRecognition()

    #讀取測試影像
    img = cv.imread("./test.jpg")
    #將影像編碼成資料流程
    img = img_object.imencode(img)

    #設定參數
    addUser_data = {"APP_ID":"12345678", "API_KEY":"12345678",
                    "SECRET_KEY":"12345678", "type":0, "userId":"lilianjie"}
    searchUser_data = {"APP_ID":"12345678", "API_KEY":"12345678",
                   "SECRET_KEY":"12345678", "type":1}
    img_object.font_path = "../../font/wqy-microhei.ttc"

    #呼叫介面進行人臉註冊
    result = img_object.inference(img, addUser_data)
    #呼叫介面進行人臉辨識
    if result[«code»] == 200:
        result = img_object.inference(img, searchUser_data)
        try:
            frame = img_object.base64_to_image(result["result_image"])
            print(result["result_data"])

            #影像顯示
```

```
            cv.imshow('frame',frame)
            while True:
                key=cv.waitKey(1)
                if key==ord('q'):
                    break
            cv.destroyAllWindows()

        except AttributeError:
            print("辨識結果影像為空！，請重新辨識！")
    else:
        print("註冊失敗！")
```

4.1.2 開發步驟與驗證

4.1.2.1 專案部署

1）硬體部署

（1）AiCam 平臺的部署。

① 準備 AiCam 平臺，並正確連接 Wi-Fi 天線、攝影機。

② 啟動 AiCam 平臺，啟動 Ubuntu 作業系統，連接區域網內的 Wi-Fi 網路，記錄 AiCam 平臺的 IP 位址，如 192.168.100.200。

（2）專案硬體裝置部署。透過以下兩種方式可以部署邊緣應用所需要的硬體裝置

方式 1：利用虛擬模擬進行硬體原型的架設，Sensor-A、Sensor-C、Sensor-D、Sensor-EH 感測器裝置的專案。

方式 2：利用 AiCam 平臺的硬體來部署專案硬體裝置，並與 AiCam 平臺建構一個感測網路。

2）專案部署

（1）執行 MobaXterm 工具，透過 SSH 登入到邊緣計算閘道。

（2）在 SSH 終端執行以下命令，建立開發專案目錄。

```
$ mkdir -p ~/aiedge-exp
```

（3）透過 SSH 將本專案開發專案程式和 aicam 專案套件（DISK-AILab/02-軟體資料 /02- 綜合應用 /aicam.zip）上傳到 ~/aiedge-exp 目錄下，並採用 unzip 命令進行解壓縮。

```
$ unzip edge_door.zip
$ unzip aicam.zip -d edge_door
```

（4）修改專案設定檔 static/edge_door/js/config.js 內的智雲帳號、百度帳號、硬體位址、邊緣服務位址等資訊，範例如下：

```
user: {
    id: '12345678',                                   // 智雲帳號
    key: '12345678',                                  // 智雲金鑰
    addr: 'ws://api.zhiyun360.com:28080',             // 智雲端服務位址
    edge_addr: 'http://192.168.100.200:4001',         // 邊緣服務位址
    baidu_id: '12345678',                             // 百度應用 ID
    baidu_apikey: '12345678',                         // 百度應用 APIKEY
    baidu_secretkey: '12345678',                      // 百度應用 SECREKEY
},
// 定義本機存放區參數（MAC 位址）
macList: {
        mac_602: '01:12:4B:00:27:22:AC:4E',           //Sensor-B 控制類感測器的位址
},
```

（5）透過 SSH 將修改後的檔案上傳到邊緣計算閘道。

3）專案執行

（1）在 SSH 終端輸入以下命令，執行開發專案。

```
$ cd ~/aiedge-exp/edge_door
$ chmod 755 start_aicam.sh
$ conda activate py36_tf114_torch15_cpu_cv345     //Ubuntu 20.04 作業系統下需要切換環境
$ ./start_aicam.sh
```

（2）在使用者端或邊緣計算閘道端開啟 Chrome 瀏覽器，輸入專案頁面位址 http://192.168.100.200:4001/static/edge_door/index.html，即可查看專案內容。

4.1.2.2 人臉門禁系統的驗證

本專案透過 AiCam 平臺的門鎖控制來模擬人臉門禁的應用場景。

1）使用 AiCam 平臺門鎖控制的人臉開關門鎖

AiCam 平臺的門鎖控制實現了人臉的註冊和辨識功能。

（1）執行 AiCam 平臺的門鎖控制功能，在其介面中選擇「人臉開關門鎖」，按一下「人臉註冊」可彈出對話方塊，將當前視訊截圖作為需要註冊的人臉物件，按照要求填寫使用者姓名（英文），按一下「註冊」按鈕將發送一次註冊請求並傳回註冊結果。人臉註冊如圖 4.6 所示。

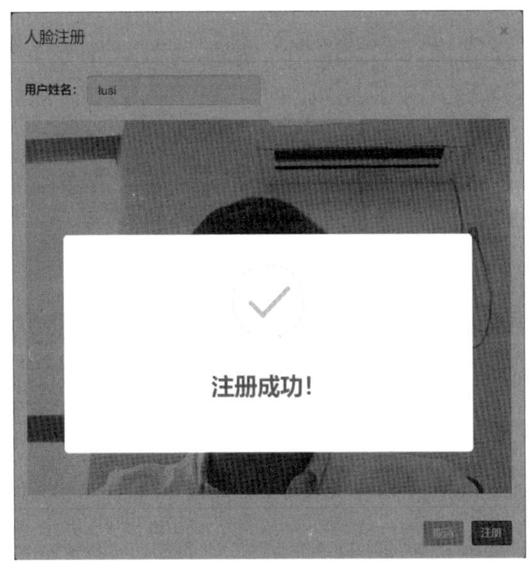

▲ 圖 4.6 人臉註冊（人臉開關門鎖）

（2）當 AiCam 平臺辨識到已註冊使用者後可開啟門鎖，並彈出訊息提示使用者名稱（成功樣式），同時門鎖開啟並在 5 s 後關閉。當辨識成功開鎖後，等待 10 s 後才再次開鎖。

成功開鎖的 AiCam 平臺介面如圖 4.7 所示。

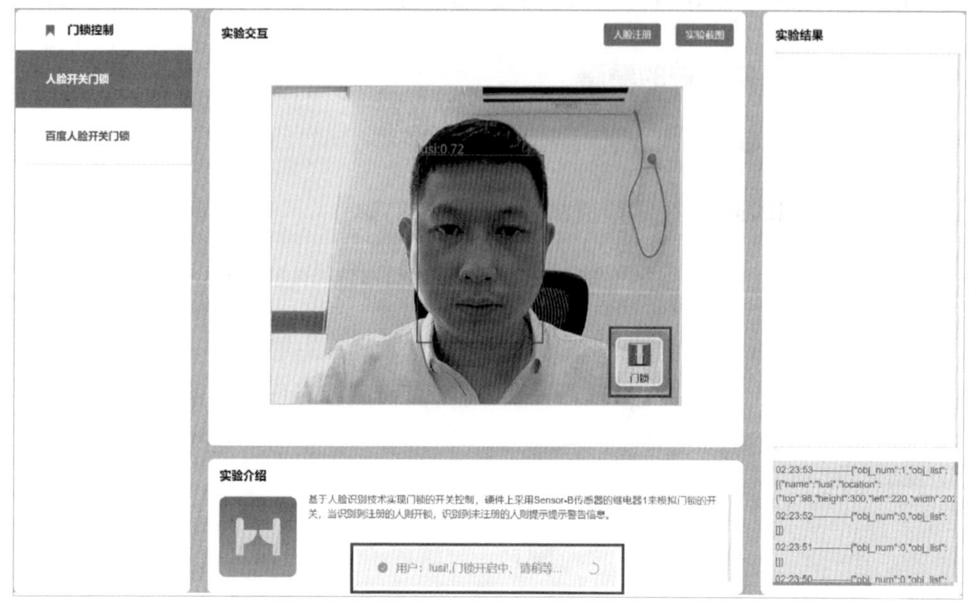

▲ 圖 4.7 成功開鎖的 AiCam 平臺介面（人臉開關門鎖）

成功開鎖的虛擬平臺介面如圖 4.8 所示。

▲ 圖 4.8 成功開鎖虛擬平臺

成功開鎖的硬體平臺如圖 4.9 所示。

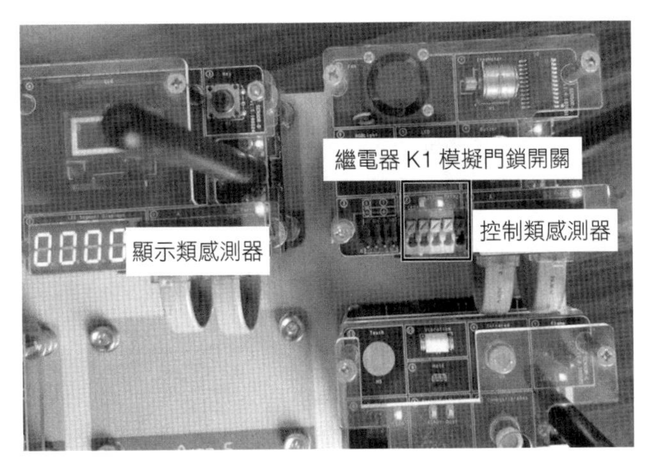

▲ 圖 4.9 成功開鎖的硬體平臺

（4）當 AiCam 平臺辨識到未註冊使用者時，將彈出訊息警告（警告樣式），並將當前視訊截圖顯示到右側清單，如圖 4.10 所示，等待 10 s 後才再次進行辨識。

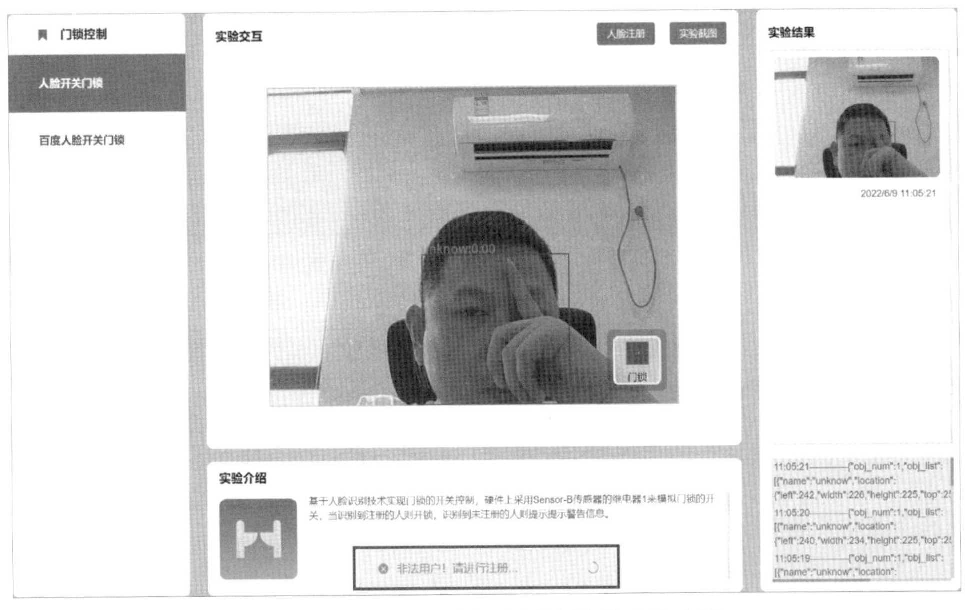

▲ 圖 4.10 檢測到未註冊使用者時的介面

2）使用 AiCam 平臺門鎖控制的百度人臉開關門鎖

百度人臉開關門鎖採用的是百度人臉辨識演算法，具有人臉註冊和辨識的功能。

（1）執行 AiCam 平臺的門鎖控制功能，在其介面中選擇「百度人臉開關門鎖」，按一下「人臉註冊」可彈出對話方塊，將當前視訊截圖作為需要註冊的人臉物件，按照要求填寫使用者姓名（英文），按一下「註冊」按鈕將發送一次註冊請求並傳回註冊結果。人臉註冊如圖 4.11 所示。

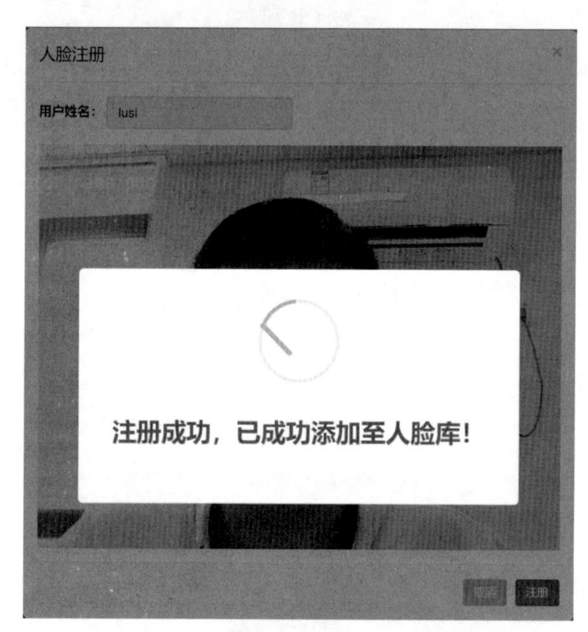

▲ 圖 4.11　人臉註冊（百度人臉開關門鎖）

（2）按一下「人臉辨識」後將發送前視訊截圖並進行一次辨識，並根據傳回結果進行相應的門鎖開關操作（將辨識到的人臉影像顯示到專案結果清單，發送門鎖開關命令，門鎖開啟後在 5 s 後自動關閉），如圖 4.12 所示。

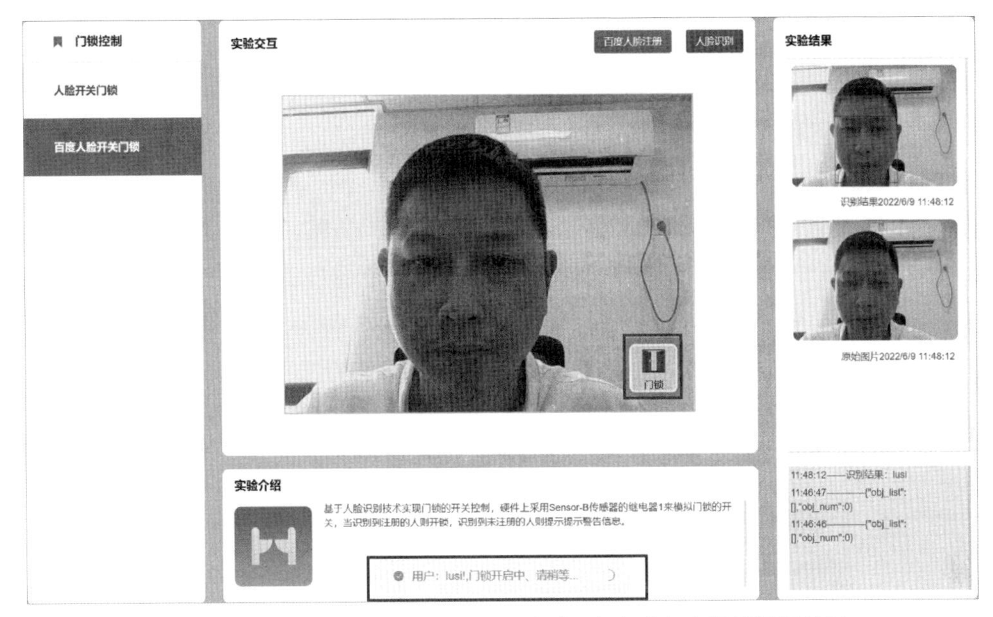

▲ 圖 4.12 成功開鎖的 AiCam 平臺介面（百度人臉開關門鎖）

4.1.3 本節小結

本節介紹了人臉門禁系統的整體框架、通訊協定、基本工作原理，以及詳細的開發步驟和硬體部署。本節透過人臉開關門鎖模擬了人臉門禁的功能。

4.1.4 思考與擴充

（1）基於深度學習人臉辨識演算法和百度人臉辨識演算法各有什麼優缺點？

（2）基於 MobileFaceNet 的人臉辨識模型有哪些主要特點？

（3）請簡述百度人臉開關門鎖的步驟。

（4）基於 AiCam 平臺的開發流程是什麼？

4.2 人體入侵監測應用程式開發

人體入侵監測是一種用於監測未經授權的人員進入特定區域的系統，該系統常用於安全領域和存取控制領域，如住宅安全系統、商業和零售業、工業設施、政府機構、醫療機構和軍事基地等，以確保只有被授權的人員才能進入受保護的區域。人體入侵監測系統通常使用各種感測器技術，如紅外感測器、超音波感測器、微波感測器和攝影機等，用來檢測人體的存在和移動。人體入侵監測系統還需要檢測區域內的任何人體運動，常用的方法包括基於規則的方法、機器學習方法和深度學習方法。深度學習技術，尤其是卷積神經網路（CNN）技術，在影像和視訊分析中獲得了顯著的進展。

本節的基礎知識如下：

- 了解基於深度學習的人體檢測技術。
- 掌握基於 YOLOv3 實現人體檢測的基本原理。
- 結合 YOLOv3 和 AiCam 平臺進行人體入侵監測應用程式開發。

4.2.1 原理分析與開發設計

深度學習是近年來人工智慧領域最為前端的技術之一，在電腦視覺、語音辨識等領域獲得了廣泛的應用，產生了一系列檢測演算法，如 R-CNN、Fast R-CNN、Faster R-CNN 和 SSD 等，但這些檢測演算法或由於精度低，或由於檢測耗時長，並不能極佳地應用到商業產品中。基於上述考慮，本節使用 YOLOv3 來解決人體檢測問題，該演算法能夠極佳地嵌入到人體入侵監測系統中。

YOLO 是在電腦視覺領域中獲得了廣泛應用的物件檢測和影像分割模型，在物件辨識、物體辨識等任務中的表現也非常出色。YOLOv3 支援全方位的視覺 AI 任務，包括檢測、分割、姿態估計、追蹤和分類。本專案基於 YOLOv3 實現人體的檢測。

人體入侵監測如圖 4.13 所示。

▲ 圖 4.13 人體入侵監測

4.2.1.1 整體框架

從邊緣計算的角度看，人體入侵監測系統可分為硬體層、邊緣層、應用層，如圖 4.14 所示。

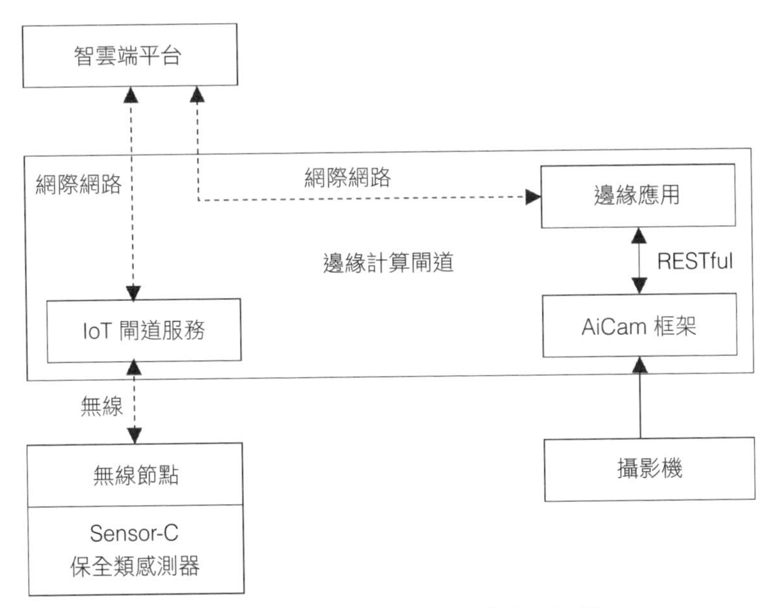

▲ 圖 4.14 人體入侵監測系統的結構

（1）硬體層：無線節點和 Sensor-C 保全類感測器組成了人體入侵監測系統的硬體層，透過 Sensor-C 保全類感測器的光柵感測器來監測是否有人入侵。

（2）邊緣層：包括邊緣計算閘道內建 IoT 閘道服務和 AiCam 框架。IoT 閘道服務負責接收和下發無線節點的資料，發送給應用端或將資料發給雲端的智雲端平台。AiCam 框架內建了演算法、模型、視訊串流推送等服務，支援應用層的邊緣計算推理任務。

（3）應用層：透過智雲介面與 IoT 的硬體層互動（預設與雲端的智雲端平台的介面互動），透過 AiCam 的 RESTful 介面與演算法層互動。

4.2.1.2 系統硬體與通訊協定設計

1）系統硬體設計

本專案既可以採用 LiteB 無線節點、Sensor-C 保全類感測器來完成硬體的架設，也可以透過虛擬模擬軟體來建立一個虛擬的硬體裝置，如圖 4.15 所示。

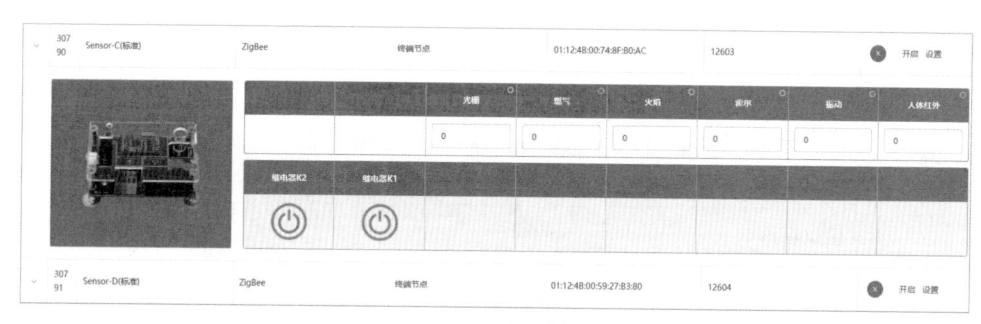

▲ 圖 4.15 虛擬硬體裝置

2）通訊協定設計

Sensor-C 保全類感測器的通訊協定如表 4.4 所示。

▼ 表 4.4 Sensor-C 保全類感測器的通訊協定

名稱	TYPE	參數	含義	許可權	說明
Sensor-C 保全類感測器	603	A0	人體紅外 / 觸控	R	人體紅外 / 觸控感測器狀態，設定值為 0 或 1，1 表示有人體活動 / 觸控動作，0 表示無人體活動 / 觸控動作

（續表）

名稱	TYPE	參數	含義	許可權	說明
Sensor-C 保全類感測器	603	A1	震動	R	震動狀態，設定值為 0 或 1，1 表示檢測到震動，0 表示未檢測到震動
		A2	霍爾	R	霍爾狀態，設定值為 0 或 1，1 表示檢測到磁場，0 表示未檢測到磁場
		A3	火焰	R	火焰狀態，設定值為 0 或 1，1 表示檢測到火焰，0 表示未檢測到火焰
		A4	瓦斯	R	瓦斯洩漏狀態，設定值為 0 或 1，1 表示檢測到瓦斯洩漏，0 表示未檢測到瓦斯洩漏
		A5	光柵	R	光柵（紅外對射）狀態值，設定值為 0 或 1，1 表示檢測到阻擋，0 表示未檢測到阻擋
		D0(OD0/CD0)	上報狀態	R/W	D0 的 Bit0 ～ Bit7 分別表示 A0 ～ A7 的上報狀態，1 表示主動上報，0 表示不上報
		D1(OD1/CD1)	繼電器	R/W	D1 的 Bit6 ～ Bit7 分別表示繼電器 K1、K2 的開關狀態，0 表示斷開，1 表示吸合
		V0	上報間隔	R/W	A0 ～ A7 和 D1 的循環上報時間間隔
		V1	語音合成資料	W	文字的 Unicode 編碼

本專案使用光柵感測器來監測是否有人入侵。當光柵感測器被遮擋時，節點會每隔3 s上傳一次光柵感測器的狀態（狀態為1）；當光柵感測器未被遮擋時，節點會每隔 30 s 上傳一次光柵感測器的狀態（狀態為 0）。光柵感測器的命令如表 4.5 所示。

▼ 表 4.5 光柵感測器的命令

發送命令	接收結果	含義
—	{A5=1/0}	光柵（紅外對射）感測器狀態

4.2.1.3 功能設計與開發

1）系統框架設計

見 4.1.1.3 節。

2）介面描述

本專案基於 AiCam 平臺開發，開發流程如下：

（1）專案配置見 4.1.1.3 節。

（2）增加模型。在 aicam 專案增加模型檔案 models/person_detection、人體檢測模型檔案 person_detector.bin\person_detector.param。

（3）增加演算法，在 aicam 專案增加人體檢測演算法檔案 algorithm/person_detection/person_ detection.py。

（4）增加應用。在 aicam 專案增加演算法專案前端應用 static/edge_grating。

3）硬體通訊設計

前端應用中的硬體控制部分透過智雲 ZCloud API 連接到硬體系統，前端應用處理範例如下：

```
getConnect()
// 服務連接
function getConnect(){                                    // 建立連接服務的函式
    rtc = new WSNRTConnect(config.user.id, config.user.key)
    rtc.setServerAddr(config.user.addr);
    rtc.connect();
    rtc.onConnect = () => {                           // 連接成功回呼函式
        online = true
        setTimeout(() => {
            if(online){
                cocoMessage.success(` 資料服務連接成功！查詢資料中 ...`)
                // 發起資料查詢
                rtc.sendMessage(config.macList.mac_603,config.sensor.mac_603.query);
            }
        }, 200);
    }
    rtc.onConnectLost = () => {                  // 資料服務掉線回呼函式
        online = false
        cocoMessage.error(` 資料服務連接失敗！請檢查網路或 IDKEY...`)
    };

    rtc.onmessageArrive = (mac, dat) => {              // 訊息處理回呼函式
        if (dat[0] == '{' && dat[dat.length - 1] == '}') {
            // 截取背景傳回的 JSON 物件（去掉 {} 符號）後，以「,」分割為陣列
            let its = dat.slice(1,-1).split(',')
            for (let i = 0; i < its.length; i++) {    // 迴圈遍歷陣列的每一個值
                let t = its[i].split("=");                  // 將每個值以「=」分割為陣列
                if (t.length != 2) continue;
                //mac_603 保全類
                if (mac == config.macList.mac_603) {
                    if(t[0] == 'A5'){                   // 光柵感測器的命令
                        console.log(' 光柵狀態：',t);
                        // 此處呼叫 AiCam 介面進行人體辨識
                    }
                }
            }
        }
    }
}
```

4）演算法互動

前端應用的演算法採用 RESTful 介面獲取處理後的視訊流，傳回 base64 編碼的結果影像和結果資料。存取 URL 位址的格式如下（IP 位址為邊緣計算閘道的位址）：

```
http://192.168.100.200:4001/stream/[algorithm_name]?camera_id=0
```

前端應用處理範例如下：

```
/******************************************************************************************
* 檔案：index.js
* 說明：主要功能是控制頁面視訊流顯示的連結切換、生成專案截圖結果等
******************************************************************************************/
if (window.location.href.indexOf('navigation') > -1) {
    $('#header').addClass('hidden');
    $('.navbar').removeClass('hidden');
}
/******************************************************************************************
* 定義本機存放區參數（ID、KEY、服務位址）
******************************************************************************************/
// 如果 configData 物件存在則賦值給 config
if (typeof configData != 'undefined') {
    config = configData
}
console.log(config);

let rtc;                              // 智雲端服務實例物件
let online = false                    // 智雲端服務是否線上
let interactionThrottle = true        // 節流，防止視訊流推送文字資料過快而頻繁觸發事件
let throttle = true                   // 節流，重複檢測到人體後的播報間隔 5 s

let linkData = [
    '/stream/index?camera_id=0',
    '/stream/person_detection?camera_id=0'
]

// 請求影像流資源
```

```
let imgData = new EventSource(linkData[0])
// 對影像流傳回的資料進行處理
imgData.onmessage = function (res) {
    let {result_image} = JSON.parse(res.data)
    $('#img_box>img').attr('src', `data:image/jpeg;base64,${result_image}`)
}

// 頁面彈窗提示使用說明
swal(
    " 使用說明 ",
    `1、光柵感測器被觸發時，右下角光柵圖示發生變化，並將初始視訊流切換為人體辨識視訊流。
     2、攝影機監測到人體後觸發彈窗警示（警告樣式），對當前視訊進行截圖並顯示到右側清單，10 s 內不再
進行警示截圖。
    3、如光柵感測器停止警示，則切換回初始視訊流。`,
    "",
    {closeOnClickOutside: false,button: " 確定 ",})

getConnect()
// 智雲端服務連接
function getConnect(){                // 建立連接服務的函式
    rtc = new WSNRTConnect(config.user.id, config.user.key)
    rtc.setServerAddr(config.user.addr);
    rtc.connect();
    rtc.onConnect = () => {                // 連接成功回呼函式
        online = true
        setTimeout(() => {
            if(online){
                cocoMessage.success(` 資料服務連接成功！查詢資料中 ...`)
                // 發起資料查詢
                rtc.sendMessage(config.macList.mac_603,config.sensor.mac_603.query);
            }
        }, 200);
    }
    rtc.onConnectLost = () => {        // 資料服務掉線回呼函式
        online = false
        cocoMessage.error(` 資料服務連接失敗！請檢查網路或 IDKEY...`)
    };

    rtc.onmessageArrive = (mac, dat) => {      // 訊息處理回呼函式
```

```
if (dat[0] == '{' && dat[dat.length - 1] == '}') {
    let its = dat.slice(1,-1).split(',')
    // 截取背景傳回的 JSON 物件（去掉 {} 符號）後，以「,」分割為陣列
    for (let i = 0; i < its.length; i++) {              // 迴圈遍歷陣列的每一個值
        let t = its[i].split("=");                      // 將每個值以「=」分割為陣列
        if (t.length != 2) continue;
        //mac_603 保全類感測器
        if (mac == config.macList.mac_603) {
            if(t[0] == 'A5'){                           // 光柵感測器
                console.log(' 光柵狀態：',t);
                if (t[1] == '1' && $('#icon').attr('src') == './img/icon-off.gif') {
                   $('#icon').attr('src','./img/icon-on.gif')
                   // 請求影像流資源
                   imgData && imgData.close()
                   imgData = new EventSource(linkData[1])
                   // 對影像流傳回的資料進行處理
                   imgData.onmessage = function (res) {
                       let {result_image} = JSON.parse(res.data)
                       $('#img_box>img').attr('src', `data:image/jpeg;base64,
                                   ${result_image}`)

                       if (interactionThrottle) {
                           interactionThrottle = false
                           let {result_data} = JSON.parse(res.data)
                           let html = `<div>${new
                                   Date().toLocaleTimeString()}————
                                   ${JSON.stringify(result_data)}</div>`
                           $('#text-list').prepend(html);
                           if (throttle && result_data.obj_num > 0) {
                               throttle = false
                               $('#result').click()
                               swal(` 非法入侵！ `, " 監測到非法入侵人員，
                                   請馬上處理！ ", "error", {button: false,
                                   timer: 2000});
                               setTimeout(() => {
                                   throttle = true
                               }, 10000);
                           }
                           setTimeout(() => {
```

```
                                        interactionThrottle = true
                                    }, 1000);
                                }
                            }
                        }else if(t[1] == '0'){
                            $('#icon').attr('src','./img/icon-off.gif')
                            // 請求影像流資源
                            imgData && imgData.close()
                            imgData = new EventSource(linkData[0])
                            // 對影像流傳回的資料進行處理
                            imgData.onmessage = function (res) {
                                let {result_image} = JSON.parse(res.data)
                                $('#img_box>img').attr('src', `data:image/jpeg;base64,
                                                  ${result_image}`)
                            }
                        }
                    }
                }
            }
        }
    }
}
$('.dropdown').hover(function (param) {
    $(this).addClass('open')
},function (param) {
    $(this).removeClass('open')
})
// 獲取當前視訊截圖，並顯示在專案結果清單中
$('#result').click(function () {
    let img = $('#img_box>img').attr('src');
    let html = `<div class="img-li">
            <div class="img-box">
            <img src="${img}" alt=""  data-toggle="modal" data-target="#myModal">
            </div>
            <div class="time">${new Date().toLocaleString()}</div>
            </div>`
    $('.list-box').prepend(html);
})
```

```
// 獲取觸發彈窗的影像，並將其顯示在彈窗介面
$('#myModal').on('show.bs.modal', function (event) {
    let img = $(event.relatedTarget).attr('src')
    $('.modal-body img').attr('src',img)
})
```

5）人體入侵監測演算法介面設計

```
##########################################################################################
# 檔案：person_detection.py
# 說明：人體入侵監測
##########################################################################################
from PIL import Image,ImageDraw,ImageFont
import numpy as np
import cv2 as cv
import os
import json
import base64
c_dir = os.path.split(os.path.realpath(__file__))[0]

class PersonDetection(object):
    def __init__(self, model_path="models/person_detection"):
        self.model_path = model_path
        self.person_model = PersonDet()
        self.person_model.init(self.model_path)

    def image_to_base64(self, img):
        image = cv.imencode('.jpg', img, [cv.IMWRITE_JPEG_QUALITY, 60])[1]
        image_encode = base64.b64encode(image).decode()
        return image_encode

    def base64_to_image(self, b64):
        img = base64.b64decode(b64.encode('utf-8'))
        img = np.asarray(bytearray(img), dtype="uint8")
        img = cv.imdecode(img, cv.IMREAD_COLOR)
        return img
    def draw_pos(self, img, objs):
        img_rgb = cv.cvtColor(img, cv.COLOR_BGR2RGB)
        pilimg = Image.fromarray(img_rgb)
```

```
        # 建立 ImageDraw 繪圖類別
        draw = ImageDraw.Draw(pilimg)
        # 設定字型
        font_size = 20
        font_path = c_dir+"/../../font/wqy-microhei.ttc"
        font_hei = ImageFont.truetype(font_path, font_size, encoding="utf-8")

        for obj in objs:
            loc = obj["location"]
            draw.rectangle((loc["left"], loc["top"], loc["left"]+loc["width"],
loc["top"]+loc["height"]),
                           outline='green',width=2)
            msg =  "%.2f"%obj["score"]
            draw.text((loc["left"], loc["top"]-font_size*1), msg, (0, 255, 0), font=font_hei)
        result = cv.cvtColor(np.array(pilimg), cv.COLOR_RGB2BGR)
        return result
    def inference(self, image, param_data):
        #code：辨識成功傳回 200
        #msg：相關提示訊息
        #origin_image：原始影像
        #result_image：處理之後的影像
        #result_data：結果資料
        return_result = {'code': 200, 'msg': None, 'origin_image': None, 'result_image': None,
                         'result_data': None}

        # 即時視頻界面：@__app.route('/stream/<action>')
        #image：攝影機即時傳遞過來的影像
        #param_data：必須為 None
        result = self.person_model.detect(image)
        result = json.loads(result)
        if result["code"] == 200 and result["result"]["obj_num"] > 0:
            r_image = self.draw_pos(image, result["result"]["obj_list"])
        else:
            r_image = image
        return_result["code"] = result["code"]
        return_result["msg"] = result["msg"]
        return_result["result_image"] = self.image_to_base64(r_image)
        return_result["result_data"] = result["result"]
        return return_result
```

```
#單元測試，如果處理類別中引用了檔案，則在單元測試中要修改檔案路徑
if __name__=='__main__':

    from persondet import PersonDet
    #建立視訊捕捉物件
    cap=cv.VideoCapture(0)
    if cap.isOpened()!=1:
        pass
    #迴圈獲取影像、處理影像、顯示影像
    while True:
        ret,img=cap.read()
        if ret==False:
            break
        #建立影像處理物件
        img_object=PersonDetection(c_dir+'/../../models/person_detection')
        #呼叫影像處理函式對影像進行加工處理
        result=img_object.inference(img,None)
        frame = img_object.base64_to_image(result["result_image"])

        #影像顯示
        cv.imshow('frame',frame)
        key=cv.waitKey(1)
        if key==ord('q'):
            break
    cap.release()
    cv.destroyAllWindows()
else :
    from .persondet import PersonDet
```

4.2.2 開發步驟與驗證

4.2.2.1 專案部署

1）硬體部署

見 4.1.2.1 節。

2）專案部署

（1）執行 MobaXterm 工具，透過 SSH 登入到邊緣計算閘道。

（2）在 SSH 終端執行以下命令，建立開發專案目錄。

```
$ mkdir -p ~/aiedge-exp
```

（3）透過 SSH 將本專案的專案程式和 aicam 專案套件上傳到 ~/aiedge-exp 目錄下，並採用 unzip 命令進行解壓縮。

```
$ unzip edge_grating.zip
$ unzip aicam.zip -d edge_grating
```

（4）修改專案設定檔 static/edge_grating/js/config.js 內的智雲帳號、硬體位址、邊緣服務位址等資訊，範例如下：

```
user: {
    id: '12345678',                                    //智雲帳號
    key: '12345678',                                   //智雲金鑰
    addr: 'ws://api.zhiyun360.com:28080',              //智雲端服務位址
    edge_addr: 'http://192.168.100.200:4001',          //邊緣服務位址
},

//定義本機存放區參數（MAC 位址）
macList: {
    mac_603: '01:12:4B:00:E5:24:1F:F1',      //Sensor-C 保全類感測器
},
```

（5）透過 SSH 將修改後的檔案上傳到邊緣計算閘道。

3）專案執行

（1）在 SSH 終端輸入以下命令，執行開發專案。

```
$ cd ~/aiedge-exp/edge_grating
$ chmod 755 start_aicam.sh
$ conda activate py36_tf114_torch15_cpu_cv345     //Ubuntu 20.04 作業系統下需要切換環境
$ ./start_aicam.sh
```

（2）在使用者端或邊緣計算閘道端開啟 Chrome 瀏覽器，輸入專案頁面位址 http://192.168.100.200:4001/static/edge_grating/index.html，即可查看專案內容。

4.2.2.2 人體入侵監測系統的驗證

本專案透過光柵感測器和人體檢測演算法來判斷是否存在非法入侵，用於模擬人體入侵監測應用。

（1）當光柵感測器被遮擋時，其狀態為 1，AiCam 平臺入侵監測介面右下角「光柵」圖示會變亮，並將初始視訊流切換為人體檢測視訊流。

（2）攝影機監測到人體後將觸發彈窗警示（警告樣式），對當前視訊進行截圖並顯示到右側清單中，10 s 內不再進行警示截圖。

（3）如果光柵感測器沒有被遮擋，則其狀態為 0，切換回初始視訊流。

（4）如果採用虛擬模擬建立的光柵感測器裝置，當設定光柵感測器狀態為 1 時表示當前處於光柵警告狀態，當攝影機視窗中出現人體時，將發出人體入侵警示並進行截圖。虛擬平臺的光柵感測器狀態如圖 4.16 所示。

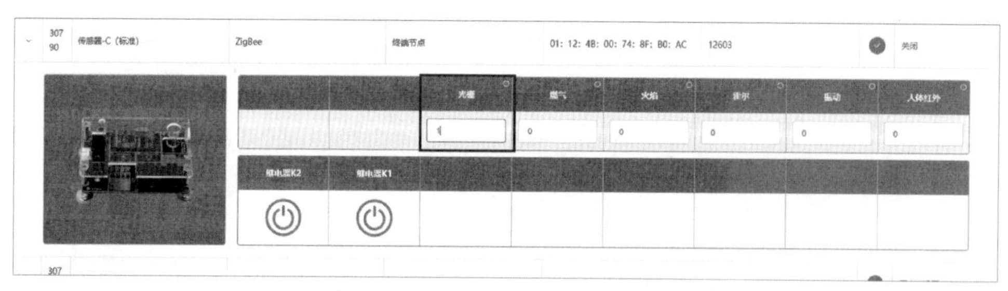

▲ 圖 4.16 虛擬平臺的光柵感測器狀態

（5）如果採用硬體裝置，當不透光的卡片遮擋光柵感測器時，則光柵感測器會上報警告（狀態為 1）。卡片遮擋光柵感測器的硬體狀態如圖 4.17 所示。

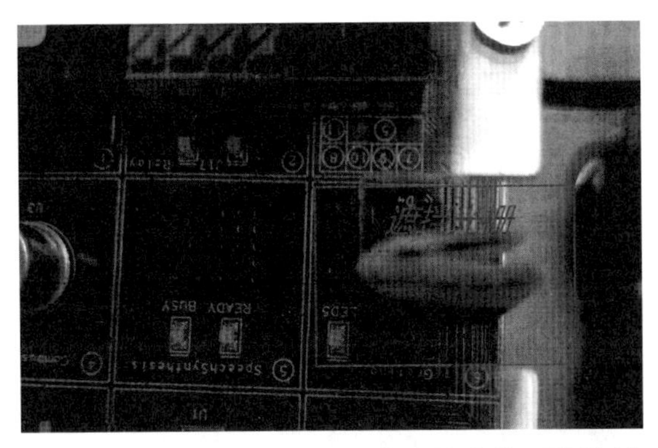

▲ 圖 4.17 卡片遮擋光柵感測器時的硬體狀態

當攝影機視窗中出現人體時，AiCam 平臺將發出人體入侵警示，並進行截圖，如圖 4.18 所示。

▲ 圖 4.18 監測到人體時的 AiCam 平臺介面

4.2.3 本節小結

本節首先介紹了 YOLOv3 在人體入侵監測系統中的應用；然後介紹了人體入侵監測系統的結構和通訊協定，描述了人體入侵監測系統的開發流程和實現方法；最後對人體入侵監測系統進行了驗證。

4.2.4 思考與擴充

（1）深度學習檢測技術有哪些？各有什麼優缺點？

（2）人體入侵監測系統的通訊協定有哪些特點？

（3）簡述人體入侵監測系統的開發步驟。

4.3 手勢開關風扇應用程式開發

手勢開關風扇是一種基於手勢辨識的控制系統，主要包括：

- 攝影機：用來捕捉使用者的手勢。
- 電腦視覺系統：用來處理攝影機捕捉的影像或視訊流，如 OpenCV 和 TensorFlow 等。
- 手勢辨識演算法：使用電腦視覺技術實現手勢辨識演算法，以辨識使用者的手勢，涉及手的位置、動作、方向和手指數量等因素。
- 風扇控制系統：用來接收來自電腦視覺系統的命令，並相應地調整風扇的轉速或開關狀態。

手勢開關風扇可以在多種情境下提供便利，可實現節能目標，以下是一些常見的用途：

- 家庭：手勢開關風扇可以用於家庭，人們可以輕鬆控制風扇的開關和轉速，無須使用遙控器或物理按鈕。
- 辦公室和商業場所：手勢開關風扇可以為員工和客戶提供更舒適的環境，減少不必要的功耗。
- 公共交通：一些公共交通工具（如火車和巴士）中，可以在座位上安裝手勢開關風扇系統，從而讓乘客自主調整風扇的轉速，提高乘坐的舒適度。

- 醫療和護理設施：手勢開關風扇可以讓病人輕鬆控制房間的溫度，提高他們的舒適感。

手勢開關風扇可以實現節能和環保，當人們離開房間或不需要風扇時，可以輕鬆關閉風扇，減少能源浪費。手勢開關風扇採用的是無接觸控制方式，是一種便捷、智慧和可持續的風扇控制方式，可以在多種環境中改善使用者體驗，並降低功耗。

本節的基礎知識如下：

- 了解基於深度學習的手勢辨識技術。
- 掌握基於 NanoDet 模型實現手勢辨識的基本原理。
- 結合 NanoDet 模型和 AiCam 平臺進行手勢辨識應用的開發。

4.3.1 原理分析與開發設計

4.3.1.1 整體框架

1）基本描述

手勢開關風扇系統透過辨識手勢來對風扇進行控制，增加了人們與風扇的互動性，為智慧家居研究中的情景化設計提供全新的想法。手勢辨識如圖 4.19 所示。

▲ 圖 4.19 手勢辨識示意圖

本專案採用深度學習演算法實現人手檢測和手勢辨識。

（1）人手檢測。YOLO、SSD、Faster R-CNN 等模型或演算法在物件辨識方面的速度較快、精度較高，但是這些模型或演算法比較大，不適合移植到行動端或嵌入式裝置。

本專案的人手檢測是基於奈米檢測網路（Nano Detecting Network，Nano Det）實現的，NanoDet 是一種 FCOS 式的單階段無錨節點（Anchor-Free）的物件辨識模型，它使用 ATSS（Adaptive Training Sample Selection）方法進行目標採樣，使用 Generalized Focal Loss 函式執行分類和邊框回歸（Box Regression），實現了高性能的物件辨識，同時保持了模型的小尺度和低計算複雜性，特別適用於嵌入式裝置和行動端應用。NanoDet 模型的結構如圖 4.20 所示。

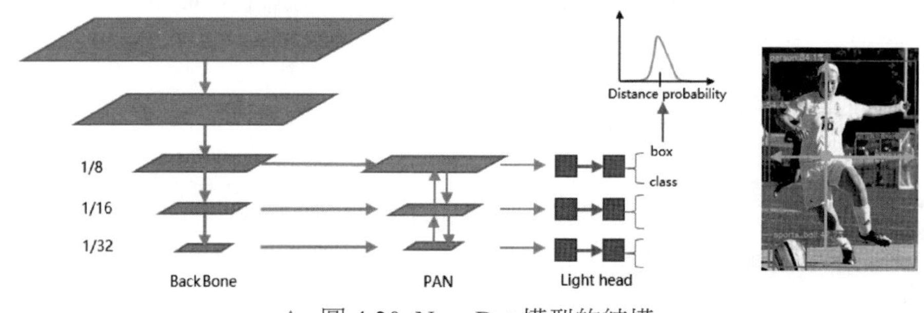

▲ 圖 4.20 NanoDet 模型的結構

NanoDet 使用了 Generalized Focal Loss 損失函式，該函式能夠去掉 FCOS 的 Centerness 分支，省去了這一分支上的大量卷積，從而減少檢測頭的計算銷耗，適合行動端的輕量化部署。Generalized Focal Loss 損失函式的結構如圖 4.21 所示。

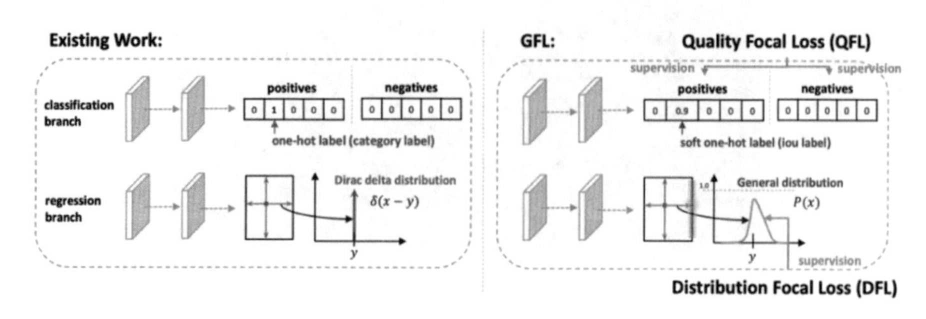

▲ 圖 4.21 Generalized Focal Loss 損失函式的結構

NanoDet 和 FCOS 物件辨識演算法系列一樣，使用共用權重的檢測頭，即對 FPN 輸出的多尺度特徵圖使用同一組卷積預測檢測框，然後每一層使用一個可學習的 Scale 值作為係數，對預測出來的框進行縮放。FCOS 的特徵圖如圖 4.22 所示。

NanoDet 的特徵提取網路選擇的是 ShuffleNetV2，並在 ShuffleNetV2 的基礎上進行了微調：首先將特徵提取網路最後一層的卷積層去掉，其次分別選擇下採樣倍數為 8、16、32 的 3 種尺度的特徵層作為特徵融合模組的輸入，最後將三種尺度的特徵層輸入 PAN 特徵融合模組得到檢測頭的輸入。

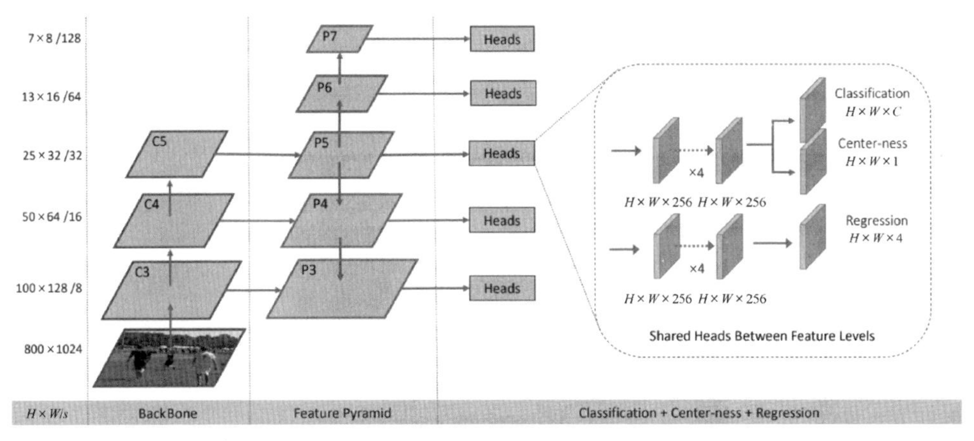

▲ 圖 4.22 FCOS 的特徵圖

（2）手勢辨識。手部姿勢估計在電腦視覺和人機互動領域有廣泛的應用，如手勢辨識、虛擬實境、擴增實境、手部追蹤等。手勢檢測可將人手的骨骼點檢測出來並連成線，將人手的結構繪製出來。從名字的角度來看，手勢辨識可以視為對人手姿態（關鍵點，如大拇指、中指、食指等）的位置估計。

HandPose 是一種用於手部姿勢估計的深度學習模型，其結構如圖 4.23 所示，主要任務是在影像或視訊中檢測和估計手部的位置和關鍵點，通常包括手掌和手指的關節位置。HandPose 模型透過檢測影像中所有的人手關鍵點，將這些關鍵點對應到不同的人手上。

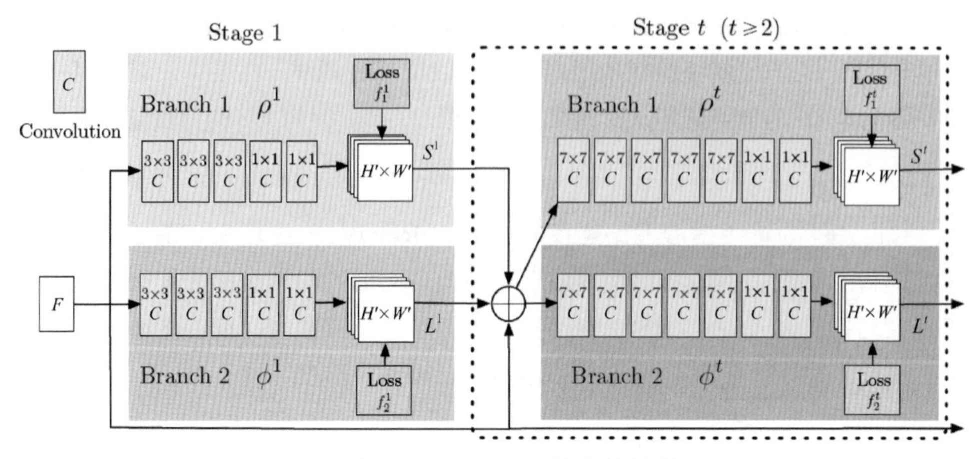

▲ 圖 4.23 HandPose 模型的結構

HandPose 基於肢體姿態模型 OpenPose 的原理，透過 PAF（Part Affinity Fields）來實現人手姿態的估計。PAF 用來描述像素在骨架中的走向，用 $L(p)$ 表示；關鍵點的回應用 $S(p)$ 表示。主體網路結構採用 VGG Pre-Train Network 作為框架，由兩個分支（Branch）分別回歸 $L(p)$ 和 $S(p)$。每一個階段（Stage）計算一次損失（Loss），之後把 L 和 S 以及原始輸入連接起來，繼續下一階段的訓練。隨著迭代次數的增加，S 能夠一定程度上區分結構的左右。損失用的 L2 範數表示，S 和 L 的真實值（Ground-Truth）需要從標註的關鍵點生成，如果某個關鍵點在標註中有缺失則不計算該關鍵點。

2）系統框架

從邊緣計算的角度看，手勢開關風扇系統可分為硬體層、邊緣層、應用層，如圖 4.24 所示。

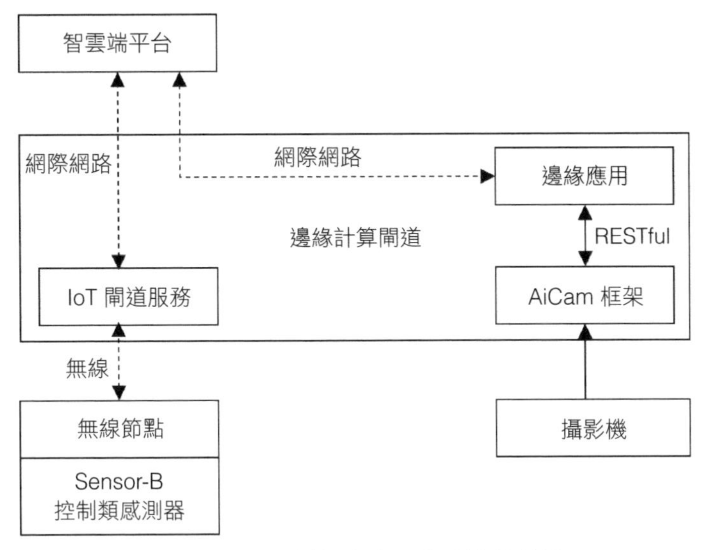

▲ 圖 4.24 手勢開關風扇系統的結構

（1）硬體層：無線節點和 Sensor-B 控制類感測器組成了手勢開關風扇系統的硬體層，透過 Sensor-B 控制類感測器的風扇來模擬實際的風扇。

（2）邊緣層：包括邊緣計算閘道內建 IoT 閘道服務和 AiCam 框架。IoT 閘道服務負責接收和下發無線節點的資料，發送給應用端或將資料發給雲端的智雲端平台。AiCam 框架內建了演算法、模型、視訊串流推送等服務，支援應用層的邊緣計算推理任務。

（3）應用層：透過智雲介面與 IoT 的硬體層互動（預設與雲端的智雲端平台的介面互動），透過 AiCam 的 RESTful 介面與演算法層互動。

4.3.1.2 系統硬體與通訊協定設計

1）系統硬體設計

本專案既可以採用 LiteB 無線節點、Sensor-B 控制類感測器來完成硬體的架設，也可以透過虛擬模擬軟體來建立一個虛擬的硬體裝置，如圖 4.25 所示。

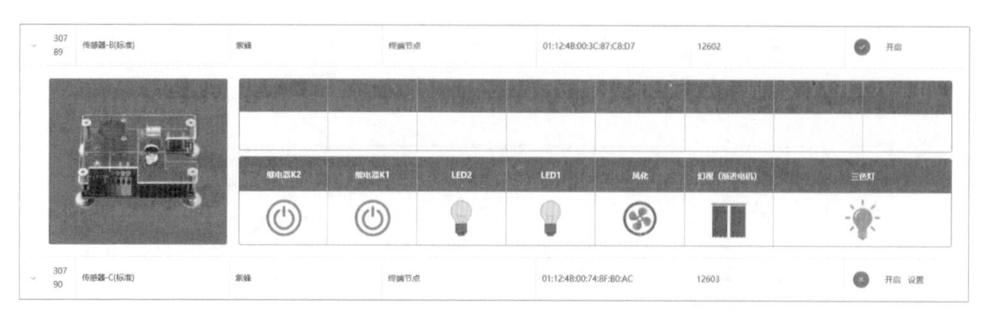

▲ 圖 4.25 虛擬硬體裝置

2）通訊協定設計

Sensor-B 控制類感測器的通訊協定見表 4.1。

本專案使用的是 Sensor-B 控制類感測器中的風扇，相關命令如表 4.6 所示。

▼ 表 4.6 風扇的命令

發送命令	接收結果	含義
{D1=?}	{D1=XX}	查詢風扇當前的開關狀態
{OD1=8,D1=?}	{D1=XX}	開啟風扇（Bit3 為 1 表示開啟風扇）
{CD1=8,D1=?}	{D1=XX}	關閉風扇（Bit3 為 0 表示關閉風扇）

4.3.1.3 功能設計與開發

1）系統框架設計

見 4.1.1.3 節。

2）介面描述

本專案基於 AiCam 平臺開發，開發流程如下：

（1）專案配置見 4.1.1.3 節。

（2）增加模型。在 aicam 專案增加模型檔案 models/handpose_detection、人手檢測模型檔案 handdet.bin/handdet.param、手勢辨識模型檔案 handpose.bin/handpose.param。

（3）增加演算法。在 aicam 專案增加手勢辨識演算法檔案 algorithm/handpose_detection/ handpose_detection.py。

（4）增加應用。在 aicam 專案增加演算法專案前端應用 static/edge_fan。

3）硬體通訊設計

前端應用中的硬體控制部分透過智雲 ZCloud API 連接到硬體系統，前端應用處理範例如下：

```
getConnect()
// 智雲端服務連接
function getConnect(){                          // 建立連接服務的函式
    rtc = new WSNRTConnect(config.user.id, config.user.key)
    rtc.setServerAddr(config.user.addr);
    rtc.connect();
    rtc.onConnect = () => {                     // 連接成功回呼函式
        online = true
        setTimeout(() => {
            if(online){
                cocoMessage.success(` 資料服務連接成功！查詢資料中 ...`)
                rtc.sendMessage(config.macList.mac_602,config.sensor.mac_602.query);  // 資料
查詢
            }
        }, 200);
    }
    rtc.onConnectLost = () => {                 // 資料服務掉線回呼函式
        online = false
        cocoMessage.error(` 資料服務連接失敗！請檢查網路或 IDKEY...`)
    };

    rtc.onmessageArrive = (mac, dat) => {   // 訊息處理回呼函式
        if (dat[0] == '{' && dat[dat.length - 1] == '}') {
            // 截取背景傳回的 JSON 物件（去掉 {} 符號）後，以「,」分割為陣列
```

```
            let its = dat.slice(1,-1).split(',')
            for (let i = 0; i < its.length; i++) {              // 迴圈遍歷陣列的每一個值
                let t = its[i].split("=");                      // 將每個值以「=」分割為陣列
                if (t.length != 2) continue;
                //mac_602 控制類感測器
                if (mac == config.macList.mac_602) {
                    console.log(' 風扇開關：',t);
                    if(t[0] == 'D1'){                  // 開關控制
                        if (t[1] & 8) {
                            $('#icon').attr('src','./img/icon-on.gif')
                        }else{
                            $('#icon').attr('src','./img/icon-off.png')
                        }
                    }
                }
            }
        }
    }
}
```

4）演算法互動

前端應用的演算法採用 RESTful 介面獲取處理後的視訊流，傳回 base64 編碼的結果影像和結果資料。存取 URL 位址的格式如下（IP 位址為邊緣計算閘道的位址）：

```
http://192.168.100.200:4001/stream/[algorithm_name]?camera_id=0
```

前端應用處理範例如下：

```
let linkData = '/stream/handpose_detection?camera_id=0'        // 視訊資源連結
// 請求影像流資源
let imgData = new EventSource(linkData)
// 對影像流傳回的資料進行處理
imgData.onmessage = function (res) {
    let {result_image} = JSON.parse(res.data)
    $('#img_box>img').attr('src', `data:image/jpeg;base64,${result_image}`)
    if (interactionThrottle) {
```

```
        interactionThrottle = false
        let {result_data} = JSON.parse(res.data)
        let html = `<div>${new Date().toLocaleTimeString()}————$
                     {JSON.stringify(result_data)}</div>`
        $('#text-list').prepend(html);
        console.log(result_data);
        // 當匹配率（result_data.obj_list[0].score）大於 95%、辨識到手勢
        // 結果（result_data.obj_list[0].name）、當前不處於節流（throttle）狀態、
        // 智雲端服務連接成功（online）後，才能進入判斷
        if (result_data.obj_num > 0 && throttle && online) {
            // 根據手勢辨識結果判斷發送的是開啟命令還是關閉命令，從而相應地控制裝置
            if (result_data.obj_list[0].name == 'one' && result_data.obj_list[0].score > 0.95
                                        && $('#icon').attr('src').indexOf('icon-on') > -1) {
                console.log(result_data.obj_list[0].name);
                throttle = false
                rtc.sendMessage(config.macList.mac_602, config.sensor.mac_602.fanClose)
                cocoMessage.success(' 辨識到手勢 1，關閉風扇！ ')
                setTimeout(() => {
                    throttle = true
                }, 5000);
            }
            if (result_data.obj_list[0].name == 'five' && result_data.obj_list[0].score > 0.95
                                        && $('#icon').attr('src').indexOf('icon-off') > -1) {
                console.log(result_data.obj_list[0].name);
                throttle = false
                rtc.sendMessage(config.macList.mac_602, config.sensor.mac_602.fanOpen)
                cocoMessage.success(' 辨識到手勢 5，開啟風扇！ ')
                setTimeout(() => {
                    throttle = true
                }, 5000);
            }
        }
        setTimeout(() => {
            interactionThrottle = true
        }, 1000);
    }
}
```

5）手勢開關風扇演算法介面設計

```
#########################################################################################
# 檔案：handpose_detection.py
# 說明：手勢辨識
#########################################################################################
from PIL import Image,ImageDraw,ImageFont
import numpy as np
import cv2 as cv
import os
import json
import base64
c_dir = os.path.split(os.path.realpath(__file__))[0]

class HandposeDetection(object):
    def __init__(self, model_path="models/handpose_detection"):
        self.model_path = model_path
        self.handpose_model = HandDetector()
        self.handpose_model.init(self.model_path)

    def image_to_base64(self, img):
        image = cv.imencode('.jpg', img, [cv.IMWRITE_JPEG_QUALITY, 60])[1]
        image_encode = base64.b64encode(image).decode()
        return image_encode

    def base64_to_image(self, b64):
        img = base64.b64decode(b64.encode('utf-8'))
        img = np.asarray(bytearray(img), dtype="uint8")
        img = cv.imdecode(img, cv.IMREAD_COLOR)
        return img

    def draw_pos(self, img, objs):
        img_rgb = cv.cvtColor(img, cv.COLOR_BGR2RGB)
        pilimg = Image.fromarray(img_rgb)
        # 建立 ImageDraw 繪圖類別
        draw = ImageDraw.Draw(pilimg)
        # 設定字型
        font_size = 20
        font_path = c_dir+"/../../font/wqy-microhei.ttc"
```

```
        font_hei = ImageFont.truetype(font_path, font_size, encoding="utf-8")

        for obj in objs:
            loc = obj["location"]
            draw.rectangle((loc["left"], loc["top"], loc["left"]+loc["width"],
loc["top"]+loc["height"]),
                           outline='green',width=2)
            msg = obj["name"]+": %.2f"%obj["score"]
            draw.text((loc["left"], loc["top"]-font_size*1), msg, (0, 255, 0), font=font_hei)

            color1 = (10, 215, 255)
            color2 = (255, 115, 55)
            color3 = (5, 255, 55)
            color4 = (25, 15, 255)
            color5 = (225, 15, 55)
            marks = obj["mark"]
            for j in range(len(marks)):
                kp = obj["mark"][j]
                draw.ellipse(((kp["x"]-4, kp["y"]-4), (kp["x"]+4,kp["y"]+4)),
                             fill=None,outline=(255,0,0),width=2)
                color = (color1,color2,color3,color4,color5)
                ii = j //4
                if  j==0 or j / 4 != ii:
                    draw.line(((marks[j]["x"],marks[j]["y"]),(marks[j+1]["x"],marks[j+1]
["y"])),
                              fill=color[ii],width=2)
                    draw.line(((marks[0]["x"],marks[0]["y"]),(marks[5]["x"],marks[5]
["y"])),
                              fill=color[1],width=2)
                    draw.line(((marks[0]["x"],marks[0]["y"]),(marks[9]["x"],marks[9]
["y"])),
                              fill=color[2],width=2)
                    draw.line(((marks[0]["x"],marks[0]["y"]),(marks[13]["x"],marks[13]
["y"])),
                              fill=color[3],width=2)
                    draw.line(((marks[0]["x"],marks[0]["y"]),(marks[17]["x"],marks[17]
["y"])),
                              fill=color[4],width=2)
        result = cv.cvtColor(np.array(pilimg), cv.COLOR_RGB2BGR)
```

```
        return result
    def inference(self, image, param_data):
        #code：辨識成功傳回 200
        #msg：相關提示訊息
        #origin_image：原始影像
        #result_image：處理之後的影像
        #result_data：結果資料
        return_result = {'code': 200, 'msg': None, 'origin_image': None,
                         'result_image': None, 'result_data': None}

        # 即時視頻界面：@__app.route('/stream/<action>')
        #image：攝影機即時傳遞過來的影像
        #param_data：必須為 None
        result = self.handpose_model.detect(image)
        result = json.loads(result)
        if result["code"] == 200 and result["result"]["obj_num"] > 0:
            r_image = self.draw_pos(image, result["result"]["obj_list"])
        else:
            r_image = image
        return_result["code"] = result["code"]
        return_result["msg"] = result["msg"]
        return_result["result_image"] = self.image_to_base64(r_image)
        return_result["result_data"] = result["result"]
        return return_result

# 單元測試，如果處理類別中引用了檔案，則在單元測試中要修改檔案路徑
if __name__=='__main__':
    from handpose import HandDetector
    # 建立視訊捕捉物件
    cap=cv.VideoCapture(0)
    if cap.isOpened()!=1:
        pass
    # 迴圈獲取影像、處理影像、顯示影像
    while True:
        ret,img=cap.read()
        if ret==False:
            break
        # 建立影像處理物件
        img_object=HandposeDetection(c_dir+'/../../models/handpose_detection')
```

```
        # 呼叫影像處理函式對影像進行加工處理
        result=img_object.inference(img,None)
        frame = img_object.base64_to_image(result["result_image"])

        # 影像顯示
        cv.imshow('frame',frame)
        key=cv.waitKey(1)
        if key==ord('q'):
            break
    cap.release()
    cv.destroyAllWindows()
else :
    from .handpose import HandDetector
```

4.3.2 開發步驟與驗證

4.3.2.1 專案部署

1）硬體部署

見 4.1.2.1 節。

2）專案部署

（1）執行 MobaXterm 工具，透過 SSH 登入到邊緣計算閘道。

（2）在 SSH 終端執行以下命令，建立開發專案目錄。

```
$ mkdir -p ~/aiedge-exp
```

（3）透過 SSH 將本專案開發專案程式和 aicam 專案套件上傳到 ~/aiedge-exp 目錄下，並採用 unzip 命令進行解壓縮。

```
$ unzip edge_fan.zip
$ unzip aicam.zip -d edge_fan
```

（4）參考 4.1.2.1 節的內容，修改專案設定檔 static/edge_fan/js/config.js 內的智雲帳號、硬體位址、邊緣服務位址等資訊。

（5）透過 SSH 將修改後的檔案上傳到邊緣計算閘道。

3）專案執行

（1）在 SSH 終端輸入以下命令執行開發專案。

```
$ cd ~/aiedge-exp/edge_fan
$ chmod 755 start_aicam.sh
$ conda activate py36_tf114_torch15_cpu_cv345    //Ubuntu 20.04 作業系統下需要切換環境
$ ./start_aicam.sh
```

（2）在使用者端或邊緣計算閘道端開啟 Chrome 瀏覽器，輸入專案頁面位址 http://192.168.100.200:4001/static/edge_fan/index.html，即可查看專案內容。

4.3.2.2 手勢開關風扇系統的驗證

本專案實現了手勢開關風扇系統，用於模擬智慧家居手勢互動的應用場景。

（1）手勢開關風扇系統具有手勢辨識功能，手勢 5 表示發送開啟風扇命令，手勢 1 表示發送關閉風扇命令，並伴有彈窗提示，5 s 內不再進行手勢辨識。

（2）AiCam 平臺中的風扇控制功能會對前視訊進行截圖並進行手勢辨識，當辨識到的手勢是 5 時開啟風扇，如圖 4.26 所示。

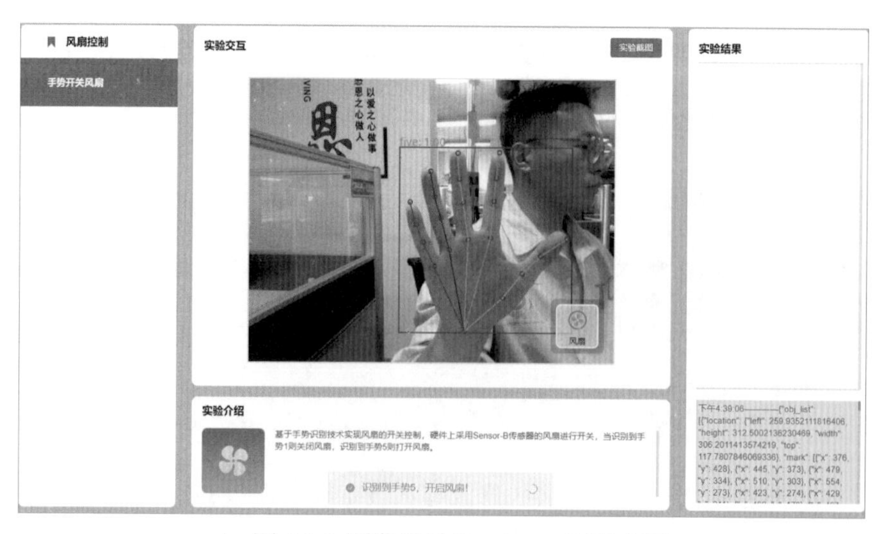

▲ 圖 4.26 開啟風扇的 AiCam 平臺介面

硬體平臺上的風扇轉動效果如圖 4.27 所示。

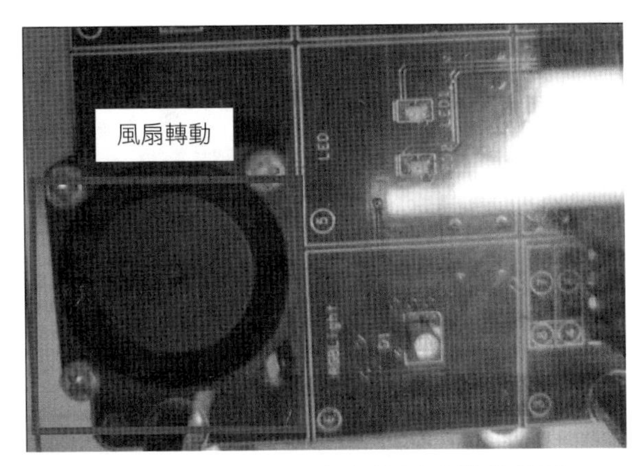

▲ 圖 4.27 硬體平臺上的風扇轉動效果

（3）當辨識到的手勢是 1 時關閉風扇，如圖 4.28 所示。

▲ 圖 4.28 關閉風扇的 AiCam 平臺介面

硬體平臺上的風扇停止效果如圖 4.29 所示。

▲ 圖 4.29 硬體平臺上的風扇停止效果

4.3.3 本節小結

本節透過手勢開關風扇系統介紹了人手檢測和手勢辨識的常用模型，以及手勢開關風扇系統的框架、通訊協定，然後借助 AiCam 平臺介紹了手勢開關風扇的硬體部署、軟體開發等開發流程。

4.3.4 思考與擴充

（1）手勢開關風扇系統可用於哪些場景？

（2）NanoDet 模型具備哪些優點？

（3）請透過具體實例說明手勢開關風扇的過程。

4.4 視覺火情監測應用程式開發

視覺火情監測是一種利用電腦視覺和影像處理技術來監測火災或火情的系統。該系統使用攝影機捕捉即時影像，透過電腦視覺和深度學習演算法分析捕捉到的即時影像，以監測火源、火勢和火情的變化。視覺火情監測系統一般包括：

- 火源檢測：系統可透過分析影像中的像素和顏色資訊來檢測潛在的火源。
- 火勢估計：一旦火源被檢測到，系統可透過比較連續影像幀之間的差異來監測火勢的變化，以估計火勢發展。
- 煙霧檢測：系統不僅可以檢測火源，還可以檢測煙霧，這對於及早發現火情並採取適當措施而言是非常重要的。
- 警示和通知：一旦系統檢測到火情或火源，它可以觸發警示並通知相關人員，如消防部門、安全人員或建築物的管理者。
- 監控和追蹤：系統可以即時監視火情並追蹤火勢的變化，這對於指導滅火工作和人員疏散而言是非常重要的。

隨著電腦視覺和深度學習技術的進步，視覺火情監測系統的準確性和性能獲得了顯著提高，被廣泛應用於多種場合，如森林、工業設施、建築物、交通隧道、油田和自然保護區等。視覺火情監測系統有助減少火災帶來的損失，提高回應速度和效率。

本節的基礎知識如下：

- 了解基於深度學習的火焰辨識技術。
- 掌握基於 YOLOv3 實現火焰辨識的基本原理。
- 結合 YOLOv3 和 AiCam 平臺進行火焰辨識應用的開發。

4.4.1 原理分析與開發設計

4.4.1.1 整體框架

基於大規模火焰資料的辨識訓練，視覺火情監測系統可即時辨識監測火情，並在辨識到火焰後發出火情警示，提醒相關部門及時查看和止損，適用於室內外多種複雜環境。火焰實況監測如圖 4.30 所示。

▲ 圖 4.30 火情實況監測

本專案採用 YOLOv3 辨識火焰，YOLOv3 是一個在電腦視覺領域獲得了廣泛應用的深度學習模型，在物件辨識、物體辨識等任務中表現非常出色。

從邊緣計算的角度看，視覺火情監測系統可分為硬體層、邊緣層、應用層，如圖 4.31 所示。

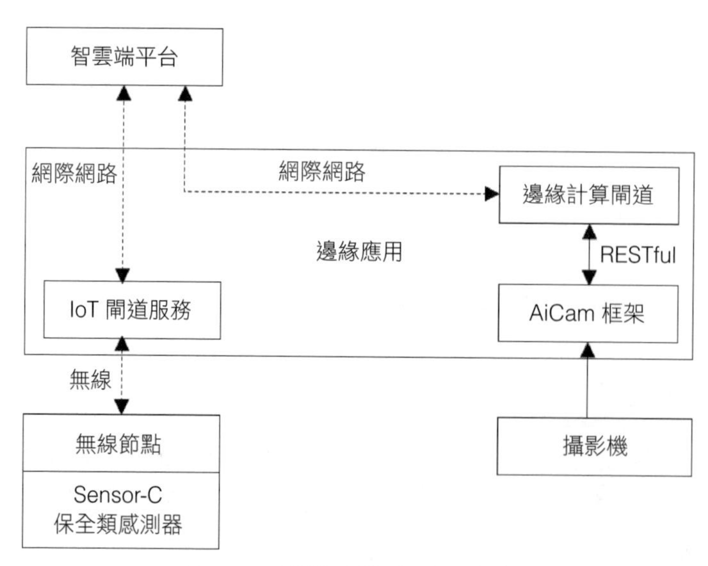

▲ 圖 4.31 視覺火情監測系統的結構

（1）硬體層：無線節點和 Sensor-C 保全類感測器組成了視覺火焰監測系統的硬體層，透過 Sensor-C 保全類感測器的火焰感測器來感應火焰狀態。

（2）邊緣層：包括邊緣計算閘道內建 IoT 閘道服務和 AiCam 框架。IoT 閘道服務負責接收和下發無線節點的資料，發送給應用端或將資料發給雲端的智雲端平台。AiCam 框架內建了演算法、模型、視訊串流推送等服務，支援應用層的邊緣計算推理任務。

（3）應用層：透過智雲介面與 IoT 的硬體層互動（預設與雲端的智雲端平台的介面互動），透過 AiCam 的 RESTful 介面與演算法層互動。

4.4.1.2 系統硬體與通訊協定設計

1）系統硬體設計

本專案既可以採用 LiteB 無線節點、Sensor-C 保全類感測器來完成硬體的架設，也可以透過虛擬模擬軟體來建立一個虛擬的硬體裝置，如圖 4.32 所示。

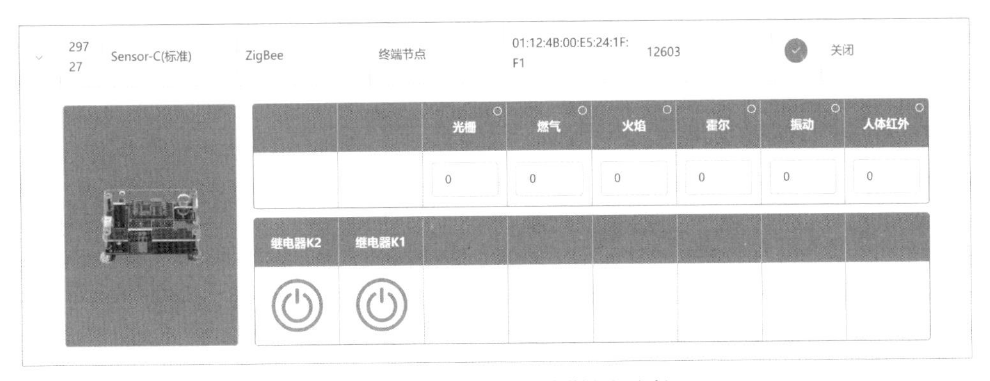

▲ 圖 4.32 虛擬硬體裝置平臺

2）通訊協定設計

Sensor-C 保全類感測器的通訊協定如表 4.4 所示。本專案使用火焰感測器來檢測火焰，當火焰感測器辨識到火焰時，會每隔 3 s 上傳一次火焰感測器的狀態（狀態為 1）；當火焰感測器未辨識到火焰時，會每隔 30 s 上傳一次火焰感測器的狀態（狀態為 0）。火焰感測器的命令如表 4.7 所示。

▼ 表 4.7 火焰感測器的命令

發送命令	接收結果	含義
-	{A3=1/0}	火焰感測器狀態

4.4.1.3 功能設計與開發

1）系統框架設計

見 4.1.1.3 節。

2）介面描述

本專案基於 AiCam 平臺開發，開發流程如下：

（1）專案配置見 4.1.1.3 節。

（2）增加模型。在 aicam 專案增加模型檔案 models/fire_detection、火焰辨識模型檔案 yolov3-tiny-fire-opt.bin/yolov3-tiny-fire-opt.param。

（3）增加演算法。在 aicam 專案增加火焰辨識演算法檔案 algorithm/fire_detection/fire_ detection.py。

（4）增加應用。在 aicam 專案增加演算法專案前端應用 static/edge_fire。

3）硬體通訊設計

前端應用中的硬體控制部分透過智雲 ZCloud API 連接到硬體系統，前端應用處理範例如下：

```
getConnect()
//智雲端服務連接
function getConnect(){                              //建立連接服務的函式
    rtc = new WSNRTConnect(config.user.id, config.user.key)
    rtc.setServerAddr(config.user.addr);
    rtc.connect();
    rtc.onConnect = () => {                         //連接成功回呼函式
        online = true
```

```
        setTimeout(() => {
            if(online){
                cocoMessage.success(` 資料服務連接成功！查詢資料中 ...`)
                // 發起資料查詢
                rtc.sendMessage(config.macList.mac_603,config.sensor.mac_603.query);
            }
        }, 200);
    }
    rtc.onConnectLost = () => {                         // 資料服務掉線回呼函式
        online = false
        cocoMessage.error(` 資料服務連接失敗！請檢查網路或 IDKEY...`)
    };

    rtc.onmessageArrive = (mac, dat) => {               // 訊息處理回呼函式
        if (dat[0] == '{' && dat[dat.length - 1] == '}') {
            // 截取背景傳回的 JSON 物件（去掉 {} 符號）後，以「,」分割為陣列
            let its = dat.slice(1,-1).split(',')
            for (let i = 0; i < its.length; i++) {      // 迴圈遍歷陣列的每一個值
                let t = its[i].split("=");              // 將每個值以「=」分割為陣列
                if (t.length != 2) continue;
                //mac_603 保全類感測器
                if (mac == config.macList.mac_603) {
                    if(t[0] == 'A3'){                   // 火焰感測器
                        console.log(' 火焰狀態：',t);
                        // 此處呼叫 AiCam 平臺的介面進行火焰辨識
                    }
                }
            }
        }
    }
}
```

4）演算法互動

前端應用的演算法採用 RESTful 介面獲取處理後的視訊流，傳回 base64 編碼的結果影像和結果資料。存取 URL 位址的格式如下（IP 位址為邊緣計算閘道的位址）：

```
http://192.168.100.200:4001/stream/[algorithm_name]?camera_id=0
```

前端應用處理範例如下：

```
let linkData = [
    '/stream/index?camera_id=0',
    '/stream/fire_detection?camera_id=0'
]
// 請求影像流資源
let imgData = new EventSource(linkData[0])
// 對影像流傳回的資料進行處理
imgData.onmessage = function (res) {
    let {result_image} = JSON.parse(res.data)
    $('#img_box>img').attr('src', `data:image/jpeg;base64,${result_image}`)
}
......
#function getConnect() 函式裡面接收到火焰警告後，呼叫火焰監測演算法
rtc.onmessageArrive = (mac, dat) => {                       // 訊息處理回呼函式
    if (dat[0] == '{' && dat[dat.length - 1] == '}') {
        // 截取背景傳回的 JSON 物件（去掉 {} 符號）後，以「,」分割為陣列
        let its = dat.slice(1,-1).split(',')
        for (let i = 0; i < its.length; i++) {          // 迴圈遍歷陣列的每一個值
            let t = its[i].split("=");                  // 將每個值以「=」分割為陣列
            if (t.length != 2) continue;
            //mac_603 保全類感測器
            if (mac == config.macList.mac_603) {
                if(t[0] == 'A3'){                       // 火焰感測器
                    console.log(' 火焰狀態：',t);
                    // 如果當前感測器的狀態為 1，且上次為 0 則進行下一步
                    if (t[1] == '1' && $('#icon').attr('src') == './img/icon-off.png') {
                        clearInterval(Timer)
                        $('#icon').attr('src','./img/icon-on.gif')

                        // 請求影像流資源
                        imgData && imgData.close()
                        imgData = new EventSource(linkData[1])
                        // 對影像流傳回的資料進行處理
                        imgData.onmessage = function (res) {
                            let {result_image} = JSON.parse(res.data)
                            $('#img_box>img').attr('src', `data:image/jpeg;base64,${result_
image}`)
```

```
    if (interactionThrottle) {
        interactionThrottle = false
        let {result_data} = JSON.parse(res.data)
        let html = `<div>${new
                 Date().toLocaleTimeString()}————
                 ${JSON.stringify(result_data)}</div>`
        console.log(result_data);
        $('#text-list').prepend(html);
        // 當辨識到火焰，且辨識率大於 50% 時，
        // 進行一次截圖 + 彈窗提示，10 s 後方可進行下一次辨識
        if (throttle && result_data.obj_num > 0 &&
                    result_data.obj_list[0].score > 0.5) {
          throttle = false
          $('#result').click()
          swal(` 火焰警示！`, "監測到火情，請馬上處理！",
              "error", {button: false,timer: 2000});
          setTimeout(() => {
              throttle = true
          }, 10000);
      }
      setTimeout(() => {
          interactionThrottle = true
      }, 1000);
  }
}
// 每 10 s 檢測一次火焰感測器的狀態是否為 0，若為 0，
// 則切換為普通視訊流
Timer = setInterval(() => {
    if ($('#icon').attr('src') == './img/icon-off.png') {
    // 請求影像流資源
    imgData && imgData.close()
    imgData = new EventSource(linkData[0])
    // 對影像流傳回的資料進行處理
    imgData.onmessage = function (res) {
        let {result_image} = JSON.parse(res.data)
        $('#img_box>img').attr('src', `data:image/jpeg;
                          base64,${result_image}`)
    }
```

```
                    }
                }, 10000);
              }
              // 如果當前感測器的狀態為 0，且上次為 1 則進行下一步
              if (t[1] == '0' && $('#icon').attr('src') == './img/icon-on.gif'){
                  $('#icon').attr('src','./img/icon-off.png')
              }
          }
      }
    }
  }
}
```

5）火情辨識演算法介面設計

```
##########################################################################################
# 檔案：fire_detection.py
# 說明：火焰辨識
##########################################################################################
from PIL import Image,ImageDraw,ImageFont
import numpy as np
import cv2 as cv
import os
import json
import base64
c_dir = os.path.split(os.path.realpath(__file__))[0]

class FireDetection(object):
    def __init__(self, model_path="models/fire_detection"):
        self.model_path = model_path
        self.fire_model = FireDet()
        self.fire_model.init(self.model_path)

    def image_to_base64(self, img):
        image = cv.imencode('.jpg', img, [cv.IMWRITE_JPEG_QUALITY, 60])[1]
        image_encode = base64.b64encode(image).decode()
        return image_encode

    def base64_to_image(self, b64):
```

```
        img = base64.b64decode(b64.encode('utf-8'))
        img = np.asarray(bytearray(img), dtype="uint8")
        img = cv.imdecode(img, cv.IMREAD_COLOR)
        return img
    def draw_pos(self, img, objs):
        img_rgb = cv.cvtColor(img, cv.COLOR_BGR2RGB)
        pilimg = Image.fromarray(img_rgb)
        # 建立 ImageDraw 繪圖類別
        draw = ImageDraw.Draw(pilimg)
        # 設定字型
        font_size = 20
        font_path = c_dir+"/../../font/wqy-microhei.ttc"
        font_hei = ImageFont.truetype(font_path, font_size, encoding="utf-8")

        for obj in objs:
            loc = obj["location"]
            draw.rectangle((loc["left"], loc["top"], loc["left"]+loc["width"],
                        loc["top"]+loc["height"]), outline='green',width=2)
            msg =  "%.2f"%obj["score"]
            draw.text((loc["left"], loc["top"]-font_size*1), msg, (0, 255, 0), font=font_hei)
        result = cv.cvtColor(np.array(pilimg), cv.COLOR_RGB2BGR)
        return result
    def inference(self, image, param_data):
        #code：辨識成功傳回 200
        #msg：相關提示訊息
        #origin_image：原始影像
        #result_image：處理之後的影像
        #result_data：結果資料
        return_result = {'code': 200, 'msg': None, 'origin_image': None,
                    'result_image': None, 'result_data': None}

        # 即時視頻界面：@__app.route('/stream/<action>')
        #image：攝影機即時傳遞過來的影像
        #param_data：必須為 None
        result = self.fire_model.detect(image)
        result = json.loads(result)
        r_image = image
        if result["code"] == 200 and result["result"]["obj_num"] > 0:
            r_image = self.draw_pos(r_image, result["result"]["obj_list"])
```

```
        return_result["code"] = result["code"]
        return_result["msg"] = result["msg"]
        return_result["result_image"] = self.image_to_base64(r_image)
        return_result["result_data"] = result["result"]
        return return_result

#單元測試，如果處理類別中引用了檔案，則在單元測試中要修改檔案路徑
if __name__=='__main__':

    from firedet import FireDet
    #建立視訊捕捉物件
    cap=cv.VideoCapture(0)
    if cap.isOpened()!=1:
        pass
    #迴圈獲取影像、處理影像、顯示影像
    while True:
        ret,img=cap.read()
        if ret==False:
            break
        #建立影像處理物件
        img_object=FireDetection(c_dir+'/../../models/fire_detection')
        #呼叫影像處理函式對影像進行加工處理
        result=img_object.inference(img,None)
        frame = img_object.base64_to_image(result["result_image"])
        #影像顯示
        cv.imshow('frame',frame)
        key=cv.waitKey(1)
        if key==ord('q'):
            break
    cap.release()
    cv.destroyAllWindows()
else :
    from .firedet import FireDet
```

4.4.2 開發步驟與驗證

4.4.2.1 專案部署

1）硬體部署

見 4.1.2.1 節。

2）專案部署

（1）執行 MobaXterm 工具，透過 SSH 登入到邊緣計算閘道。

（2）在 SSH 終端執行以下命令，建立開發專案目錄。

```
$ mkdir -p ~/aiedge-exp
```

（3）透過 SSH 將本專案開發專案程式和 aicam 專案套件上傳到 ~/aiedge-exp 目錄下，並採用 unzip 命令進行解壓縮。

```
$ unzip edge_fire.zip
$ unzip aicam.zip -d edge_fire
```

（4）參考 4.1.2.1 節的內容，修改專案設定檔 static/edge_fan/js\config.js 內的智雲帳號、硬體位址、邊緣服務位址等資訊。

（5）透過 SSH 將修改後的檔案上傳到邊緣計算閘道。

3）專案執行

（1）在 SSH 終端輸入以下命令執行開發專案：

```
$ cd ~/aiedge-exp/edge_fire
$ chmod 755 start_aicam.sh
$ conda activate py36_tf114_torch15_cpu_cv345    //Ubuntu 20.04 作業系統下需要切換環境
$ ./start_aicam.sh
```

（2）在使用者端或邊緣計算閘道端開啟 Chrome 瀏覽器，輸入專案頁面位址 http://192.168.100.200:4001/static/edge_fire/index.html，即可查看專案內容。

4.4.2.2 視覺火情監測系統的驗證

本專案透過火焰感測器和火焰辨識演算法實現了對火情的監測，用於模擬視覺火情監測的應用場景。

（1）火焰感測器的狀態為 1（辨識到火焰）時警告被觸發，AiCam 平臺火情監測介面右下角火焰圖示會變亮，並將初始視訊流切換為火焰監測視訊流。

（2）當攝影機監測到火焰（可透過手機開啟火災影像來模擬）時將觸發彈窗警示（警告樣式），並對當前視訊進行截圖顯示到右側清單，10 s 內不再進行警示截圖。攝影機監測到火焰時的 AiCam 平臺介面如圖 4.33 所示。

（3）在火焰監測視訊流狀態下，AiCam 平臺每 10 s 判斷一次火焰感測器是否處於觸發狀態，如果感測器火焰警告停止（狀態為 0），則切換回初始視訊流。

（4）如果採用虛擬模擬建立的火焰感測器裝置，則先將火焰感測器的狀態設定為 1，表示當前處於火焰警告狀態；再將火災影像放置到攝影機視窗內，當監測到火焰且置信度大於或等於 0.5 時，將發出火焰警示並截圖。虛擬平臺的火焰感測器狀態如圖 4.34 所示。

▲ 圖 4.33 攝影機監測到火焰時的 AiCam 平臺介面

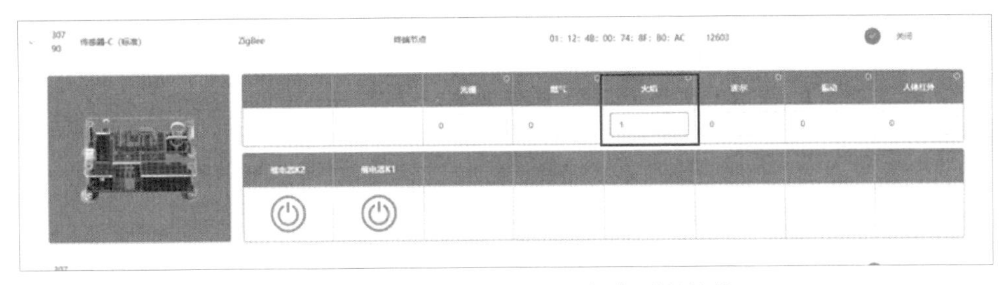

▲ 圖 4.34 虛擬平臺的火焰感測器狀態

如果採用真實的硬體裝置，將打火機打著火靠近火焰感測器時，當火焰感測器監測到火焰後會上報火焰感測器警告（狀態為 1），如圖 4.35 所示。

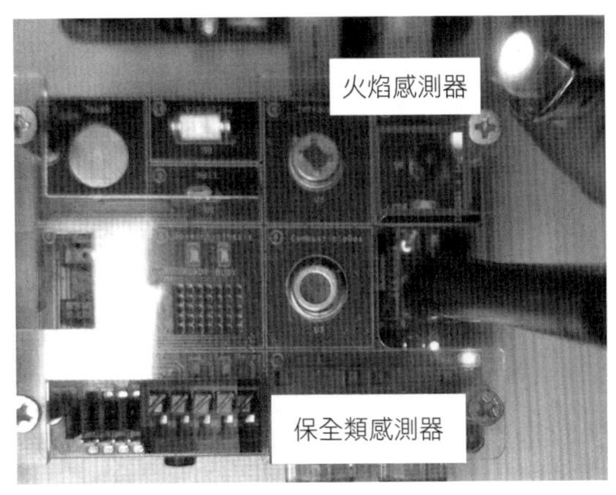

▲ 圖 4.35 監測到火焰時的硬體平臺狀態

4.4.3 本節小結

本節採用火焰感測器和火焰辨識演算法實現了視覺火情監測系統，詳細描述了系統框架、通訊協定，並進行了驗證。

4.4.4 思考與擴充

（1）視覺火情監測系統與人臉門禁系統有哪些共同之處？

（2）視覺火情監測系統需要的硬體裝置有哪些？

（3）當火焰感測器的狀態為 1 時表示其處於什麼狀態？

4.5 視覺車牌辨識應用程式開發

車牌辨識是一種電腦視覺技術，用於自動辨識和提取車輛上的車牌號碼，其步驟是影像擷取→影像前置處理→車牌定位→字元分割→字元辨識。車牌辨識技術可以應用於多種場景，如交通管理、停車場管理、安全監控、物流和運輸、道路收費系統、停車追蹤、智慧交通燈控制、犯罪調查、社區門禁系統和停車限時監控。車牌辨識如圖 4.36 所示。

▲ 圖 4.36 CNN 車牌辨識現場

車牌辨識技術經歷了幾十年的發展，從最早的基於規則的方法到現在的基於深度學習和人工智慧的方法。以下是車牌辨識技術的主要發展歷程：

早期的車牌辨識技術（1970 年代到 1990 年代）：早期的車牌辨識技術主要依賴於規則和傳統的電腦視覺技術，包括字元範本匹配、邊緣檢測、顏色分割和形狀分析等。

基於範本匹配的車牌辨識技術（1990 年代到 2000 年代）：這個時期的車牌辨識技術逐漸採用了範本匹配和特徵提取的方法，範本匹配方法是透過比較影像中的字元與預先定義的字元範本來辨識車牌的。

基於機器學習的車牌辨識技術（2000 年代到 2010 年代）：隨著機器學習技術的發展，車牌辨識開始採用基於機器學習的方法，如支援向量機（SVM）和隨機森林等。

深度學習時代（2010 年代至今）：深度學習技術的崛起徹底改變了車牌辨識技術，卷積神經網路（CNN）和循環神經網路（RNN）等深度學習框架被廣泛用於車牌辨識技術。

車牌辨識技術經歷了長期的發展過程，從最初的基本方法到現在的基於深度學習和多模態融合的技術，不斷提高了準確性、速度和堅固性，使其在多種應用場景中得到廣泛應用。

本節的基礎知識如下：

- 了解基於深度學習的車牌辨識技術。
- 掌握基於 LPRNet 模型實現車牌辨識的基本原理。
- 結合 LPRNet 模型和 AiCam 平臺進行車牌辨識應用的開發。

4.5.1 原理分析與開發設計

4.5.1.1 整體框架

1）車牌辨識概述

車牌辨識系統是指能夠檢測受控路面的車輛並自動提取車輛牌照資訊進行處理的技術，是現代智慧交通系統中的重要組成部分之一。車牌辨識技術以數位影像處理、模式辨識、電腦視覺等技術為基礎，首先透過攝影機提取車牌視訊影像，對提取的每一幀影像，利用高效視訊檢測技術對車牌進行定位和追蹤，從中自動提取車牌影像；然後經過車牌精確定位、切分和辨識等模組準確地自動分割和辨識字元，得到車牌的全部字元資訊和顏色資訊。透過一些後續處理手段可以實現停車場收費管理、交通流量控制。視覺車牌辨識系統的框架如圖 4.37 所示。

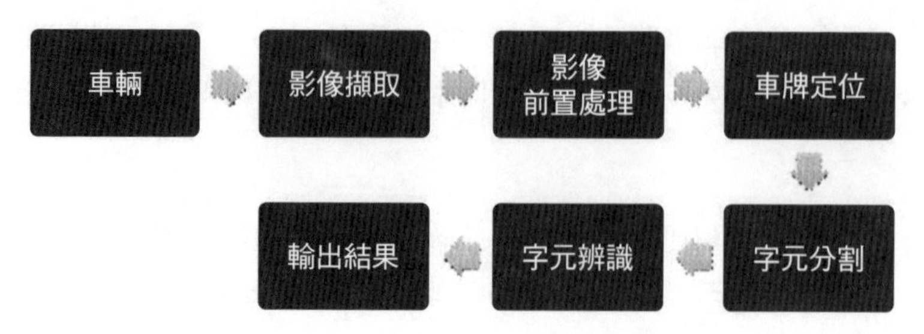

▲ 圖 4.37 視覺車牌辨識系統的框架

車牌辨識中常用的深度學習演算法或模型如下：

（1）卷積神經網路（CNN）模型：CNN 是車牌辨識中常用的深度學習模型，該模型可用於車牌檢測和字元辨識。一般來說 CNN 用於提取車牌影像中的特徵，後續使用其他方法來辨識車牌中的字元。CNN 模型的結構如圖 4.38 所示。

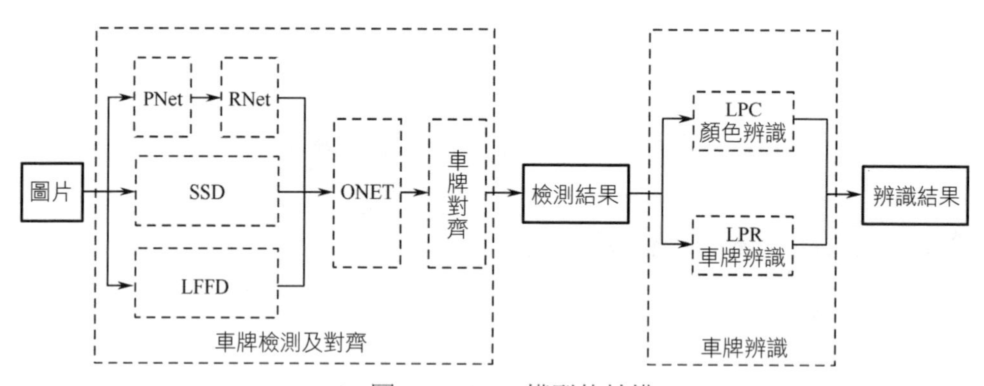

▲ 圖 4.38 CNN 模型的結構

（2）YOLO（You Only Look Once）系列模型：YOLO 是物件辨識的系列模型，常用於車牌檢測。YOLO 系列模型可以同時辨識多個目標，並且速度很快，適用於即時車牌辨識。

（3）R-CNN 模型（如 Faster R-CNN 和 Mask R-CNN）：這些模型是用於物件辨識的深度學習模型，可用於車牌檢測，不僅能提供較高的檢測精度，還可以辨識車牌的位置。

（4）CRNN（Convolutional Recurrent Neural Network）模型：是一種將卷積神經網路和循環神經網路結合起來的模型，用於點對點的車牌字元辨識。該模型可以接收整個車牌影像並輸出字元序列。

（5）CTC（Connectionist Temporal Classification，連接時序分類）模型：CTC 是一種用於序列辨識任務的深度學習模型，常用於車牌字元辨識。該模型可以將輸入影像對應到字元序列，無須明確的字元定位資訊。

（6）自注意力模型（如 Transformer）：自注意力模型已經被應用於車牌字元辨識，該模型可以在辨識字元時關注輸入影像的不同部分，具有較高的車牌辨識性能。

（7）深度生成對抗網路（GAN）模型：GAN 模型可用於生成車牌影像，以擴充訓練資料，改善了車牌辨識的性能。

（8）卷積循環神經網路（CRNN）模型：CRNN 模型結合了卷積層和循環層，用於點對點的車牌辨識，可以同時檢測車牌位置和辨識字元。

本專案基於 LPRNet 模型實現車牌的辨識。LPRNet 模型是一個行動端導向的準商業級車牌辨識庫，以 NCNN 為推理後端，以 DNN 模型為核心，支援多種車牌檢測演算法，可辨識車牌資訊和車牌顏色。LPRNet 模型的特點如下：

- 超輕量：核心函式庫只依賴 NCNN，支援模型量化。
- 多檢測：支援 SSD、MTCNN、LFFD 等物件辨識演算法。
- 精度高：LFFD 物件辨識演算法在 CCPD 中的檢測精度可達到 98.9%，車牌辨識率達到 99.95%，綜合辨識率超過 99%。
- 易使用：只需要 10 行程式即可完成車牌辨識。
- 易擴充：可快速擴充各類檢測演算法。

LPRNet（License Plate Recognition Network）是一個用於車牌辨識的深度學習模型，主要用於辨識和提取車牌上的字元。在特徵提取網路方面，LPRNet 模型使用輕量化的卷積神經網路，訓練集階段的損失函式是 CTC Loss，對中文車牌的辨識準確率達到了 95%。LPRNet 模型在辨識中文車牌時有以下幾大優點：

（1）LPRNet 模型不需要預先對字元進行分割，可辨識可變長度的車牌，特別是對於字元差異性比較大的車牌，可以實現點對點的檢測辨識。

（2）LPRNet 模型以卷積神經網路為基礎，沒有採用循環卷積神經網路，使得網路結構更加輕量化，並且能夠在各種嵌入式裝置上。

（3））LPRNet 模型可在光照條件惡劣、拍攝角度扭曲等環境下對車牌進行辨識，具有優良的堅固性與泛化性。

LPRNet 模型的結構如圖 4.39 所示，包含輸入影像、CBR（Convolution, Batch Normalization, Rectified Linear Unit）、MaxPool、AvgPool、Small basic block、concat 和 container。其中，輸入影像的尺寸為 94×24×3；CBR 由一個卷積層、批次歸一化層和啟動函式 ReLU 組成；MaxPool 是 3 維最大池化操作，三維核心尺寸分別為 1、3 和 3，步進值均為 1；AvgPool 是二維平均池化操作；Small basic block 由 4 個卷積層和 3 個啟動函式 ReLU 組成；concat 以通道維度拼接多個特徵圖；container 包含 1 個卷積核心為 1×1 的卷積層。

LPRNet 採用堆疊的卷積層作為特徵提取網路，以原始的 RGB 影像作為輸入。為了更進一步地融合多層特徵資訊，LPRNet 模型對 4 個不同尺度的特徵圖進行融合，進而在利用高層次的細粒度特徵資訊的同時與淺層的全域影像資訊相結合。淺層特徵圖包含更多的影像全域特徵資訊且有較高的解析度。

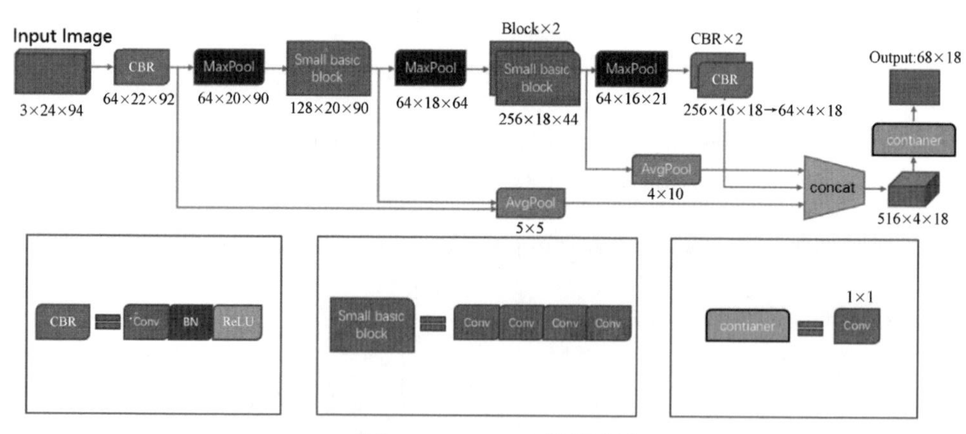

▲ 圖 4.39 LPRNet 模型框架

1）系統框架

從邊緣計算的角度看，視覺車牌辨識系統可分為硬體層、邊緣層、應用層，如圖 4.40 所示。

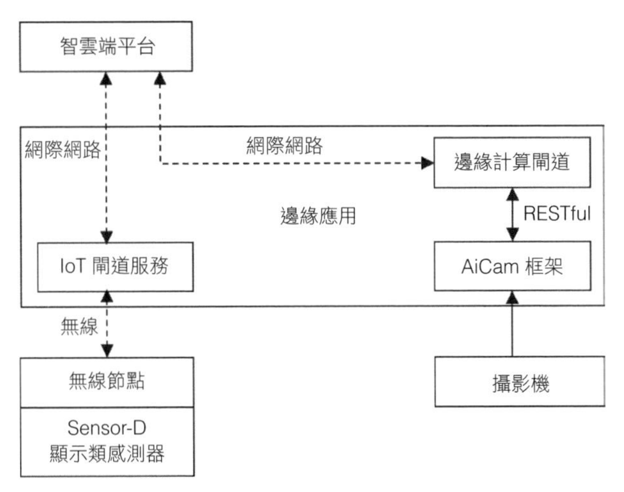

▲ 圖 4.40 視覺車牌辨識系統的結構

（1）硬體層：無線節點和 Sensor-D 顯示類感測器組成了視覺車牌辨識系統的硬體層，透過 Sensor-D 顯示類感測器顯示檢測到的車牌號碼。

（2）邊緣層：包括邊緣計算閘道內建 IoT 閘道服務和 AiCam 框架。IoT 閘道服務負責接收和下發無線節點的資料，發送給應用端或將資料發給雲端的智雲端平台。AiCam 框架內建了演算法、模型、視訊串流推送等服務，支援應用層的邊緣計算推理任務。

（3）應用層：透過智雲介面與 IoT 的硬體層互動（預設與雲端的智雲端平台的介面互動），透過 AiCam 的 RESTful 介面與演算法層互動。

4.5.1.2 系統硬體與通訊協定設計

1）系統硬體設計

本專案可以採用 LiteB 無線節點、Sensor-D 顯示類感測器來完成硬體的架設。

2）通訊協定設計

Sensor-D 顯示類感測器的通訊協定如表 4.8 所示。

▼ 表 4.8 Sensor-D 顯示類感測器的通訊協定

名稱	TYPE	參數	含義	許可權	說明
Sensor-D 顯示類感測器	604	五向開關狀態	A0	R	觸發上報，1 表示上（UP）、2 表示左（LEFT）、3 表示下（DOWN）、4 表示右（RIGHT）、5 表示中心（CENTER）
		OLED 背光開關	D1(OD1/CD1)	R/W	D1 的 Bit0 代表 LCD 的背光開關狀態，1 表示開啟背光開關，0 表示關閉背光開關
		數位管背光開關	D1(OD1/CD1)	R/W	D1 的 Bit1 代表數位管的背光開關狀態，1 表示開啟背光開關，0 表示關閉背光開關
		上報間隔	V0	R/W	A0 值的循環上報時間間隔
		車牌 / 儀表	V1	R/W	車牌號碼 / 儀表值
		車位數	V2	R/W	停車場空閒車位數
		模式設定	V3	R/W	1 表示停車模式，2 表示抄表模式

本專案使用 Sensor-D 顯示類感測器來顯示車牌號碼，當 Sensor-D 顯示類感測器處於停車模式時，等待應用層傳輸過來的車牌影像，當接收到車牌影像後，可顯示車牌號碼並維持 5 s。

4.5.1.3 功能設計與開發

1）系統框架設計

見 4.1.1.3 節。

2）介面描述

本專案基於 AiCam 平臺開發，開發流程如下：

（1）專案配置見 4.1.1.3 節。

（2）增加模型。在 aicam 專案增加模型檔案 models/plate_recognition、車牌檢測模型檔案 det3.bin/det3.param、車牌對齊模型檔案 lffd.bin/lffd.param、顏色辨識模型檔案 lpc.bin/lpc.param、車牌辨識模型檔案 lpr.bin/lpr.param。

（3）增加演算法。在 aicam 專案增加基於深度學習的車牌辨識演算法 algorithm/plate_ recognition/plate_recognition.py。

（4）增加應用。在 aicam 專案增加演算法專案前端應用 static/edge_plate。

3）硬體通訊設計

前端應用中的硬體控制部分透過智雲 ZCloud API 連接到硬體系統，前端應用處理範例如下：

```
getConnect()
//智雲端服務連接
function getConnect(){                          //建立連接服務的函式
    rtc = new WSNRTConnect(config.user.id, config.user.key)
    rtc.setServerAddr(config.user.addr);
    rtc.connect();
    rtc.onConnect = () => {                     //連接成功回呼函式
        online = true
        setTimeout(() => {
            if(online){
                cocoMessage.success(`資料服務連接成功！查詢資料中...`)
                rtc.sendMessage(config.macList.mac_604, '{V3=1,V3=?}');    //發起資料查詢
            }
        }, 200);
    }
    rtc.onConnectLost = () => {                                   //資料服務掉線回呼函式
        online = false
        cocoMessage.error(`資料服務連接失敗！請檢查網路或IDKEY...`)
```

```
    };
}
```

4）演算法互動

前端應用的演算法採用 RESTful 介面獲取處理後的視訊流，傳回 base64 編碼的結果影像和結果資料。存取 URL 位址的格式以下（IP 位址為邊緣計算閘道的位址）：

```
http://192.168.100.200:4001/stream/[algorithm_name]?camera_id=0
```

前端應用處理範例如下：

```
let linkData = '/stream/plate_recognition?camera_id=0'                    // 視訊資源連結
// 請求影像流資源
let imgData = new EventSource(linkData)
// 對影像流傳回的資料進行處理
imgData.onmessage = function (res) {
    let {result_image} = JSON.parse(res.data)
    $('#img_box>img').attr('src', `data:image/jpeg;base64,${result_image}`)

    let {result_data} = JSON.parse(res.data)
    // 將辨識到的車牌資訊顯示到文字專案結果顯示框（設定增加間隔不得小於 1 s）
    if (result_data && interactionThrottle) {
        interactionThrottle = false
        let html = `<div>${new Date().toLocaleTimeString()}————
                    ${JSON.stringify(result_data)}</div>`
                    $('#text-list').prepend(html);
        //result_data.obj_list[0].plate_no 表示車牌，throttle 表示命令發送暫停時間（如 8 s），
        //online 表示智雲端服務連接成功後才能進入判斷
        console.log(result_data,count);
        if(result_data.obj_num > 0 && throttle && online){
            // 設定每個辨識到的車牌為物件屬性名稱，初始值為 1。當辨識次數累計達到 5 次後，
            // 發送更新車牌顯示命令，並清空計數
            if(count[result_data.obj_list[0].plate_no]){
                count[result_data.obj_list[0].plate_no] += 1
                if(count[result_data.obj_list[0].plate_no] == 5){
                    throttle = false
```

```
                    console.log('車牌為>>>>>>', result_data.obj_list[0].plate_no);
                    swal(`辨識到車牌：${result_data.obj_list[0].plate_no}`, " ",
                                    "success", {button: false,timer: 3000});
                    //對辨識到的車牌首位中文字進行編碼
                    for (let index in provice_dict) {
                        if(result_data.obj_list[0].plate_no[0] == index){
                            result_data.obj_list[0].plate_no = result_data.obj_list[0].
plate_no.
                                replace(result_data.obj_list[0].plate_no[0],provice_
dict[index])
                        }
                    }
                    rtc.sendMessage(config.macList.mac_604,
                                `{V1=${result_data.obj_list[0].plate_no},V1=?}`)
                    console.log(config.macList.mac_604,
                                `{V1=${result_data.obj_list[0].plate_no},V1=?}`);
                    count = {}
                    setTimeout(() => {
                        throttle = true
                    }, 8000);
                }
            }else{
                count[result_data.obj_list[0].plate_no] = 1
            }
        }
        setTimeout(() => {
            interactionThrottle = true
        }, 1000);
    }
}
```

5）車牌辨識演算法介面設計

```
#####################################################################################
# 檔案：plate_recognition.py
# 說明：車牌辨識
#####################################################################################
from PIL import Image,ImageDraw,ImageFont
import numpy as np
```

```
import cv2 as cv
import os
import json
import base64
c_dir = os.path.split(os.path.realpath(__file__))[0]

load = False
class PlateRecognition(object):
    def __init__(self, model_path="models/plate_recognition/"):
        global load
        if load:
            model_path="./"
        self.plate_model = PlateRecognizer()
        self.plate_model.init(model_path)
        load = True

    def image_to_base64(self, img):
        image = cv.imencode('.jpg', img, [cv.IMWRITE_JPEG_QUALITY, 60])[1]
        image_encode = base64.b64encode(image).decode()
        return image_encode

    def base64_to_image(self, b64):
        img = base64.b64decode(b64.encode('utf-8'))
        img = np.asarray(bytearray(img), dtype="uint8")
        img = cv.imdecode(img, cv.IMREAD_COLOR)
        return img

    def draw_pos(self, img, objs):
        img_rgb = cv.cvtColor(img, cv.COLOR_BGR2RGB)
        pilimg = Image.fromarray(img_rgb)
        # 建立 ImageDraw 繪圖類別
        draw = ImageDraw.Draw(pilimg)
        # 設定字型
        font_size = 20
        font_path = c_dir+"/../../font/wqy-microhei.ttc"
        font_hei = ImageFont.truetype(font_path, font_size, encoding="utf-8")

        for obj in objs:
            loc = obj["location"]
```

```
            draw.rectangle((loc["left"], loc["top"], loc["left"]+loc["width"],
                        loc["top"]+loc["height"]), outline='green',width=2)
            msg = obj["plate_no"]+" : %.2f"%obj["score"]
            draw.text((loc["left"], loc["top"]-font_size*1), msg, (0, 255, 0), font=font_hei)
        result = cv.cvtColor(np.array(pilimg), cv.COLOR_RGB2BGR)
        return result

    def inference(self, image, param_data):
        #code：辨識成功傳回 200
        #msg：相關提示訊息
        #origin_image：原始影像
        #result_image：處理之後的影像
        #result_data：結果資料
        return_result = {'code': 200, 'msg': None, 'origin_image': None,
                    'result_image': None, 'result_data': None}

        # 即時視頻界面：@__app.route('/stream/<action>')
        #image：攝影機即時傳遞過來的影像
        #param_data：必須為 None
        result = self.plate_model.plate_recognize(image)
        result = json.loads(result)
        r_image = image

        if result["code"] == 200 and result["result"]["obj_num"] > 0:
            r_image = self.draw_pos(r_image, result["result"]["obj_list"])

        return_result["code"] = result["code"]
        return_result["msg"] = result["msg"]
        return_result["result_image"] = self.image_to_base64(r_image)
        return_result["result_data"] = result["result"]
        return return_result

# 單元測試，如果處理類別中引用了檔案，則在單元測試中要修改檔案路徑
if __name__=='__main__':

    from plateRecognize import PlateRecognizer
    # 建立影像處理物件
    img_object=PlateRecognition(c_dir+'/../../models/plate_recognition')
```

```
    cap=cv.VideoCapture(0)
    if cap.isOpened()!=1:
        pass
    #迴圈獲取影像、處理影像、顯示影像
    while True:
        ret,img=cap.read()
        if ret==False:
            break
        #呼叫影像處理函式對影像進行加工處理
        result=img_object.inference(img,None)
        frame = img_object.base64_to_image(result["result_image"])

        #影像顯示
        cv.imshow('frame',frame)

        key=cv.waitKey(1)
        if key==ord('q'):
            break
    cv.destroyAllWindows()
else :
    from .plateRecognize import PlateRecognizer
```

4.5.2 開發步驟與驗證

4.5.2.1 專案部署

1）硬體部署

見 4.1.2.1 節。

2）專案部署

（1）執行 MobaXterm 工具，透過 SSH 登入到邊緣計算閘道。

（2）在 SSH 終端執行以下命令，建立開發專案目錄。

```
$ mkdir -p ~/aiedge-exp
```

（3）透過 SSH 將本專案開發專案程式和 aicam 專案套件上傳到 ~/aiedge-exp 目錄下，並採用 unzip 命令進行解壓縮。

```
$ unzip edge_plate.zip
$ unzip aicam.zip -d edge_plate
```

（4）參考 4.1.2.1 節的內容，修改專案設定檔 static/edge_fan/js/config.js 內的智雲帳號、硬體位址、邊緣服務位址等資訊。

（5）透過 SSH 將修改後的檔案上傳到邊緣計算閘道。

3）專案執行

（1）在 SSH 終端輸入以下命令執行開發專案：

```
$ cd ~/aiedge-exp/edge_plate
$ chmod 755 start_aicam.sh
$ conda activate py36_tf114_torch15_cpu_cv345    //Ubuntu 20.04 作業系統下需要切換環境
$ ./start_aicam.sh
// 開始執行指令稿
* Serving Flask app "start_aicam" (lazy loading)
* Environment: production
   WARNING: Do not use the development server in a production environment.
   Use a production WSGI server instead.
* Debug mode: off
* Running on http://0.0.0.0:4001/ (Press CTRL+C to quit)
```

（2）在使用者端或邊緣計算閘道端開啟 Chrome 瀏覽器，輸入專案頁面位址 http://192.168.100.200:4001/static/edge_plate/index.html，即可查看專案內容。

4.5.2.2 視覺車牌辨識系統的驗證

本專案實現了車牌辨識功能，並將辨識到的車牌號碼顯示在 LCD 螢幕上，用於模擬停車場的應用場景。

（1）視覺車牌辨識系統可以即時監測車牌，並將辨識到的車牌資訊顯示在 LCD 上。

（2）為了提高辨識的準確度，視覺車牌辨識系統將對所有辨識到的車牌進行統計，單一車牌辨識次數達到 5 次後將該車牌號碼發往 Sensor-D 顯示類感測器進行更新，並重置所有的車牌計數，在 8 s 內不再進行車牌辨識。

（3）將測試樣圖放置在攝影機的視窗內，視覺車牌辨識系統會在即時視訊流中將車牌框出來，並且將辨識的車牌內容顯示出來，在成功辨識 5 次後會彈窗提示辨識到的車牌資訊。車牌辨識成功時的 AiCam 平臺介面如圖 4.41 所示。

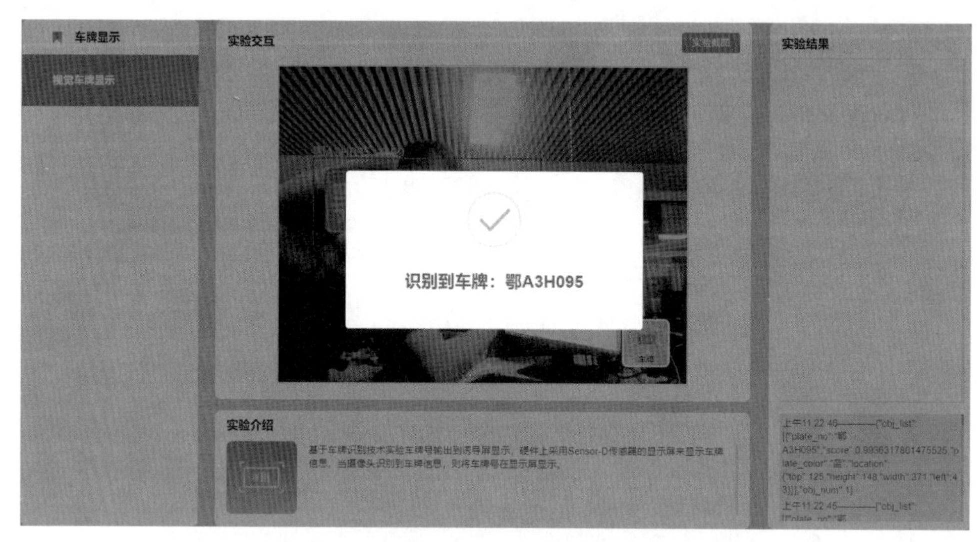

▲ 圖 4.41 車牌辨識成功時的 AiCam 平臺介面

（4）辨識到的車牌號碼將顯示在 LCD，如圖 4.42 所示。

▲ 圖 4.42 LCD 上顯示的車牌

4.5.3 本節小結

本節基於 LPRNet 模型和 AiCam 平臺實現了視覺車牌辨識系統，首先介紹了車牌辨識技術的相關知識和 LPRNet 模型；然後詳細介紹了視覺車牌辨識系統的系統框架和通訊協定，並基於 AiCam 平臺完成了車牌辨識演算法的介面，最後對視覺車牌辨識系統進行了驗證。

4.5.4 思考與擴充

（1）LPRNet 模型支援哪幾種物件辨識演算法？

（2）Sensor-D 顯示類感測器的通訊協定包含哪些內容？

（3）簡述視覺車牌辨識系統的開發步驟。

4.6 視覺智慧抄表應用程式開發

智慧抄表是一種利用先進的技術和裝置來即時監測和記錄各種儀器（如電錶、水錶、瓦斯表），在多個領域中都有廣泛的應用，其主要目的是即時監測和管理資源的使用情況，提高效率、降低成本、減少浪費，並提供更好的資料分析和決策支援。

智慧抄表的主要用途如下：

（1）電力抄表：監測電能的消耗，幫助電力公司更準確地資費，減少電力盜用。

（2）水錶和瓦斯表：即時監測水和天然氣的用量，減少浪費，提高用水和用氣效率。

（3）樓宇自動化：智慧抄表可以與樓宇管理系統集成，實現對照明、空調、供暖等能源的智慧控制，提高能源使用率。

（4）漏水檢測：透過即時監測水錶資料，能夠迅速監測到漏水事件並減少損失。

（5）能源管理：在工廠和生產環境中，智慧抄表可用於監測裝置的能源消耗，最佳化生產過程，降低能源成本。

本節的基礎知識如下：

- 了解數字辨識技術在智慧抄表中的應用。
- 掌握基於百度數字辨識演算法實現數字辨識的基本原理。
- 結合百度數字辨識演算法和 AiCam 平臺進行視覺智慧抄表的應用程式開發。

4.6.1 原理分析與開發設計

4.6.1.1 整體框架

視覺智慧抄表系統通常由多個元件組成，這些元件協作工作以實現遠端抄表、資料管理和監控。舉例來說，智慧水錶遠端抄表系統如圖 4.43 所示。

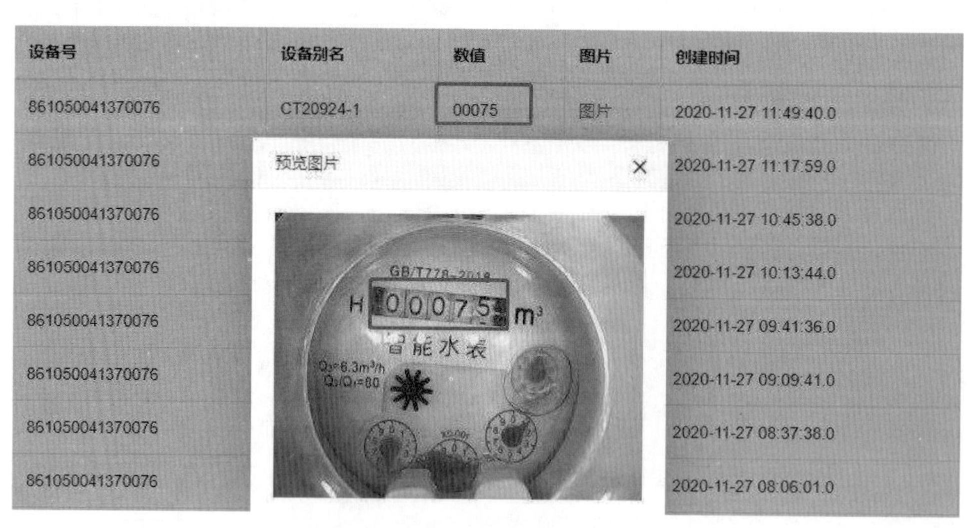

▲ 圖 4.43 智慧水錶遠端抄表系統

智慧水錶遠端抄表系統的一般包括：

（1）智慧水錶：智慧水錶是智慧水錶遠端抄表系統的核心元件。智慧水錶配備了數位電子計量技術，能夠即時記錄用水資料，並與通訊模組連接，將資料傳輸到遠端伺服器。

（2）通訊模組：智慧水錶遠端抄表系統配備了通訊模組，通常是無線通訊模組，如 GSM、3G、4G、LoRaWAN 或 NB-IoT 等模組。通訊模組可將智慧水錶的資料傳輸到遠端伺服器。

（3）資料獲取器：用於收集來自多個智慧水錶的資料。資料獲取器可以是現場裝置或雲端裝置，負責接收、儲存和處理來自智慧水錶的資料。

（4）遠端伺服器：遠端伺服器是智慧水錶遠端抄表系統的核心背景，接收從資料獲取器傳輸的資料，並提供儲存、處理和分析功能。遠端伺服器還負責與使用者介面和其他系統進行整合通訊。

上述元件共同組成了智慧水錶遠端抄表系統的框架，使系統能夠實現遠端監控、管理和最佳化用水情況，提高效率並降低資源浪費。智慧水錶遠端抄表系統的具體設計和功能可以根據需求和應用場景進行訂製。

本節採用百度數字辨識演算法實現儀表數字的讀取。百度數字辨識演算法可對影像中的數字進行提取和辨識，自動過濾非數字內容，僅傳回數字內容及其位置資訊，辨識準確率超過 99%。

透過百度數字辨識演算法，還可以對快遞面單、物流單據、外賣小票中的電話號碼進行辨識和提取，大幅提升收貨人資訊的輸入效率，方便進行收件通知，同時還可辨識純數字形式的快遞三段碼，有效提升快件的分揀速度。

從邊緣計算的角度看，視覺智慧抄表系統可分為硬體層、邊緣層、應用層，如圖 4.44 所示。

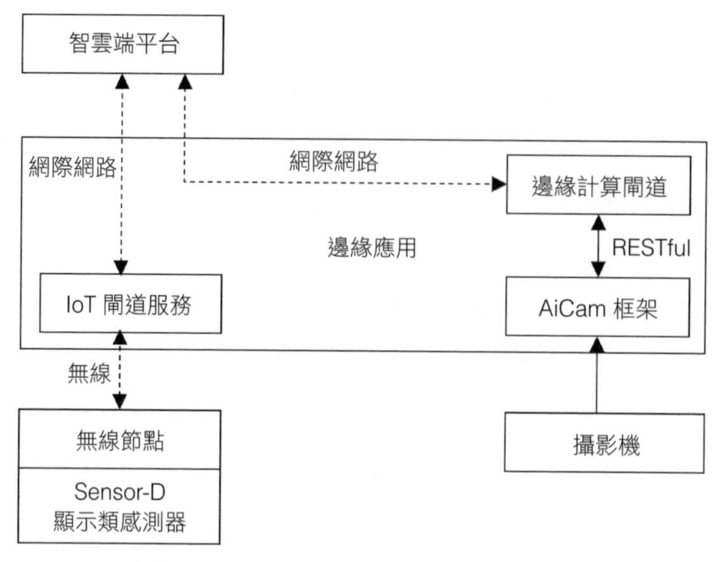

▲ 圖 4.44 智慧水電抄表系統邊緣應用結構圖

（1）硬體層：無線節點和 Sensor-D 顯示類感測器組成了視覺智慧抄表系統的硬體層，透過 Sensor-D 顯示類感測器的抄表模式顯示儀表的數字，在截取影像中的數字部分後呼叫百度數字辨識演算法進行辨識，並將辨識的結果傳回到應用端。

（2）邊緣層：包括邊緣計算閘道內建 IoT 閘道服務和 AiCam 框架。IoT 閘道服務負責接收和下發無線節點的資料，發送給應用端或將資料發給雲端的智雲端平台。AiCam 框架內建了演算法、模型、視訊串流推送等服務，支援應用層的邊緣計算推理任務。

（3）應用層：透過智雲介面與 IoT 的硬體層互動（預設與雲端的智雲端平台的介面互動），透過 AiCam 的 RESTful 介面與演算法層互動。

4.6.1.2 系統硬體與通訊協定設計

1）系統硬體設計

本專案可以採用 LiteB 無線節點（見圖 4.45）和 Sensor-D 顯示類感測器（見圖 2.16）完成硬體的架設。

▲ 圖 4.45 LiteB 無線節點

2）通訊協定設計

Sensor-D 顯示類感測器的通訊協定可參考表 4.8。本專案使用 Sensor-D 顯示類感測器來模擬儀表的顯示，當 Sensor-D 顯示類感測器處於抄表模式時，其中的 LCD 會顯示儀表的資料，並每隔 30 s 更新一次並進行上報。

4.6.1.3 功能設計與開發

1）系統框架設計

見 4.1.1.3 節。

2）介面描述

本專案基於 AiCam 平臺開發，開發流程如下：

（1）專案配置見 4.1.1.3 節。

（2）增加演算法。在 aicam 專案增加百度數字辨識演算法檔案 algorithm/baidu_meter_ recognition/baidu_meter_recognition.py。

（3）增加應用。在 aicam 專案增加演算法專案前端應用 static/edge_meter。

3）硬體通訊設計

前端應用中的硬體控制部分透過智雲 ZCloud API 連接到硬體系統，前端應用處理範例如下：

```
getConnect()
//智雲端服務連接
function getConnect(){                              //建立連接服務的函式
    rtc = new WSNRTConnect(config.user.id, config.user.key)
    rtc.setServerAddr(config.user.addr);
    rtc.connect();
    rtc.onConnect = () => {                         //連接成功回呼函式
        online = true
        setTimeout(() => {
            if(online){
                cocoMessage.success(`資料服務連接成功！查詢資料中...`)
                rtc.sendMessage(config.macList.mac_604, '{V3=2,V3=?}');  //設定為抄表模式
            }
        }, 200);
    }
    rtc.onConnectLost = () => {                                  //資料服務掉線回呼函式
        online = false
        cocoMessage.error(`資料服務連接失敗！請檢查網路或IDKEY...`)
    };
    rtc.onmessageArrive = (mac, dat) => {                        //訊息處理回呼函式
        if (dat[0] == '{' && dat[dat.length - 1] == '}') {
            //截取背景傳回的JSON物件（去掉{}符號）後，以「,」分割為陣列
            let its = dat.slice(1,-1).split(',')
            for (let i = 0; i < its.length; i++) {              //迴圈遍歷陣列的每一個值
                let t = its[i].split("=");                 //將每個值以「=」分割為陣列
                if (t.length != 2) continue;
                //mac_604顯示類感測器
                if (mac == config.macList.mac_604) {
                    console.log('抄表：',t);
                    if(t[0] == 'V1'){                      //抄表
                        $('.label').text(t[1])
                    }
                }
            }
```

```
        }
    }
}
```

4）演算法互動

透過 Ajax 介面將前端應用中截取的影像，以及包含百度帳號資訊的資料傳遞給百度數字辨識演算法進行數字辨識。Ajax 介面的參數如表 4.9 所示。

▼ 表 4.9 Ajax 介面的參數

參數	範例
url	"/file/baidu_meter_recognition?camera_id=0"
method	'POST'
processData	false
contentType	false
dataType	'json'

data	let img = $('.camera>img').attr('src')
	let blob = dataURItoBlob(img)
	var formData = new FormData();
	formData.append('file_name',blob,'image.png');
	formData.append('param_data', JSON.stringify({"APP_ID":config.user.baidu_id,
	"API_KEY":config.user.baidu_apikey, "SECRET_KEY":config.user.baidu_secretkey}));

success	function(res){} 內容： return_result = {'code': 200, 'msg': None, 'origin_image': None, 'result_image': None, 'result_data': None} 範例： code/msg：200 表示辨識成功、404 表示辨識失敗、500 表示介面呼叫失敗。 origin_image/result_image：原始影像 / 結果影像。 result_data：演算法傳回的儀表數位資訊

前端應用將待辨識的儀表影像傳遞給百度數字辨識演算法進行數字辨識，並傳回原始影像、結果影像、結果資料，相關程式如下：

```
setInterval(() => {
    getMeter()
}, 15000);

// 按一下發起專案結果請求、並對傳回的結果進行相應的處理
function  getMeter() {
    let img = $('#img_box>img').attr('src')
    let blob = dataURItoBlob(img)
    swal(' 辨識資料中，請稍等 ...',' ',"success",{button: false,timer: 2000});
    var formData = new FormData();
    formData.append('file_name',blob,'image.png');
    formData.append('param_data', JSON.stringify({"APP_ID":config.user.baidu_id,
                "API_KEY":config.user.baidu_apikey,
                "SECRET_KEY":config.user.baidu_secretkey}));
    $.ajax({
        url: '/file/baidu_meter_recognition',
        method: 'POST',
        processData: false,                 // 必需的
        contentType: false,                 // 必需的
        dataType: 'json',
        data: formData,
        success: function(result) {
            console.log(result);
            if(result.code==200) {
                swal({
                    icon: "success",
                    title: " 辨識成功 ",
                    text: " 已成功辨識！ ",
                    button: false,
                    timer: 2000
                });
                let img = 'data:image/jpeg;base64,' + result.origin_image;
                let html = `<div class="img-li">
                        <div  class="img-box">
                        <img src="${img}" alt=""  data-toggle="modal" data-
target="#myModal">
```

```
                </div>
                <div class="time"> 原始影像 <span></span><span>${new
                               Date().toLocaleString()}</span></div>
                </div>`
        $('.list-box').prepend(html);

        let img1 = 'data:image/jpeg;base64,' + result.result_image;
        let html1 = `<div class="img-li">
                  <div  class="img-box">
                  <img src="${img1}" alt=""  data-toggle="modal"
                                    data-target="#myModal">
                  </div>
                  <div class="time"> 辨識結果 <span></span><span>${new
                               Date().toLocaleString()}</span></div>
                  </div>`
        $('.list-box').prepend(html1);
        // 將辨識到的儀表資訊著色到頁面上
        let text = result.result_data.words_result[0].words
        let html2 = `<div>${new Date().toLocaleTimeString()}——辨識結果：${text}</div>`
                   $('#text-list').prepend(html2);
        swal(` 辨識到抄表資料 :${text}`,' ',"success",{button: false,timer: 2000});
    }else if(result.code==404){
        swal({
            icon: "error",
            title: " 辨識失敗 ",
            text: result.msg,
            button: false,
            timer: 2000
        });
        let img = 'data:image/jpeg;base64,' + result.origin_image;
        let html = `<div class="img-li">
                  <div  class="img-box">
                  <img src="${img}" alt=""  data-toggle="modal" data-
target="#myModal">
                  </div>
                  <div class="time"> 原始影像 <span></span><span>${new
                                Date().toLocaleString()}</span></div>
                  </div>`
         $('.list-box').prepend(html);
```

```
            }else{
                swal({
                    icon: "error",
                      title: " 辨識失敗 ",
                    text: result.msg,
                    button: false,
                    timer: 2000
                });
            }
            // 請求影像流資源
            imgData.close()
            imgData = new EventSource(linkData)
            // 對影像流傳回的資料進行處理
            imgData.onmessage = function (res) {
                let {result_image} = JSON.parse(res.data)
                $('#img_box>img').attr('src', `data:image/jpeg;base64,${result_image}`)
            }
        }, error: function(error){
            console.log(error);
            swal(' 呼叫介面失敗 ',' ',"error",{button: false,timer: 2000});
            // 請求影像流資源
            imgData.close()
            imgData = new EventSource(linkData)
            // 對影像流傳回的資料進行處理
            imgData.onmessage = function (res) {
                let {result_image} = JSON.parse(res.data)
                $('#img_box>img').attr('src', `data:image/jpeg;base64,${result_image}`)
            }
        }
    });
}
```

5）視覺智慧抄表演算法介面設計

```
##################################################################################
# 檔案：baidu_numbers_detect.py
# 說明：呼叫百度數字辨識演算法
##################################################################################
from PIL import Image, ImageDraw, ImageFont
```

```
import numpy as np
import cv2 as cv
import os,sys,time
import json
import base64
from aip import AipOcr

class BaiduMeterRecognition(object):
    def __init__(self, font_path="font/wqy-microhei.ttc"):
        self.font_path = font_path
        self.lower_blue = np.array([80,89,218])              # 顯示幕藍色範圍低設定值
        self.upper_blue = np.array([96,255,255])             # 顯示幕藍色範圍高設定值
        self.x = 0
        self.y = 0
    def imencode(self,image_np):
        # 將 JPG 格式的影像編碼為資料流程
        data = cv.imencode('.jpg', image_np)[1]
        return data

    def image_to_base64(self, img):
        image = cv.imencode('.jpg', img, [cv.IMWRITE_JPEG_QUALITY, 60])[1]
        image_encode = base64.b64encode(image).decode()
        return image_encode

    def base64_to_image(self, b64):
        img = base64.b64decode(b64.encode('utf-8'))
        img = np.asarray(bytearray(img), dtype="uint8")
        img = cv.imdecode(img, cv.IMREAD_COLOR)
        return img

    def contours_area(cnt):
        # 計算 countour 的面積
        (x, y, w, h) = cv.boundingRect(cnt)
        return w * h

    def cropGreenImg(self, image):
        # 獲取顯示幕藍色字元區域並傳回截圖
        hsv_img = cv.cvtColor(image, cv.COLOR_BGR2HSV)
```

```
        mask_green = cv.inRange(hsv_img, self.lower_blue, self.upper_blue) #根據藍色範圍篩選
        mask_green = cv.medianBlur(mask_green, 7)                          #中值濾波
        mask_green, contours, hierarchy = cv.findContours(mask_green, cv.RETR_EXTERNAL,
                                     cv.CHAIN_APPROX_NONE)     #獲取輪廓
        if len(contours) == 1:
            (x, y, w, h) = cv.boundingRect(contours[0])
            self.x = x
            self.y=y
            return image[y:y+h, x:x+w]
        elif len(contours) > 1:       #如果輪廓數大於 1 個，則獲取最大面積的輪廓
            max_cnt = max(contours, key=lambda cnt: self.contours_area(cnt))
            (x, y, w, h) = cv.boundingRect(max_cnt)
            self.x = x
            self.y = y
            return image[y:y+h, x:x+w]
        else:
            #沒有找到顯示幕藍色區域，傳回空白物件
            print('No blue rectangle found on LED display.')
            return []

    def inference(self, image, param_data):
        #code：辨識成功傳回 200
        #msg：相關提示訊息
        #origin_image：原始影像
        #result_image：處理之後的影像
        #result_data：結果資料
        return_result = {'code': 200, 'msg': None, 'origin_image': None,
                    'result_image': None, 'result_data': None}

        #應用請求介面：@__app.route('/file/<action>', methods=["POST"])
        #image：應用傳遞過來的資料（根據實際應用可能為影像、音訊、視訊、文字）
        #param_data：應用傳遞過來的參數，不能為空
        if param_data != None:
            #讀取應用傳遞過來的影像
            image = np.asarray(bytearray(image), dtype="uint8")
            image = cv.imdecode(image, cv.IMREAD_COLOR)
            #獲取顯示幕藍色字元區域
            cropImg = self.cropGreenImg(image)
            #判斷是否找到顯示幕，如果沒有找到，則直接傳回錯誤資訊
```

```
if len(cropImg)==0:
    return_result['code'] = 500
    return_result['msg'] = "沒有檢測到顯示幕！"
    return return_result

#影像資料格式的壓縮，方便網路傳輸。
img = self.imencode(cropImg)

#呼叫百度數字辨識演算法，透過下面的使用者金鑰連接百度伺服器
#APP_ID：百度應用 ID
#API_KEY：百度 API_KEY
#SECRET_KEY：百度使用者金鑰
client = AipOcr(param_data['APP_ID'], param_data['API_KEY'],
         param_data['SECRET_ KEY'])

#配置可選參數
options={}
#small：定位單字元位置
options['recognize_granularity']='small'

#帶有參數進行數字辨識
response=client.numbers(img, options)

#應用部分
if "error_msg" in response:
    if response['error_msg']!='SUCCESS':
        return_result["code"] = 500
        return_result["msg"] = "數字辨識介面呼叫失敗！"
        return_result["result_data"] = response
        return return_result
if response['words_result_num'] == 0:
    return_result["code"] = 404
    return_result["msg"] = "沒有檢測到數字！"
    return_result["origin_image"] = self.image_to_base64(image)
    return_result["result_data"] = response
    return return_result
if response['words_result_num']>0:
    #影像輸入
    img_rgb = cv.cvtColor(image, cv.COLOR_BGR2RGB)   #影像色彩格式轉換
```

```
        pilimg = Image.fromarray(img_rgb)   # 使用 PIL 讀取影像像素陣列
        draw = ImageDraw.Draw(pilimg)
        # 設定字型
        font_size = 20
        font_hei = ImageFont.truetype(self.font_path, font_size, encoding="utf-8")
        # 取資料
        words_result=response['words_result']
        for m in words_result:
            loc=m['location']                  # 文字位置
            words=m['words']                   # 文字資料
            # 使用紅色字型和方框標註文字資訊
            draw.rectangle((int(loc["left"]) + self.x, int(loc["top"]) + self.y,
                        (int(loc["left"]) + self.x + int(loc["width"])),
                        (int(loc["top"]) + self.y + int(loc["height"]))),
            outline='red',width=1)
            chars=m['chars']
            for n in chars:
                loc=n['location']              # 字元位置
                char=n['char']                 # 字元資料
                # 使用紅色字型和方框標註字元資訊
                draw.rectangle((int(loc["left"])+self.x, int(loc["top"])+self.y,
                            (int(loc["left"])+self.x + int(loc["width"])),
                            (int(loc["top"])+self.y + int(loc["height"]))),
                outline='red',width=1)
                draw.text((loc["left"]+self.x, loc["top"]+self.y-font_size-2),
                         char,fill= 'red', font=font_hei)
        # 輸出影像
        result = cv.cvtColor(np.array(pilimg), cv.COLOR_RGB2BGR)
        return_result[«code»] = 200
        return_result[«msg»] = " 數字辨識成功！ "
        return_result["origin_image"] = self.image_to_base64(image)
        return_result["result_image"] = self.image_to_base64(result)
        return_result["result_data"] = response
    else:
        return_result["code"] = 500
        return_result["msg"] = " 百度介面呼叫失敗！ "
        return_result["result_data"] = response

# 即時視頻界面：@__app.route('/stream/<action>')
```

```
        #image：攝影機即時傳遞過來的影像
        #param_data：必須為 None
        else:
            return_result["result_image"] = self.image_to_base64(image)

        return return_result

# 單元測試，如果處理類別中引用了檔案，則在單元測試中要修改檔案路徑
if __name__=='__main__':
    # 建立影像處理物件
    img_object = BaiduMeterRecognition()

    # 讀取測試影像
    img = cv.imread("./test.jpg")
    # 將影像編碼成資料流程
    img = img_object.imencode(img)

    # 設定參數
    param_data = {"APP_ID":"123456", "API_KEY":"123456", "SECRET_KEY":"123456"}
    img_object.font_path = "../../font/wqy-microhei.ttc"

    # 呼叫百度數字辨識演算法介面處理影像並傳回結果
    result = img_object.inference(img, param_data)
    if result["code"] == 200:
        frame = img_object.base64_to_image(result["result_image"])
        print(result["result_data"])

        # 影像顯示
        cv.imshow('frame',frame)
        while True:
            key=cv.waitKey(1)
            if key==ord('q'):
                break
        cv.destroyAllWindows()
    else:
        print(" 辨識失敗！ ")
```

4.6.2 開發步驟與驗證

4.6.2.1 專案部署

1）硬體部署

見 4.1.2.1 節。

2）專案部署

（1）執行 MobaXterm 工具，透過 SSH 登入到邊緣計算閘道。

（2）在 SSH 終端執行以下命令，建立開發專案目錄。

```
$ mkdir -p ~/aiedge-exp
```

（3）透過 SSH 將本專案開發專案程式和 aicam 專案套件上傳到 ~/aiedge-exp 目錄下，並採用 unzip 命令進行解壓縮。

```
$ unzip edge_meter.zip
$ unzip aicam.zip -d edge_meter
```

（4）參考 4.1.2.1 節的內容，修改專案設定檔 static/edge_fan/js/config.js 內的智雲帳號、硬體位址、邊緣服務位址等資訊。

（5）透過 SSH 將修改後的檔案上傳到邊緣計算閘道。

3）專案執行

（1）在 SSH 終端輸入以下命令執行開發專案。

```
$ cd ~/aiedge-exp/edge_meter
$ chmod 755 start_aicam.sh
$ conda activate py36_tf114_torch15_cpu_cv345     //Ubuntu 20.04 作業系統下需要切換環境
$ ./start_aicam.sh
```

（2）在使用者端或邊緣計算閘道端開啟 Chrome 瀏覽器，輸入專案頁面位址 http://192.168.100.200:4001/static/edge_meter/index.html，即可查看專案內容。

4.6.2.2 視覺智慧抄表系統的驗證

本專案實現了儀表數字辨識功能，透過讀取 Sensor-D 顯示類感測器顯示的儀表數字，呼叫百度數字辨識演算法進行儀表數字的辨識並進行記錄，用於模擬視覺智慧抄表系統的應用。

（1）將攝影機對準 Sensor-D 顯示類感測器中的顯示幕，如圖 4.46 所示。

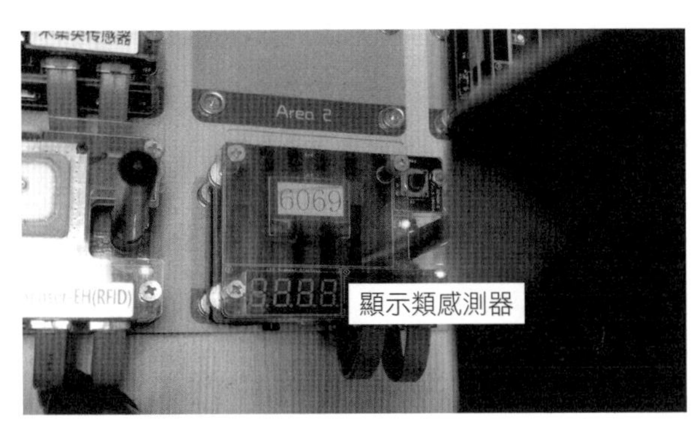

▲ 圖 4.46 Sensor-D 顯示類感測器顯示幕

（2）AiCam 平臺每隔 15 s 抓取一次攝影機拍攝的影像並進行檢測辨識，透過彈窗顯示辨識到的文字結果，並將影像結果顯示到右側的清單中。成功辨識儀表數字時的 AiCam 平臺介面如圖 4.47 所示。

▲ 圖 4.47 成功辨識儀表數字時的 AiCam 平臺介面

（3）在 AiCam 平臺智慧抄表介面的右下角圖示中會顯示 Sensor-D 顯示類感測器上傳的儀表數字。

（4）視覺智慧抄表系統將辨識到的儀表數字即時傳遞到應用端並進行顯示，如圖 4.48 所示。

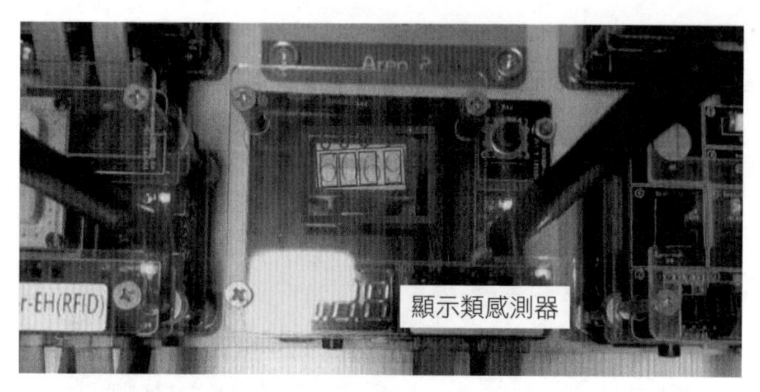

▲ 圖 4.48 應用端顯示的儀表數字

4.6.3 本節小結

本節基於百度數字辨識演算法和 AiCam 平臺實現了視覺智慧抄表系統，首先介紹了智慧水錶遠端抄表系統和百度數字辨識演算法的相關內容；然後介紹了視覺智慧抄表系統的系統框架、通訊協定，並完成了介面設計；接著借助 AiCam 平臺進行了硬體部署、專案部署、專案執行等開發流程；最後，透過具體案例對影像中的數字進行提取和辨識，自動過濾非數字內容，實現了儀表數字的即時辨識，並在 AiCam 平臺介面和應用端對辨識結果進行了顯示。

4.6.4 思考與擴充

（1）智慧水錶遠端抄表系統有哪些特點？

（2）百度數字辨識演算法有哪些優點？

（3）簡述視覺智慧抄表系統的開發流程。

4.7 語音窗簾控制應用程式開發

語音窗簾控制是一種智慧家居技術，可透過語音來控制窗簾。語音窗簾控制具有多種實用用途，可以增強家居的智慧性和便利性。語音辨識的目的是將一段語音轉換成文字，其過程蘊含著複雜的演算法和邏輯。語音辨識的工作原理如圖 4.49 所示，其中的前置處理主要包括預加重、加窗分幀和端點檢測。

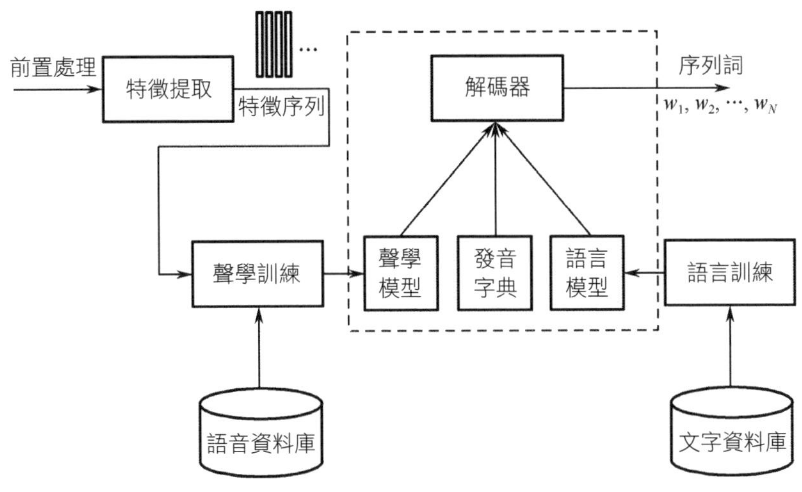

▲ 圖 4.49 語音辨識的工作原理

本節的基礎知識如下：

- 了解語音辨識技術在智慧家居系統中的應用。
- 掌握基於百度語音辨識介面進行語音辨識的基本原理。
- 結合百度語音辨識介面和 AiCam 平臺進行語音窗簾控制系統的開發。

4.7.1 原理分析與開發設計

4.7.1.1 整體框架

1）語音辨識技術概述

語音辨識技術（也稱為語音辨識或自動語音辨識，ASR）是一種能夠將語音轉化為文字或命令的技術，該技術在醫療保健、客戶服務、智慧家居、汽車、教育、娛樂、殘疾人輔助工具等多個領域都有廣泛的應用。語音辨識技術的關鍵技術如下：

（1）音訊訊號處理：語音辨識系統首先接收音訊訊號，然後對其進行處理，包括去除雜訊、分割音訊流、提取聲音特徵等。

（2）聲學模型：聲學模型是語音辨識系統的核心組成部分，用於辨識語音中的不同音素和語音單元。聲學模型通常基於大量的訓練資料來進行訓練，以便準確地辨識不同說話者的語音。

（3）語言模型：語音辨識系統使用語言模型來根據上下文確定最可能的文字輸出，語言模型考慮了語言的結構、詞彙、語法和語境，以提高辨識的準確性。

（4）聲學特徵提取：語音訊號通常需要轉化為聲學特徵，如梅爾頻率倒譜系數（MFCC）或聲譜圖，以便進行分析和辨識。

語音辨識技術中的常用深度學習模型包括：

（1）深度神經網路（DNN）模型：常用於聲學建模，將聲學特徵（如梅爾頻率倒譜系數）映射到音素或音素狀態。DNN 模型在提高語音辨識性能方面獲得了顯著的進展。

（2）卷積神經網路（CNN）模型：CNN 模型可用於處理聲學特徵的卷積層，以捕捉時間和頻率上的局部特徵。CNN 模型通常與其他深度學習模型結合使用，用於提取聲學特徵。

（3）循環神經網路（RNN）模型：RNN 模型在語音辨識中用於建模時間序列特徵，RNN 模型的變形〔如長短時記憶網路（LSTM）和門控循環單元（GRU）等〕特別適合處理變長序列資料（如語音）。

（4）CTC（Connectionist Temporal Classification，連接時序分類）模型：CTC 是一種深度學習模型，可將語音訊號映射到文字序列，且不需要對齊資訊，常用於語音辨識的點對點建模。

（5）深度轉換器（Deep Transform）模型：這是一種深度學習模型，用於語音特徵的轉換，以改善語音辨識性能。Deep Transform 模型可將聲學特徵映射到更有助辨識的表示。

（6）注意力機制（Attention Mechanism）：注意力機制被廣泛用於語音辨識性能的改進，特別是在點對點模型中，它能夠使模型在辨識時關注輸入的不同部分，以提高性能。

（7）自注意力機制（Self-Attention Mechanism）：Transformer 模型中使用的自注意力機制已經在語音辨識中獲得了顯著的成功，該機制允許模型考慮輸入序列的不同部分，從而提高性能。

（8）深度生成對抗網路（GAN）：GAN 模型可用於生成更真實的語音資料，這對於資料增強和域適應在語音辨識中的應用非常有用。

本專案採用百度語音辨識介面實現語音窗簾控制系統。百度語音辨識（標準版）介面可以將 60 s 內的語音精準地辨識為文字（Android、iOS、Linux SDK 支持超過 60 s 的語音辨識），可用於手機語音輸入、智慧語音互動、語音指令、語音搜索等語音互動場景。語音辨識技術在智慧家居中的應用場景如圖 4.50 所示。

▲ 圖 4.50 語音辨識技術在智慧家居中的應用場景

2）語音窗簾控制系統的結構

從邊緣計算的角度看，語音窗簾控制系統可分為硬體層、邊緣層、應用層，如圖 4.51 所示。

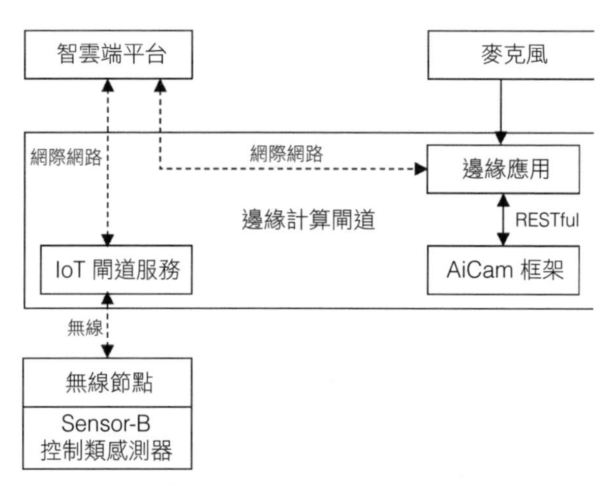

▲ 圖 4.51 語音窗簾控制系統的結構

（1）硬體層：無線節點和 Sensor-B 控制類感測器組成了語音窗簾控制系統的硬體層，透過 Sensor-B 控制類感測器的步進馬達來模擬窗簾的開和關。

（2）邊緣層：包括邊緣計算閘道內建 IoT 閘道服務和 AiCam 框架。IoT 閘道服務負責接收和下發無線節點的資料，發送給應用端或將資料發給雲端的智雲端平台。AiCam 框架內建了演算法、模型、視訊串流推送等服務，支援應用層的邊緣計算推理任務。

（3）應用層：透過智雲介面與 IoT 的硬體層互動（預設與雲端的智雲端平台的介面互動），透過 AiCam 的 RESTful 介面與演算法層互動。

4.7.1.2 系統硬體與通訊協定設計

1）系統硬體設計

本專案可以採用 LiteB 無線節點、Sensor-B 控制類感測器來完成硬體的架設，也可以透過虛擬模擬軟體來建立一個虛擬的硬體裝置，如圖 4.52 所示。

▲ 圖 4.52 虛擬硬體平臺

2）通訊協定設計

Sensor-B 控制類感測器的通訊協定如表 4.1 所示。本專案使用步進馬達來模擬窗簾的開和關，步進馬達的命令如表 4.10 所示。

▼ 表 4.10 步進馬達的命令

發送命令	接收結果	含義
{D1=?}	{D1=XX}	查詢步進窗簾（步進馬達轉動）狀態
{OD1=4,D1=?}	{D1=XX}	開啟窗簾（Bit2 為 1 表示步進馬達反轉）
{CD1=4,D1=?}	{D1=XX}	關閉窗簾（Bit2 為 0 表示步進馬達正轉）

4.7.1.3 功能設計與開發

1）系統框架設計

見 4.1.1.3 節。

2）介面描述

本專案基於 AiCam 平臺開發，開發流程如下：

（1）專案配置見 4.1.1.3 節。

（2）增加演算法。在 aicam 專案增加百度語音辨識介面檔案 algorithm/baidu_speech_ recognition/baidu_speech_recognition.py。

（3）增加應用。在 aicam 專案增加演算法專案前端應用 static/edge_curtain。

3）硬體通訊設計

前端應用中的硬體控制部分透過智雲 ZCloud API 連接到硬體系統，前端應用處理範例如下：

```
getConnect()
// 智雲端服務連接
function getConnect(){                              // 建立連接服務的函式
    rtc = new WSNRTConnect(config.user.id, config.user.key)
    rtc.setServerAddr(config.user.addr);
    rtc.connect();
    rtc.onConnect = () => {                         // 連接成功回呼函式
        online = true
        setTimeout(() => {
            if(online){
                cocoMessage.success(` 資料服務連接成功！查詢資料中 ...`)
                rtc.sendMessage(config.macList.mac_602,config.sensor.mac_602.query);
// 發起資料查詢
            }
        }, 200);
```

```
    }
    rtc.onConnectLost = () => {                              // 資料服務掉線回呼函式
        online = false
        cocoMessage.error(` 資料服務連接失敗！請檢查網路或 IDKEY...`)
    };

    rtc.onmessageArrive = (mac, dat) => {                    // 訊息處理回呼函式
        if (dat[0] == '{' && dat[dat.length - 1] == '}') {
            // 截取背景傳回的 JSON 物件（去掉 {} 符號）後，以「,」分割為陣列
            let its = dat.slice(1,-1).split(',')
            for (let i = 0; i < its.length; i++) {           // 迴圈遍歷陣列的每一個值
                let t = its[i].split("=");                   // 將每個值以「=」分割為陣列
                if (t.length != 2) continue;
                //mac_602 控制類感測器
                if (mac == config.macList.mac_602) {
                    if (t[0] == 'D1') {                      // 開關控制
                        console.log(' 窗簾開關：', t);
                        if (t[1] & 4) {
                            // 判斷接收到的命令後是否跟上次一樣，若是則不做任何操作，
                            // 否則執行動畫效果
                            // 透過當前顯示的圖示可得到上次命令結果
                            if ($('#icon1').attr('src').indexOf('curtain-off') > -1) {
                                $('#icon1').attr('src', './img/curtain-on.gif')
                            }
                        } else {
                            // 判斷接收到的命令後是否跟上次一樣，是則不做任何操作，
                            // 否則執行動畫效果
                            // 透過當前顯示的圖示可得到上次命令結果
                            if ($('#icon1').attr('src').indexOf('curtain-on') > -1) {
                                $('#icon1').attr('src', './img/curtain-off.gif')
                            }
                        }
                    }
                }
            }
        }
    }
}
```

4）演算法互動

前端應用透過應用端的麥克風裝置進行錄音，透過 Ajax 介面將音訊資料傳遞給百度語音辨識介面進行語音辨識。百度語音辨識介面的參數如表 4.11 所示。

▼ 表 4.11 百度語音辨識介面的參數

參數	範例
url	"/file/baidu_speech_recognition"
method	'POST'
processData	false
contentType	false
dataType	'json'

data	let config = configData
	let blob = recorder.getWAVBlob();
	let formData = new FormData();
	formData.set('file_name',blob,'audio.wav');
	formData.append('param_data', JSON.stringify({"APP_ID":config.user.baidu_id, "API_KEY":config.user.baidu_apikey, "SECRET_KEY":config.user.baidu_secretkey}));

success	function(res){} 內容： return_result = {'code': 200, 'msg': None, 'origin_image': None, 'result_image': None, 'result_data': None} 範例： code/msg：200 表示語音辨識成功、500 表示語音辨識失敗。 result_data：傳回語音辨識後的文字內容

按一下 AiCam 平臺窗簾控制介面中的「錄音」按鈕可進行錄音，前端應用將錄製的音訊資料發送到演算法層進行辨識，並在轉為文字後顯示在頁面中。前端應用範例如下：

```
$('#interaction .item').click(function () {
    if ($(this).find('.label').text() == '錄音') {
        $(this).find('img').attr('src', './img/microphone-on.gif')
        $(this).find('.label').text('錄音中...')
        recorder.start().then(() => {
            //開始錄音
        }, (error) => {
            //出錯了
            console.log(`${error.name} : ${error.message}`);
        });
    } else {
        $(this).find('img').attr('src', './img/microphone-off.png')
        $(this).find('.label').text('錄音')
        recorder.stop();
        let blob = recorder.getWAVBlob();
        console.log(blob);
        let formData = new FormData();
        formData.set('file_name',blob,'audio.wav');
        formData.append('param_data', JSON.stringify({"APP_ID":config.user.baidu_id,
                    "API_KEY":config.user.baidu_apikey,
                    "SECRET_KEY":config.user.baidu_secretkey}));
        $.ajax({
            url: '/file/baidu_speech_recognition',
            method: 'POST',
            processData: false,                         //必需的
            contentType: false,                         //必需的
            dataType: 'json',
            data: formData,
            headers: { 'X-CSRFToken': getCookie('csrftoken') },
            success: function(result) {
                console.log(result);
                if(result.code == 200){
                    if(result.result_data.indexOf('開啟窗簾') > -1){
                        cocoMessage.success(`辨識到開啟窗簾語音字樣！窗簾開啟中...`)
                        rtc.sendMessage(config.macList.mac_602,config.sensor.mac_602.
curtainOpen);
                    }
                    if(result.result_data.indexOf('關閉窗簾') > -1){
                        cocoMessage.success(`辨識到關閉窗簾語音字樣！窗簾關閉中...`)
```

```
                        rtc.sendMessage(config.macList.mac_602,config.sensor.mac_602.
curtainClose);
                    }
                    let html = `<div class="msg"><div>${result.result_data}</div></div>`
                    $('#message_box').append(html)
                    $('#message_box').scrollTop($('#message_box')[0].scrollHeight);
                }else{
                    swal({
                        icon: "error",
                        title: " 辨識失敗 ",
                        text: result.msg,
                        button: false,
                        timer: 2000
                    });
                }
            },
            error: function(error){
                console.log(error);
                swal({
                    icon: "error",
                    title: " 辨識失敗 ",
                    text: '',
                    button: false,
                    timer: 2000
                });
            }
        });
    }
})
```

5）語音窗簾控制演算法介面設計

```
#########################################################################################
# 檔案：baidu_speech_recognition.py
# 說明：百度語音辨識介面
#########################################################################################
import os
import wave
import numpy as np
```

```
from aip import AipSpeech
import ffmpeg
import tempfile

class BaiduSpeechRecognition(object):
    def __init__(self):
        pass
    def __check_wav_file(self,filePath):
        # 讀取 wav 檔案
        wave_file = wave.open(filePath, 'r')
        # 獲取檔案的每秒顯示畫面和通道
        frame_rate = wave_file.getframerate()
        channels = wave_file.getnchannels()
        wave_file.close()
        if frame_rate == 16000 and channels == 1:
            return True
            #feature_path=filePath
        else:
            return False
    def inference(self, wave_data, param_data):
        #code：辨識成功傳回 200
        #msg：相關提示訊息
        #origin_image：原始影像
        #result_image：處理之後的影像
        #result_data：結果資料
        return_result = {'code': 200, 'msg': None, 'origin_image': None,
                         'result_image': None, 'result_data': None}

        # 應用請求介面：@__app.route('/file/<action>', methods=["POST"])
        #wave_data：應用傳遞過來的資料（根據實際應用可能為影像、音訊、視訊、文字）：
        # 語音資料，格式為 wav，取樣速率為 16000
        #param_data：應用傳遞過來的參數，不能為空
        if param_data != None:
            fd, path = tempfile.mkstemp()
            try:
                with os.fdopen(fd, 'wb') as tmp:
                    tmp.write(wave_data)
                if not self.__check_wav_file(path):
                    fd2, path2 = tempfile.mkstemp()
```

```
                ffmpeg.input(path).output(path2, ar=16000).run()
                os.remove(path)
                path = path2
            f = open(path, "rb")
            f.seek(4096)
            pcm_data = f.read()
            f.close()
        finally:
            os.remove(path)

        # 呼叫百度語音辨識介面，透過以下使用者金鑰連接百度伺服器
        #APP_ID：百度應用 ID
        #API_KEY：百度 API_KEY
        #SECRET_KEY：百度使用者金鑰
        client = AipSpeech(param_data['APP_ID'], param_data['API_KEY'], param_data['SECRET_KEY'])
        # 語音檔案的格式為 pcm，取樣速率為 16000，dev_pid 為普通話（純中文辨識），辨識本地檔案
        response = client.asr(pcm_data,'pcm', 16000, {'dev_pid': 1537,})

        # 處理伺服器傳回結果
        if response['err_msg']=='success.':
            return_result["code"] = 200
            return_result["msg"] = "語音辨識成功！"
            return_result["result_data"] = response['result'][0]
        else:
            return_result["code"] = 500
            return_result["msg"] = response['err_msg']
        return return_result

# 單元測試，如果處理類別中引用了檔案，則在單元測試中要修改檔案路徑
if __name__=='__main__':
    # 建立音訊處理物件
    test = BaiduSpeechRecognition()
    param_data = {"APP_ID":"12345678", "API_KEY":"12345678", "SECRET_KEY":"12345678"}
    with open("./test.wav", "rb") as f:
        wdat = f.read()
        result = test.inference(wdat, param_data)
    print(result["result_data"])
```

4.7.2 開發步驟與驗證

4.7.2.1 專案部署

1）硬體部署

見 4.1.2.1 節。

2）專案部署

（1）執行 MobaXterm 工具，透過 SSH 登入到邊緣計算閘道。

（2）在 SSH 終端執行以下命令，建立開發專案目錄。

```
$ mkdir -p ~/aiedge-exp
```

（3）透過 SSH 將本專案開發專案程式和 aicam 專案套件上傳到 ~/aiedge-exp 目錄下，並採用 unzip 命令進行解壓縮。

```
$ unzip edge_curtain.zip
$ unzip aicam.zip -d edge_curtain
```

（4）參考 4.1.2.1 節的內容，修改專案設定檔 static/edge_curtain/js/config.js 內的智雲帳號、硬體位址、邊緣服務位址等資訊。

（5）透過 SSH 將修改後的檔案上傳到邊緣計算閘道。

3）專案執行

（1）在 SSH 終端輸入以下命令執行開發專案：

```
$ cd ~/aiedge-exp/edge_curtain
$ chmod 755 start_aicam.sh
$ conda activate py36_tf114_torch15_cpu_cv345    //Ubuntu 20.04 作業系統下需要切換環境
$ ./start_aicam.sh
```

（2）在使用者端或邊緣計算閘道端開啟 Chrome 瀏覽器，輸入專案頁面位址 https://192.168.100.200:1446/static/edge_curtain/index.html，即可查看專案內容。

4.7.2.2 語音窗簾控制系統的驗證

本專案透過百度語音辨識介面實現了語音窗簾控制系統，用於模擬語音辨識技術在智慧家居中的應用場景。

按一下 AiCam 平臺窗簾控制介面右下角的「錄音」按鈕進行語音錄製，再次按一下「錄音」按鈕可結束語音錄製並對錄製的語音進行辨識，辨識的結果以文字的形式顯示在實驗互動區，如「開啟窗簾」。成功開啟窗簾時的 AiCam 平臺介面如圖 4.53 所示。

如果辨識到「開啟窗簾」則發送開啟窗簾命令，如果辨識到「關閉窗簾」則發送關閉窗簾命令，並伴有相應動畫效果及彈窗提示。語音窗簾控制系統在接收到命令時，會與上一次的命令進行比較，如果兩次的命令一樣，則不做動畫效果處理。

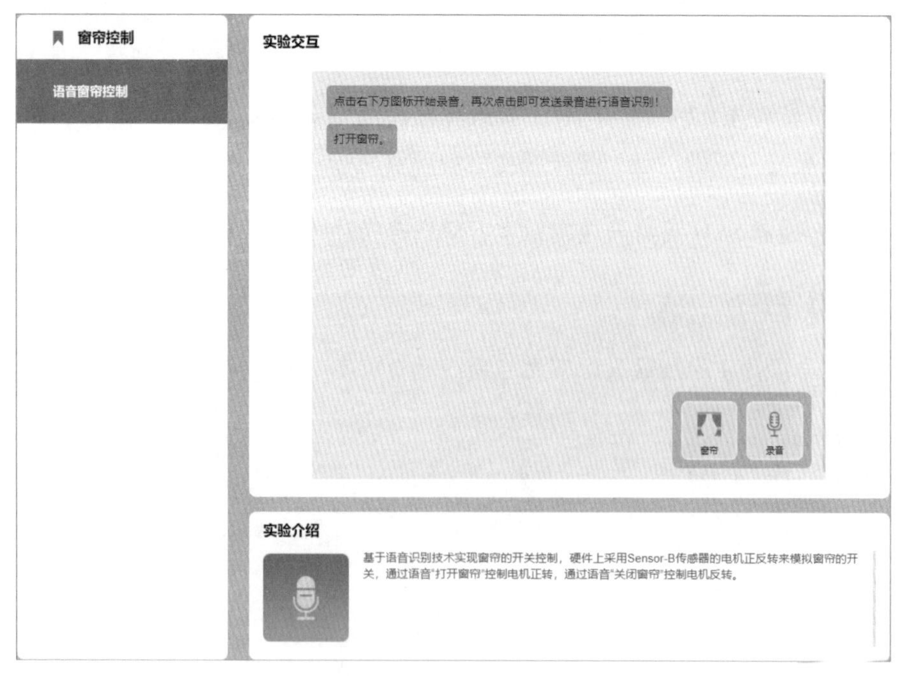

▲ 圖 4.53 成功開啟窗簾時的 AiCam 平臺介面

步進馬達模擬窗簾開關的效果如圖 4.54 所示。

▲ 圖 4.54 步進馬達模擬窗簾開關的效果

4.7.3 本節小結

本節透過百度語音辨識介面和 AiCam 平臺實現了語音窗簾控制系統，首先介紹了語音辨識技術在智慧家居領域中的應用；然後介紹了語音窗簾控制系統的系統框架、通訊協定，並完成了系統框架設計、語音窗簾控制演算法介面設計；接著借助 AiCam 平臺描述了硬體部署、專案部署、專案執行等開發流程；最後對語音窗簾控制系統進行了驗證。

4.7.4 思考與擴充

（1）百度語音辨識介面有哪些特點？

（2）語音窗簾控制系統與視覺智慧抄表系統的邊緣應用場景有什麼區別和聯繫？

（3）簡述語音窗簾控制系統的開發流程？

4.8 語音環境播報應用程式開發

語音合成技術也被稱為文字到語音（Text-to-Speech，TTS）合成，其主要功能是將文字轉換成自然流暢的語音。語音合成技術具有廣泛的應用領域，以下是一些常見的用途：

（1）語音幫手和虛擬幫手：語音合成技術可用於建立語音幫手和虛擬幫手，如 Siri、Google Assistant 和 Alexa，使這些幫手能夠與使用者自然地互動並提供語音回應。

（2）自動電話應答系統：企業客服中心可以透過語音合成技術來建構自動電話應答系統，用來回答常見的問題，提供給使用者指導。

（3）有聲讀物：透過語音合成技術可將電子書、新聞文章和其他文字內容轉為有聲讀物，以幫助視覺障礙者和那些在駕駛車輛或鍛煉時希望聽書的人。

（4）語音瀏覽：透過語音合成技術，GPS 可以提供駕駛和步行導航指示，使駕駛者和行人能夠在不分心的情況下獲取導航資訊。

（5）教育應用：透過語音合成技術可以建立教育內容，包括線上課程、教科書的有聲版本，以及輔助閱讀和學習的工具。

（6）語音輔助技術：包括螢幕閱讀器、語音放大器和語音命令系統，可以幫助視覺障礙者和身體障礙者使用電腦和行動裝置。

（7）媒體和娛樂：在電影、電視和視訊遊戲中，虛擬角色和故事情節通常可以使用語音合成技術來實現聲音效果。

本節的基礎知識如下：

- 了解語音合成技術在語音環境播報中的應用。
- 掌握基於百度語音合成介面實現語音環境播報系統的基本原理。
- 結合百度語音合成介面和 AiCam 平臺進行語音環境播報系統的開發。

4.8.1 原理分析與開發設計

4.8.1.1 整體框架

1）語音合成技術概述

語音合成技術中的常用深度學習模型如下：

（1）WaveNet：WaveNet 是由 DeepMind 開發的深度學習模型，可用於語音合成。WaveNet 模型基於深度卷積神經網路，能夠生成高品質的語音波形，非常逼近人類語音。WaveNet 模型透過逐樣本生成聲音波形，具有出色的聲音品質，但計算成本較高。

（2）Tacotron：Tacotron 是一種基於序列到序列的模型，可將文字轉為聲音的聲學特徵。在語音合成中，通常先使用 Tacotron 模型生成聲學特徵，再使用聲學模型生成聲音波形。Tacotron 模型的重要版本是 Tacotron 2，它可以和 WaveNet 模型結合起來生成高品質的語音。

（3）Transformer TTS：這是一種基於 Transformer 框架的語音合成模型，該模型使用自注意力機制來處理輸入的文字，先將文字轉為聲學特徵序列，再使用聲學模型生成語音。Transformer TTS 在語音合成領域獲得了巨大的成功。

（4）Deep Voice：Deep Voice 採用深度卷積神經網路和循環神經網路，用於建模聲學特徵，被廣泛用於點對點的語音合成。

（5）FastSpeech：FastSpeech 是一種用於文字到語音合成的模型。FastSpeech 模型使用了 Transformer 模型框架，能夠更快速地生成聲學特徵序列，可直接生成聲學特徵，繞過中間的文字到音素或音素到聲學特徵的轉換步驟。

（6）Parallel WaveGAN：Parallel WaveGAN 是一種生成對抗網路（GAN）模型的變形，專門用於生成高品質的語音波形。Parallel WaveGAN 模型可以與不同的聲學模型結合使用，以產生自然流暢的語音。

（7）Hifi-GAN：Hifi-GAN 是一種 GAN 模型的變形，旨在生成高保真度的語音波形。Hifi-GAN 模型透過訓練生成器和判別器提高了生成的語音的真實感。

上述的深度學習模型代表了現代語音合成領域的前端技術，能夠生成高品質、自然流暢的語音，適用於各種應用，如虛擬幫手、有聲讀物、語音瀏覽、廣告等。隨著深度學習技術的不斷進步，語音合成技術也在不斷發展和改進。

本專案採用百度語音合成介面實現語音環境播報系統。百度語音合成介面是基於 HTTP 請求的 REST API 介面，能夠將文字轉為可以播放的音訊檔案，適用於語音互動、語音內容分析、智慧硬體、客服中心智慧客服等多種場景。語音合成應用範例如圖 4.55 所示。

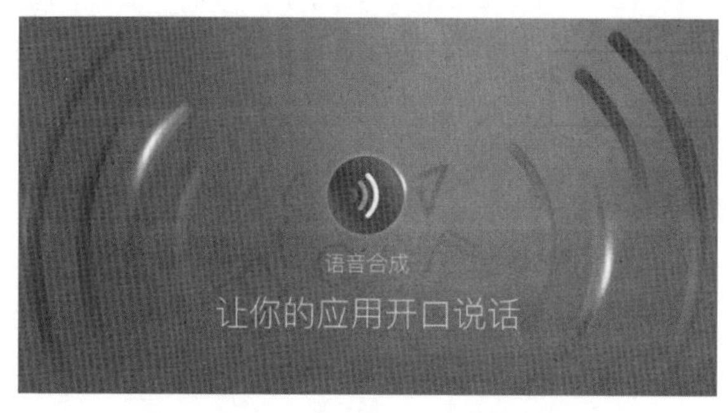

▲ 圖 4.55 語音合成應用範例

2）語音環境播報系統的結構

從邊緣計算的角度看，語音環境播報系統可分為硬體層、邊緣層、應用層，如圖 4.56 所示。

（1）硬體層：無線節點和 Sensor-A 擷取類感測器組成了語音環境播報系統的硬體層，透過 Sensor-A 擷取類感測器獲取當前環境的溫 / 濕度資訊，並傳回應用端進行顯示。

（2）邊緣層：包括邊緣計算閘道內建 IoT 閘道服務和 AiCam 框架。IoT 閘道服務負責接收和下發無線節點的資料，發送給應用端或將資料發給雲端的智雲端平台。AiCam 框架內建了演算法、模型、視訊串流推送等服務，支援應用層的邊緣計算推理任務。

（3）應用層：透過智雲介面與 IoT 的硬體層互動（預設與雲端的智雲端平台的介面互動），透過 AiCam 的 RESTful 介面與演算法層互動。

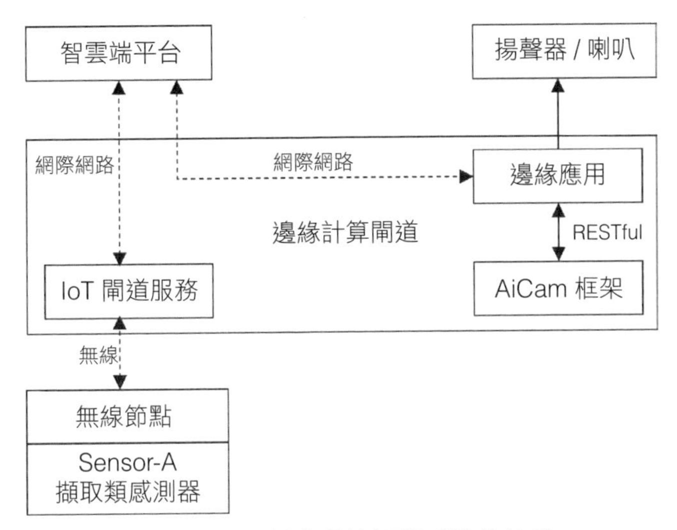

▲ 圖 4.56 語音環境播報系統的結構

4.8.1.2 系統硬體與通訊協定設計

1）系統硬體設計

本專案可以採用 LiteB 無線節點、Sensor-A 擷取類感測器來完成硬體的架設，也可以透過虛擬模擬軟體來建立一個虛擬的硬體裝置如圖 4.57 所示。

▲ 圖 4.57 虛擬硬體平臺

2）通訊協定設計

Sensor-A 擷取類感測器的通訊協定如表 4.12 所示。

▼ 表 4.12 Sensor-A 擷取類感測器的通訊協定

節點名稱	TYPE	參數	屬性	許可權	說明
Sensor-A 擷取類感測器	601	A0	溫度	R	溫度值，浮點型，精度為 0.1，範圍為 -40.0 ～ 105.0，單位為℃
		A1	濕度	R	濕度值，浮點型，精度為 0.1，範圍為 0 ～ 100.0，單位為 %RH
		A2	光照度	R	光照度值，浮點型，精度為 0.1，範圍為 0 ～ 65535.0，單位為 lx
		A3	空氣品質	R	空氣品質值，表徵空氣污染程度，整數，範圍為 0 ～ 20000，單位為 ppm
		A4	大氣壓力	R	大氣壓力值，浮點型，精度為 0.1，範圍為 800.0 ～ 1200.0，單位為 hPa
		A5	跌倒狀態	R	透過三軸感測器計算出跌倒狀態，0 表示未跌倒，1 表示跌倒
		A6	距離	R	距離值，浮點型，精度為 0.1，範圍為 10.0 ～ 80.0，單位為 cm
		D0(OD0/CD0)	上報狀態	R/W	D0 的 Bit0 ～ Bit7 分別代表 A0 ～ A7 的上報狀態，1 表示主動上報，0 表示不上報
		D1(OD1/CD1)	繼電器	R/W	D1 的 Bit6 ～ Bit7 分別代表繼電器 K1、K2 的開關狀態，0 表示斷開，1 表示吸合
		V0	上報時間間隔	R/W	A0 ～ A7 和 D1 的主動上報時間間隔，預設為 30，單位為 s

本專案採用 Sensor-A 擷取類感測器獲得當前環境的濕度和溫度，每隔 30 s 更新一次濕度和溫度，並顯示在應用端。Sensor-A 擷取類感測器的命令如表 4.13 所示。

▼ 表 4.13 Sensor-A 擷取類感測器的命令

發送命令	接收結果	含義
—	{A0=XXX,A1=XXX}	溫度為 A0，濕度為 A1

4.8.1.3 功能設計與開發

1）系統框架設計

見 4.1.1.3 節。

2）介面描述

本專案基於 AiCam 平臺開發，開發流程如下：

（1）專案配置見 4.1.1.3 節。

（2）增加演算法。在 aicam 專案增加百度語音合成介面 algorithm/baidu_speech_synthesis/ baidu_ speech_synthesis.py。

（3）增加應用。在 aicam 專案增加演算法專案前端應用 static/edge_env。

3）硬體通訊設計

前端應用中的硬體控制部分透過智雲 ZCloud API 連接到硬體系統，前端應用處理範例如下：

```
getConnect()
// 智雲端服務連接
function getConnect(){                                   // 建立連接服務的函式
    rtc = new WSNRTConnect(config.user.id, config.user.key)
    rtc.setServerAddr(config.user.addr);
    rtc.connect();
```

```
rtc.onConnect = () => {                    // 連接成功回呼函式
    online = true
    setTimeout(() => {
        if(online){
            cocoMessage.success(` 資料服務連接成功！查詢資料中 ...`)
            // 發起資料查詢
            rtc.sendMessage(config.macList.mac_601,config.sensor.mac_601.query);
        }
    }, 200);
}
rtc.onConnectLost = () => {                                 // 資料服務掉線回呼函式
    online = false
    cocoMessage.error(` 資料服務連接失敗！請檢查網路或 IDKEY...`)
};

rtc.onmessageArrive = (mac, dat) => {                       // 訊息處理回呼函式
    if (dat[0] == '{' && dat[dat.length - 1] == '}') {
        // 截取背景傳回的 JSON 物件（去掉 {} 符號）後，以「,」分割為陣列
        let its = dat.slice(1,-1).split(',')
        for (let i = 0; i < its.length; i++) {              // 迴圈遍歷陣列的每一個值
            let t = its[i].split("=");                      // 將每個值以「=」分割為陣列
            if (t.length != 2) continue;
            //mac_601 擷取類感測器
            if (mac == config.macList.mac_601) {
                if (t[0] == 'A0') {                         // 溫度
                    console.log(' 溫度：',t);
                    $('.item span:eq(0)').text(t[1])
                    temperature = t[1]
                }
                if (t[0] == 'A1') { // 濕度
                    console.log(' 濕度：',t);
                    $('.item span:eq(1)').text(t[1])
                    humidity = t[1]
                    setTimeout(() => {
                        if(audioAuthority && temperature){
                            audioAuthority = false
                            let html = `<div class="msg"><div>${new
                                Date().toLocaleTimeString()}——
```

```
                                        當前溫度 ${temperature}℃，
                                        濕度 ${humidity}%</div></div>`
                                $('#message_box').append(html)
                                $('#message_box').scrollTop($('#message_box')[0].
scrollHeight);
                                voiceSynthesis(` 當前溫度 ${temperature} 度，
                                            濕度百分之 ${humidity}`)
                                setTimeout(() => {
                                    audioAuthority = true
                                }, 5000);
                            }
                        }, 500);
                    }
                }
            }
        }
    }
}
```

4）演算法互動

前端應用透過 Ajax 介面將需要進行語音合成的文字傳遞給百度語音合成介面，百度語音合成介面的參數如表 4.14 所示。

▼ 表 4.14 百度語音合成介面的參數

參 數	範 例
url	"/file/baidu_speech_synthesis"
method	'POST'
processData	false
contentType	false
dataType	'json'

（續表）

參數	範例

data	let config = configData
	let text = $(this).parent('.item').find('textarea').val()
	let blob = new Blob([text],{
	type:'text/plain'
	});
	// 語音合成播報
	let formData = new FormData();
	formData.append('file_name',blob,'text.txt');
	formData.append('param_data', JSON.stringify({"APP_ID":config.user.baidu_id, "API_KEY":config.user.baidu_apikey, "SECRET_KEY":config.user.baidu_secretkey}));

success	function(res){} 內容： return_result = {'code': 200, 'msg': None, 'origin_image': None, 'result_image': None, 'result_data': None} 範例： code/msg：200 表示語音合成成功，500 表示語音合成失敗。 result_data：傳回 base64 編碼的音訊檔案

前端應用範例如下：

```
// 語音合成功能，val 表示需要進行語音合成的文字
function voiceSynthesis(val) {
    let blob = new Blob([val], {
        type: 'text/plain'
    });
    // 語音合成播報
    let formData = new FormData();
    formData.append('file_name', blob, 'text.txt');
    formData.append('param_data', JSON.stringify({
        "APP_ID": config.user.baidu_id,
        "API_KEY": config.user.baidu_apikey,
        "SECRET_KEY": config.user.baidu_secretkey
```

```
    }));
    $.ajax({
        url: '/file/baidu_speech_synthesis',
        method: 'POST',
        processData: false,              // 必需的
        contentType: false,              // 必需的
        dataType: 'json',
        data: formData,
        headers: {
            'X-CSRFToken': getCookie('csrftoken')
        },
        success: function (result) {
            console.log(result);
            if (result.code == 200) {
                // 播放合成的音訊
                let mp3 = new Audio(`data:audio/x-wav;base64,${result.result_data}`)
                mp3.play()
            } else {
                swal({
                    title: "介面呼叫失敗！",
                    icon: "error",
                    text: " ",
                    timer: 1000,
                    button: false
                });
            }
        },
        error: function (error) {
            console.log(error);
            swal({
                title: "介面呼叫失敗！",
                icon: "error",
                text: " ",
                timer: 1000,
                button: false
            });
        }
    })
}
```

5）語音環境播報演算法介面設計

```
###########################################################################################
#檔案：baidu_speech_synthesis.py
#說明：百度語音合成
###########################################################################################
import cv2 as cv
import base64
import os,sys,time
import wave
import numpy as np
from aip import AipSpeech

class BaiduSpeechSynthesis(object):
    def __init__(self):
        pass

    def image_to_base64(self, img):
        image = cv.imencode('.jpg', img, [cv.IMWRITE_JPEG_QUALITY, 60])[1]
        image_encode = base64.b64encode(image).decode()
        return image_encode

    def base64_to_image(self, b64):
        img = base64.b64decode(b64.encode('utf-8'))
        img = np.asarray(bytearray(img), dtype="uint8")
        img = cv.imdecode(img, cv.IMREAD_COLOR)
        return img

    def inference(self, content, param_data):
        #code：辨識成功傳回 200
        #msg：相關提示訊息
        #origin_image：原始影像
        #result_image：處理之後的影像
        #result_data：結果資料
        return_result = {'code': 200, 'msg': None, 'origin_image': None,
                  'result_image': None, 'result_data': None}

        #應用請求介面：@__app.route('/file/<action>', methods=["POST"])
        #content：應用傳遞過來的資料（根據實際應用可能為影像、音訊、視訊、文字）
```

```
        #param_data：應用傳遞過來的參數，不能為空
        if param_data != None:
            #呼叫百度語音合成介面，透過以下使用者金鑰連接百度伺服器
            #APP_ID：百度應用 ID
            #API_KEY：百度 API_KEY
            #SECRET_KEY：百度使用者金鑰
            client = AipSpeech(param_data['APP_ID'], param_data['API_KEY'],
                        param_data['SECRET_KEY'])

            #配置可選參數
            options={}
            options['spd']=5                    #語速，設定值 0～9，預設為 5（中語速）
            options['pit']=5                    #音調，設定值 0～9，預設為 5（中語調）
            options['vol']=5                    #音量，設定值 0～15，預設為 5（中音量）
            options['per']=0                    #發音人選擇
            #呼叫百度語音合成介面
            response = client.synthesis(content, 'zh', 1, options)

            if not isinstance(response, dict):
                return_result["code"] = 200
                return_result["msg"] = "語音合成成功！"
                return_result["result_data"] = base64.b64encode(response).decode()
            else:
                return_result["code"] = 500
                return_result["msg"] = "語音合成失敗！"
                return_result["result_data"] = response
            return return_result

#單元測試，如果處理類別中引用了檔案，則在單元測試中要修改檔案路徑
if __name__=='__main__':
    test = BaiduSpeechSynthesis()
    param_data={"APP_ID":"12345678", "API_KEY":"12345678", "SECRET_KEY":"12345678"}
    with open("test.txt", "rb") as f:
        txt = f.read()
        result=test.inference(txt, param_data)
        if result["code"] == 200:
            dat = base64.b64decode(result["result_data"].encode('utf-8'))
            with open("out.mp3", "wb") as fw:
                fw.write(dat)
        print(result)
```

4.8.2 開發步驟與驗證

4.8.2.1 專案部署

1）硬體部署

見 4.1.2.1 節。

2）專案部署

（1）執行 MobaXterm 工具，透過 SSH 登入到邊緣計算閘道。

（2）在 SSH 終端執行以下命令，建立開發專案目錄。

```
$ mkdir -p ~/aiedge-exp
```

（3）透過 SSH 將本專案開發專案程式和 aicam 專案套件上傳到 ~/aiedge-exp 目錄下，並採用 unzip 命令進行解壓縮。

```
$ unzip edge_env.zip
$ unzip aicam.zip -d edge_env
```

（4）參考 4.1.2.1 節的內容，修改專案設定檔 static/edge_env/js/config.js 內的智雲帳號、硬體位址、邊緣服務位址等資訊。

（5）透過 SSH 將修改後的檔案上傳到邊緣計算閘道。

3）專案執行

（1）在 SSH 終端輸入以下命令執行開發專案。

```
$ cd ~/aiedge-exp/edge_env
$ chmod 755 start_aicam.sh
$ conda activate py36_tf114_torch15_cpu_cv345    //Ubuntu 20.04 作業系統下需要切換環境
$ ./start_aicam.sh
```

（2）在使用者端或邊緣計算閘道端開啟 Chrome 瀏覽器，輸入專案頁面位址 http://192.168.100.200:4001/static/edge_env/index.html，即可查看專案內容。

4.8.2.2 語音環境播報系統的驗證

本專案透過百度語音合成介面和 AiCam 平臺實現了語音環境播報系統，可以即時播報感測器擷取的溫度和濕度資訊。

按一下 AiCam 平臺環境播報介面右下角的圖示，可以顯示由 Sensor-A 擷取類感測器上報的溫 / 濕度資訊，生成由當前時間 + 溫 / 濕度資訊組成的文字並顯示在實驗互動區。語音環境播報系統在接收到新的溫 / 濕度資訊時，都會對其進行語音合成並進行播報，播報的最小間隔時間為 5 s。

語音環境播報系統執行時期的 AiCam 平臺介面如圖 4.58 所示。

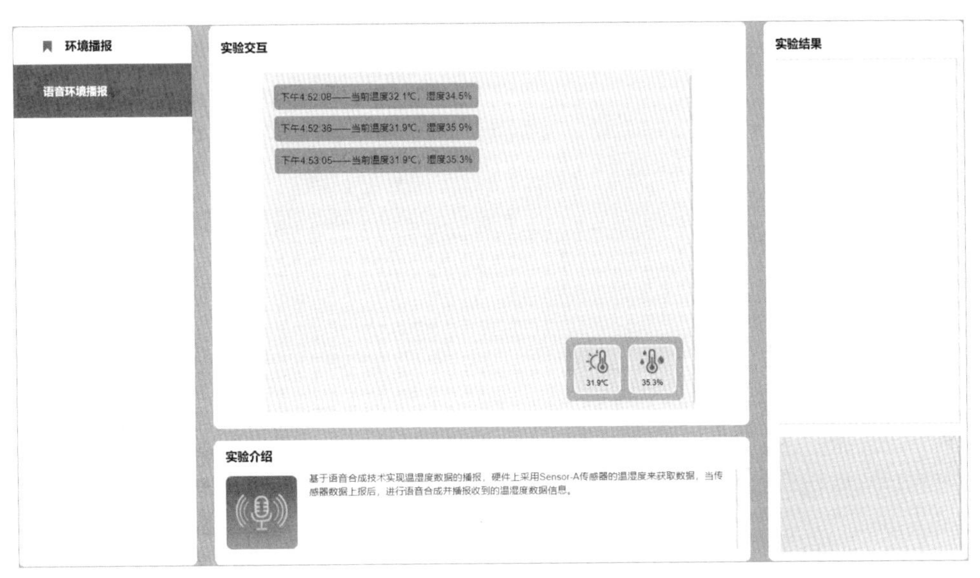

▲ 圖 4.58 語音環境播報系統執行時期的 AiCam 平臺介面

4.8.3 本節小結

本節透過百度語音合成介面和 AiCam 平臺實現了語音環境播報系統，首先介紹了語音合成技術的相關內容；然後介紹了語音環境播報系統的系統框架、通訊協定，並完成了系統框架設計、語音環境播報演算法介面設計；接著借助

AiCam 平臺描述了硬體部署、專案部署、專案執行等開發流程；最後對語音環境播報系統進行了驗證。

4.8.4 思考與擴充

（1）簡述語音環境播報系統的結構。

（2）實現語音合成的 Sensor-A 擷取類感測器的通訊協定包括哪些內容？

（3）簡述語音環境播報系統的開發過程。

MEMO

MEMO

5 邊緣計算與人工智慧綜合應用程式開發

本章介紹邊緣計算與人工智慧的綜合應用程式開發，包括 2 個開發案例。

（1）智慧家居系統設計與開發：主要內容包括智慧家居系統的應用場景、需求分析，智慧家居系統的框架設計、功能設計、硬體設計，基於手勢辨識、語音辨識和入侵偵測的智慧家居系統開發。

（2）輔助駕駛系統設計與開發：主要內容包括輔助駕駛系統的應用場景、需求分析，輔助駕駛系統的框架設計、功能設計、硬體設計，基於手勢辨識、語音辨識和駕駛行為檢測的輔助駕駛系統開發。

5.1 智慧家居系統設計與開發

智慧家居（Smart Home）系統利用綜合佈線技術、網路通訊技術、安全防範技術、自動控制技術、音視訊技術提高住宅生活品質、安全性、便利性和能源效率的家庭環境，透過聯網的裝置和感測器來實現遠端控制和自動化，讓家庭中的各種裝置能夠更智慧地協作工作。智慧家居系統具有遠端控制、聲控和語音幫手、智慧照明、智慧家電、安全和監控、能源管理、娛樂系統、健康和醫療監測等功能，目標是提高家庭的生活品質、舒適性和便利性，同時提高能源效率和安全性，在日常生活中發揮著越來越重要的作用。

本節的基礎知識如下：

- 了解智慧家居系統的應用場景、需求分析。
- 掌握智慧家居系統的框架設計、功能設計、硬體設計。
- 掌握基於手勢辨識、語音辨識和入侵偵測的智慧家居系統開發。

5.1.1 原理分析與開發設計

5.1.1.1 智慧家居系統框架分析

1）需求分析

智慧家居系統以住宅為平臺，集系統、結構、服務、管理、控制於一體，利用先進的網路通訊技術、電力自動化技術、電腦技術、無線電技術，將與家居生活有關的各種裝置有機地結合起來，透過網路化的綜合管理家中裝置，來創造一個優質、高效、舒適、安全、便利、節能、健康、環保的居住生活環境空間。智慧家居系統具有多種功能，旨在提高家庭生活的便捷性、舒適性、安全性和節能性。智慧家居系統的主要需求如下：

（1）遠端控制：智慧家居系統允許使用者透過智慧型手機、平板電腦或電腦遠端監控和控制家庭裝置，如照明、溫度、安全攝影機等，讓使用者能夠隨時隨地管理家庭。

（2）自動化控制：定時和計畫任務，使用者可以設定時間表和計畫，自動控制家庭裝置的操作，如可以在特定時間開啟或關閉燈光、加熱系統或電器裝置，以提高能源效率；情景模式，使用者可以建立自訂情景，如「回家模式」或「離家模式」，從而觸發多個裝置的聯動操作，以滿足特定需求，如自動解鎖門、調整溫度、開啟安全系統等。

（3）節能管理：智慧能源監測，智慧家居系統可以監測能源的使用情況，提供即時資料和報告，幫助使用者了解並減少能源浪費；能源最佳化，根據家庭成員的活動和習慣，智慧家居系統可以自動調整照明、加熱和冷卻系統，以減少能源消耗。

（4）安全和監控：安全警示，智慧家居系統可以透過行動應用程式或電子郵件發送安全警示，如入侵偵測、火警或漏水警示；即時監控，使用者可以即時監控家庭，透過智慧攝影機或門窗感測器來檢測異常活動。

（5）娛樂和多媒體：智慧音響，可以提供音樂、新聞、天氣預報等資訊，還可以與語音幫手互動；智慧電視和媒體中心，可以讓使用者存取各種串流媒體服務和娛樂內容。

（6）健康和健身：智慧健康裝置，如智慧體重秤、心跳監測器等，可以追蹤使用者的健康資料，並將其上傳到雲端供分析和查看；睡眠監測，智慧床墊或睡眠追蹤器可以監測睡眠品質並提供建議。

（7）語音控制：智慧家居裝置通常支援語音控制，使用者可以使用語音幫手（如 Amazon Alexa、Google Assistant、Apple Siri）來執行各種任務，如調整裝置、獲取資訊等。

（8）通訊：智慧家居系統可以用於家庭通訊，如留言、通話、視訊會議等。

智慧家居系統可以根據使用者的需求和偏好進行個性化的配置，提供更智慧、便捷和舒適的家庭生活體驗，有助提高家庭的能源效率和安全性，降低營運成本，同時提供更多的娛樂和健康管理選擇。隨著智慧家居系統的發展，市場消費群眾已經形成了對智慧家居單品的穩定需求。從最早的 Wi-Fi 聯網控制到如今的手勢互動、語音辨識，智慧家居系統的互動性能也在逐步提升，在人工智慧的世界裡，滑鼠鍵盤轉變成觸控、語音、手勢、視覺等，多模態人機互動技術正在彼此融合。

2）智慧家居系統框架

從邊緣計算的角度看，智慧家居系統可分為硬體層、邊緣層、應用層，如圖 5.1 所示。

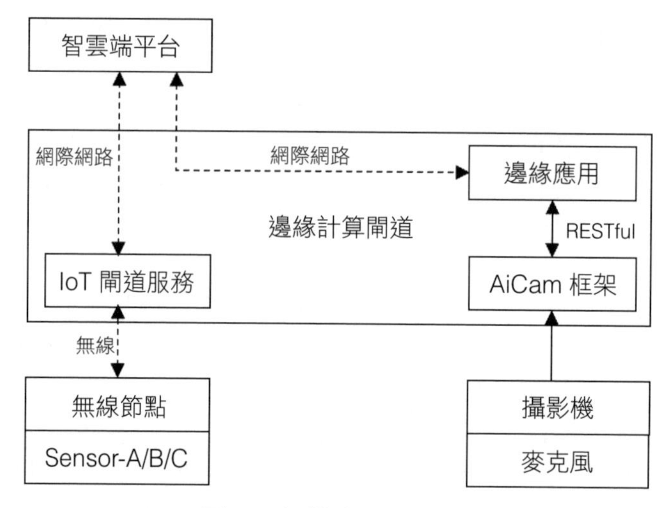

▲ 圖 5.1 智慧家居系統的結構

（1）硬體層：無線節點和 Sensor-A 擷取類感測器、Sensor-B 控制類感測器、Sensor-C 保全類感測器組成了智慧家居系統的硬體層，透過光柵感測器監測是否開啟手勢辨識等 AI 智慧互動操作。

（2）邊緣層：包括邊緣計算閘道內建 IoT 閘道服務和 AiCam 框架。IoT 閘道服務負責接收和下發無線節點的資料，發送給應用端或將資料發給雲端的智雲端平台。AiCam 框架內建了演算法、模型、視訊串流推送等服務，支援應用層的邊緣計算推理任務。

（3）應用層：透過智雲介面與 IoT 的硬體層互動（預設與雲端的智雲端平台的介面互動），透過 AiCam 的 RESTful 介面與演算法層互動。

5.1.1.2 系統功能設計

1）功能模組設計

本專案的智慧家居系統可分為以下功能模組：

（1）智慧物聯：即時顯示家居環境、設施、保全等資料，支援模式設定。

（2）手勢互動：透過攝影機進行手勢開關家居裝置的操作，在本專案中，手勢 1 表示對窗簾進行開關操作、手勢 2 表示對燈光進行開關操作、手勢 3 表示對風扇進行開關操作、手勢 4 表示對空調進行開關操作、手勢 5 表示對加濕器進行開關操作。

（3）語音互動：按一下智慧家居系統首頁右下角的麥克風圖示可錄製語音，錄製結束後可透過麥克風對家居裝置進行語音控制操作，可透過語音命令來開啟 / 關閉窗簾、燈光、風扇、空調、和加濕器。

（4）入侵監測：當光柵感測器警告時，呼叫人體檢測模型進行人體檢測，並每隔 15 s 拍照一次。

（5）應用設定：對專案的智雲帳號、裝置位址、百度帳號進行設定。

2）功能框架設計

智慧家居系統的功能框架如圖 5.2 所示。

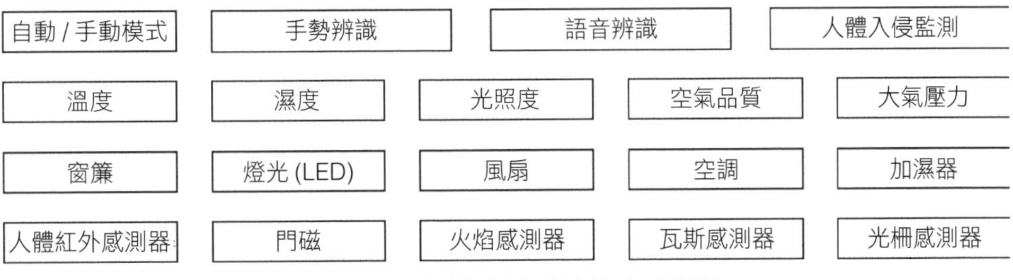

▲ 圖 5.2 智慧家居系統的功能框架

5.1.1.3 硬體與通訊設計

1）硬體設計

智慧家居系統既可以採用 LiteB 無線節點、Sensor-A 擷取類感測器、Sensor-B 控制類感測器、Sensor-C 保全類感測器來完成一套智慧家居系統硬體的架設，也可以透過虛擬模擬軟體來建立一個智慧家居系統專案，並增加對應的感測器。

（1）Sensor-A 擷取類感測器：主要用於檢測溫 / 濕度、光照度、空氣品質、大氣壓力、跌倒狀態等資訊，其虛擬控制平臺如圖 5.3 所示。

（2）Sensor-B 控制類感測器：主要用於控制窗簾（步進馬達）、風扇、LED 燈、空調（繼電器 K1）、加濕器（繼電器 K2），其虛擬控制平臺如圖 5.4 所示。

▲ 圖 5.3 Sensor-A 擷取類感測器的虛擬控制平臺

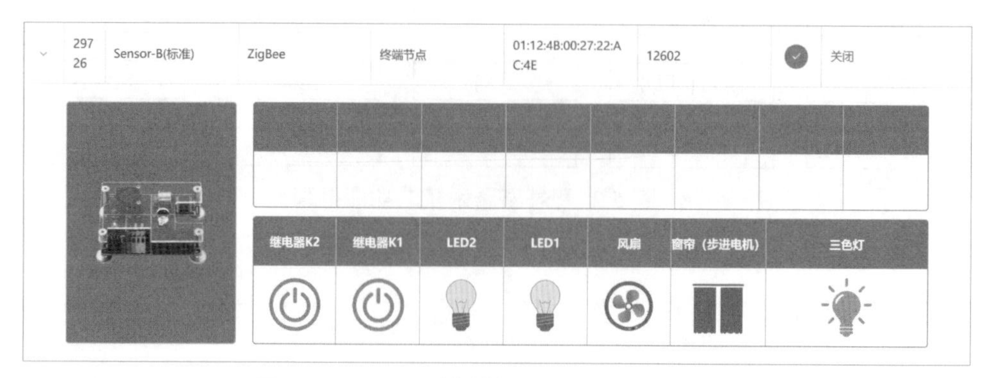

▲ 圖 5.4 Sensor-B 控制類感測器的虛擬控制平臺

（3）Sensor-C 保全類感測器：主要用於檢測瓦斯、火焰、光柵、門磁（霍爾）、人體紅外等資訊，其虛擬控制平臺如圖 5.5 所示。

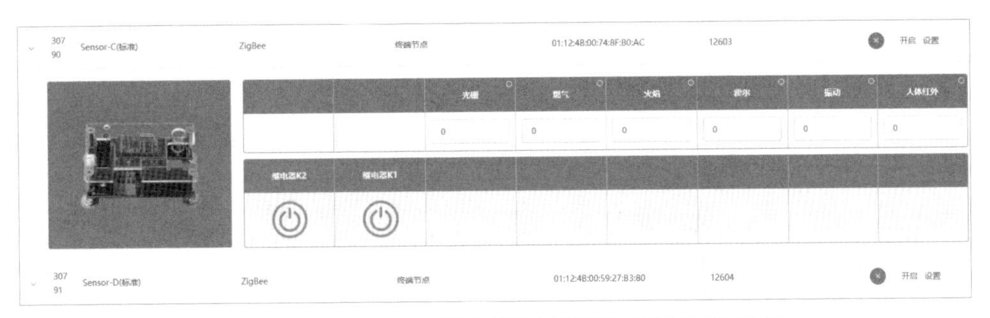

▲ 圖 5.5 Sensor-C 保全類感測器的虛擬控制平臺

2）通訊協定設計

智慧家居系統的通訊協定如表 5.1 所示。

▼ 表 5.1 智慧家居系統的通訊協定

節點名稱	TYPE	參 數	屬 性	許可權	說　明
Sensor-A 擷取類感測器	601	A0	溫度	R	溫度值，浮點型，精度為 0.1，範圍為－40.0～105.0，單位為℃
		A1	濕度	R	濕度值，浮點型，精度為 0.1，範圍為 0～100.0，單位為 %RH
		A2	光照度	R	光照度值，浮點型，精度為 0.1，範圍為 0～65535.0，單位為 lx
		A3	空氣品質	R	空氣品質值，表徵空氣污染程度，整數，範圍為 0～20000，單位為 ppm
		A4	大氣壓力	R	大氣壓力值，浮點型，精度為 0.1，範圍為 800.0～1200.0，單位為 hPa

節點名稱	TYPE	參 數	屬 性	許可權	說 明
Sensor-A 擷取類感測器	601	A5	跌倒狀態	R	透過三軸感測器計算出跌倒狀態，0 表示未跌倒，1 表示跌倒
		A6	距離	R	距離值，浮點型，精度為 0.1，範圍為 10.0 ～ 80.0，單位為 cm
		D0(OD0/CD0)	上報狀態	R/W	D0 的 Bit0 ～ Bit7 分別代表 A0 ～ A7 的上報狀態，1 表示主動上報，0 表示不上報
		D1(OD1/CD1)	繼電器	R/W	D1 的 Bit6 ～ Bit7 分別代表繼電器 K1、K2 的開關狀態，0 表示斷開，1 表示吸合
		V0	上報時間間隔	R/W	A0 ～ A7 和 D1 的主動上報時間間隔，預設為 30，單位為 s
Sensor-B 控制類感測器	602	D1(OD1/CD1)	RGB	R/W	D1 的 Bit0 ～ Bit1 代表 RGB 三色燈的顏色狀態，00 表示關、01 表示紅色、10 表示綠色、11 表示藍色
		D1(OD1/CD1)	步進馬達	R/W	D1 的 Bit2 表示步進馬達的正反轉動狀態，0 表示正轉、1 表示反轉
		D1(OD1/CD1)	風扇 / 蜂鳴器	R/W	D1 的 Bit3 表示風扇 / 蜂鳴器的開關狀態，0 表示關閉，1 表示開啟
		D1(OD1/CD1)	LED	R/W	D1 的 Bit4 ～ Bit5 表示 LED1、LED2 的開關狀態 ,0 表示關閉，1 表示開啟
		D1(OD1/CD1)	繼電器	R/W	D1 的 Bit6 ～ Bit7 表示繼電器 K1、K2 的開關狀態，0 表示斷開，1 表示吸合
		V0	上報間隔	R/W	A0 ～ A7 和 D1 的循環上報時間間隔

節點名稱	TYPE	參數	屬性	許可權	說　明
Sensor-C 保全類感測器	603	A0	人體紅外 / 觸控	R	人體紅外 / 觸控感測器狀態，設定值為 0 或 1，1 表示有人體活動 / 觸控動作，0 表示無人體活動 / 觸控動作
		A1	震動	R	震動狀態，設定值為 0 或 1，1 表示檢測到震動，0 表示未檢測到震動
		A2	霍爾	R	霍爾狀態，設定值為 0 或 1，1 表示檢測到磁場，0 表示未檢測到磁場
		A3	火焰	R	火焰狀態，設定值為 0 或 1，1 表示檢測到火焰，0 表示未檢測到火焰
		A4	瓦斯	R	瓦斯洩漏狀態，設定值為 0 或 1，1 表示檢測到瓦斯洩漏，0 表示未檢測到瓦斯洩漏
		A5	光柵	R	光柵（紅外對射）狀態值，設定值為 0 或 1，1 表示檢測到阻擋，0 表示未檢測到阻擋
		D0(OD0/CD0)	上報狀態	R/W	D0 的 Bit0 ～ Bit7 分別表示 A0 ～ A7 的上報狀態，1 表示主動上報，0 表示不上報
		D1(OD1/CD1)	繼電器	R/W	D1 的 Bit6 ～ Bit7 分別表示繼電器 K1、K2 的開關狀態，0 表示斷開，1 表示吸合
		V0	上報間隔	R/W	A0 ～ A7 和 D1 的循環上報時間間隔

3）硬體通訊設計

前端應用中的硬體控制部分透過智雲 ZCloud API 連接到硬體系統，前端應用處理範例如下：

```
/*******************************************************************************
* 名稱：getConnect()
* 功能：建立即時連接服務，監聽資料並進行處理
*******************************************************************************/
getConnect(){                                              // 建立連接服務的函式
    rtc = new WSNRTConnect(this.config.user.id, this.config.user.key)
    rtc.setServerAddr(this.config.user.addr);
    rtc.connect();
    rtc.onConnect = () => {          // 連接成功回呼函式
        this.onlineBtn = '斷開'
        setTimeout(() => {
            if(this.onlineBtn == '斷開'){
                cocoMessage.success(`資料服務連接成功！查詢資料中...`)
                // 發起資料查詢
                rtc.sendMessage(this.config.macList.mac_601, this.config.sensor.mac_601.
query);
                rtc.sendMessage(this.config.macList.mac_602, this.config.sensor.mac_602.
query);
                rtc.sendMessage(this.config.macList.mac_603, this.config.sensor.mac_603.
query);
            }
        }, 200);
    }
    rtc.onConnectLost = () => {                                // 資料服務掉線回呼函式
        this.onlineBtn = '連接'
        cocoMessage.error(`資料服務連接失敗！請檢查網路或 IDKEY...`)
    };
    rtc.onmessageArrive = (mac, dat) => {                      // 訊息處理回呼函式
        if (dat[0] == '{' && dat[dat.length - 1] == '}') {
            // 截取背景傳回的 JSON 物件（去掉 {} 符號）後，以「,」分割為陣列
            let its = dat.slice(1,-1).split(',')
            for (let i = 0; i < its.length; i++) {             // 迴圈遍歷陣列的每一個值
                let t = its[i].split("=");                     // 將每個值以「=」分割為陣列
                if (t.length != 2) continue;
                //mac_601 擷取類感測器
                if (mac == this.config.macList.mac_601) {
                    this.onlineState[0] = true
                    if(t[0] == 'A0'){                  // 溫度
                        this.data601[0].value = t[1]
```

```
                    //自動模式下根據設定的設定值進行判斷
                    if(this.modelVal == '自動模式'){
                        if(t[1] > this.data601[0].range[1] && this.data602[3].value ==
'已關閉'){
                            rtc.sendMessage(this.config.macList.mac_602,
`{OD1=64,D1=?}`);
                            cocoMessage.success("當前溫度較高！空調開啟中...");
                        }
                        if(t[1] < this.data601[0].range[0] && this.data602[3].value !=
'已關閉'){
                            rtc.sendMessage(this.config.macList.mac_602,
`{CD1=64,D1=?}`);
                            cocoMessage.success("當前溫度較低！空調關閉中...");
                        }
                    }
                }
                if(t[0] == 'A1'){                   //濕度
                    this.data601[1].value = t[1]
                    //自動模式下根據設定的設定值進行判斷
                    if(this.modelVal == '自動模式'){
                        if(t[1] < this.data601[1].range[0] && this.data602[4].value ==
'已關閉'){
                            rtc.sendMessage(this.config.macList.mac_602,
`{OD1=128,D1=?}`);
                            cocoMessage.success("當前濕度較低！加濕器開啟中...");
                        }
                        if(t[1] > this.data601[1].range[1] && this.data602[4].value !=
'已關閉'){
                            rtc.sendMessage(this.config.macList.mac_602,
`{CD1=128,D1=?}`);
                            cocoMessage.success("當前濕度較高！加濕器關閉中...");
                        }
                    }
                }
                if(t[0] == 'A2'){                   //光照度
                    this.data601[2].value = t[1]
                    //自動模式下根據設定的設定值進行判斷
                    if(this.modelVal == '自動模式'){
                        if(t[1] < this.data601[2].range[0] && this.data602[0].value ==
'已關閉'){
```

```
                        rtc.sendMessage(this.config.macList.mac_602,`{OD1=4,D1=?}`);
                        cocoMessage.success("當前光照度較低！窗簾開啟中...");
                    }
                    if(t[1] > this.data601[2].range[1] && this.data602[0].value != 
'已關閉'){
                        rtc.sendMessage(this.config.macList.mac_602,`{CD1=4,D1=?}`);
                        cocoMessage.success("當前光照度較高！窗簾關閉中...");
                    }
                }
            }
            if(t[0] == 'A3'){        //TVOC（Total Volatile Organic Compounds）
                this.data601[3].value = t[1]
                //自動模式下根據設定的設定值進行判斷
                if(this.modelVal == '自動模式'){
                    if(t[1] > this.data601[3].range[1] && this.data602[1].value == 
'已關閉'){
                        rtc.sendMessage(this.config.macList.mac_602,`{OD1=8,D1=?}`);
                        cocoMessage.success("當前空氣品質較差！風扇開啟中...");
                    }
                    if(t[1] < this.data601[3].range[0] && this.data602[1].value != 
'已關閉'){
                        rtc.sendMessage(this.config.macList.mac_602,`{CD1=8,D1=?}`);
                        cocoMessage.success("當前空氣品質正常！風扇關閉中...");
                    }
                }
            }
            if(t[0] == 'A4'){                  //大氣壓力
                this.data601[4].value = t[1]
            }
        }
        //mac_602控制類感測器
        if (mac == this.config.macList.mac_602) {
            this.onlineState[1] = true
            if(t[0] == 'D1'){
                if(t[1] & 4){                  //窗簾
                    this.data602[0].value = '已開啟'
                }else{
                    this.data602[0].value = '已關閉'
                }
```

```
            if(t[1] & 8){                  //風扇
                this.data602[1].value = '已開啟'
            }else{
                this.data602[1].value = '已關閉'
            }
            if(t[1] & 16 || t[1] & 32){
                if(t[1] & 16){             //LED
                    this.data602[2].value = '一級燈'
                }
                if(t[1] & 32){             //LED燈
                    this.data602[2].value = '二級燈'
                }
            }else{
                this.data602[2].value = '已關閉'
            }
            if(t[1] & 64){                 //空調
                this.data602[3].value = '已開啟'
            }else{
                this.data602[3].value = '已關閉'
            }
            if(t[1] & 128){                //加濕器
                this.data602[4].value = '已開啟'
            }else{
                this.data602[4].value = '已關閉'
            }
        }
    }
    //mac_603保全類感測器
    if (mac == this.config.macList.mac_603) {
        this.onlineState[2] = true
        if(t[0] == 'A0'){                  //人體紅外
            this.data603[0].value = t[1]
        }
        if(t[0] == 'A2'){                  //霍爾（門磁）
            this.data603[1].value = t[1]
        }
        if(t[0] == 'A3'){                  //火焰
            this.data603[2].value = t[1]
```

```
                    }
                    if(t[0] == 'A4'){                   // 瓦斯
                        this.data603[3].value = t[1]
                    }
                    if(t[0] == 'A5'){                   // 光柵
                        // 觸發光柵且上次為 0 時切換到人體辨識檢測
                        if(this.data603[4].value == 0 && t[1] == 1){
                            this.getVisitor()
                        }
                        // 光柵為 0 時且上次為觸發狀態時切換到手勢辨識檢測
                        if(this.data603[4].value == 1 && t[1] == 0){
                            this.gestureRecognition()
                        }
                        this.data603[4].value = t[1]
                    }
                }

            }
        }
    }
}

/*********************************************************************************
* 名稱：control()
* 功能：根據按一下的模組發送相應查詢、控制命令
* 參數：type 表示對應的 MAC 節點， val 表示對應該模組協定參數
*********************************************************************************/
control(type,val) {
    if(this.onlineBtn == '斷開'){
        if(type == '601'){
            rtc.sendMessage(this.config.macList.mac_601,`{${val}=?}`);      // 發起資料查詢
        }
        if(type == '602'){
            if(this.modelVal == '自動模式'){
                return swal("請切換為手動模式後操作！"," ","error",{button: false,timer:
2000});
            }
            if(val == 2){
                if(this.data602[val].value == '已關閉'){
```

```
                //發送開關命令
                rtc.sendMessage(this.config.macList.mac_602,`{OD1=16,D1=?}`);
            }
            if(this.data602[val].value == '一級燈'){
                //發送開關命令
                rtc.sendMessage(this.config.macList.mac_602,`{OD1=32,D1=?}`);
            }
            if(this.data602[val].value == '二級燈'){
                //發送開關命令
                rtc.sendMessage(this.config.macList.mac_602,`{CD1=48,D1=?}`);
            }
        }else{
            if(this.data602[val].value == '已關閉'){
                //發送開關命令
                rtc.sendMessage(this.config.macList.mac_602,`{OD1=${
                                this.data602[val].val},D1=?}`);
            }else{
                //發送開關命令
                rtc.sendMessage(this.config.macList.mac_602,`{CD1=${
                                this.data602[val].val},D1=?}`);
            }
        }
    }
    if(type == '603'){
        rtc.sendMessage(this.config.macList.mac_603,`{${val}=?}`);    //發起資料查詢
    }
}else{
    swal("資料服務尚未連接！請檢查網路或IDKEY...","","error",{button: false,timer:
2000});
}
```

5.1.1.4 專案開發

1）系統框架設計

見 4.1.1.3 節。

2）開發流程

本專案基於 AiCam 平臺開發，開發流程如下：

（1）專案配置見 4.1.1.3 節。

（2）增加模型。智慧家居系統用到了手勢辨識、人體檢測深度學習模型，需要在 aicam 專案增加人手檢測模型檔案 models/handpose_detection/handdet.bin、handdet.param、手勢辨識模型檔案 models/handpose_detection/handpose.bin、handpose.param、人體檢測模型檔案 models/person_detection/person_detector.bin、person_detector.param。

（3）增加演算法。智慧家居系統用到了手勢辨識、人體檢測、百度語音辨識等演算法，需要在 aicam 專案增加手勢辨識演算法檔案 algorithm/handpose_detection/handpose_detection.py、人體檢測演算法檔案 algorithm/person_detection/person_detection.py、語音辨識演算法檔案 algorithm/ baidu_speech_recognition/baidu_speech_recognition.py。

（4）增加應用。在 aicam 專案增加演算法專案前端應用 static/edge_smarthome。

3）演算法互動

（1）手勢辨識演算法。智慧家居系統採用 EventSource 介面獲取處理後的視訊流，透過即時推理介面呼叫手勢辨識演算法來辨識視訊流中的手勢，傳回 base64 編碼的結果影像和結果資料。前端應用處理範例如下：

```
/*******************************************************************************
* 名稱：gestureRecognition
* 功能：辨識視訊流中的手勢，對辨識到的手勢次數進行累計，當累計到 4 次時向相應的家居
        裝置發送開關命令
*******************************************************************************/
gestureRecognition(){
    imgData && imgData.close()
    let resultThrottle = true                // 將節流結果辨識間隔時間設定為 500 ms
    let gestureThrottle = true               // 將節流閘門開啟間隔時間設定為 8 s
```

```
    let count = {}                              // 累計手勢辨識次數
    // 請求視訊資源
    imgData = new EventSource(this.config.user.edge_addr + this.linkData[1])
    console.log(this.config.user.edge_addr + this.linkData[1]);
    // 對視訊資源傳回的資料進行處理
    imgData.onmessage = res => {
        let {result_image} = JSON.parse(res.data)
        this.homeVideoSrc =  `data:image/jpeg;base64,${result_image}`
        let {result_data} = JSON.parse(res.data)
        // 對辨識到的手勢進行判斷（設定間隔不得小於 500 ms 一次）
        if(result_data && resultThrottle){
            resultThrottle = false
            if (result_data.obj_num > 0 && result_data.obj_list[0].score > 0.70 &&
gestureThrottle) {
                console.log(result_data);
                // 設定變數並賦值辨識到的手勢
                let gesture = result_data.obj_list[0].name
                // 為辨識到的每個手勢設定為物件屬性名稱，初始值為 1。當某個手勢的辨識
                // 次數累計達到 4 次時，向相應的家居裝置發送開關命令，並清除計數
                if(count[gesture]){
                    count[gesture] += 1
                    if(count[gesture] == 4 && gesture != 'undefined'){
                        count = {}
                        gestureThrottle = false
                        let index
                        if(gesture == 'one')  index = 1
                        if(gesture == 'two')  index = 2
                        if(gesture == 'three')  index = 3
                        if(gesture == 'four')  index = 4
                        if(gesture == 'five')  index = 5
                        console.log(gesture,index);
                        if(index){
                            if(this.data602[index-1].value == '已關閉'){
                                rtc.sendMessage(this.config.macList.mac_602,`{OD1=${
                                    this.data602[index-1].val},D1=?}`);     // 發送開關命令
                                swal(`辨識到手勢 ${index}, 開啟 ${
                                    this.data602[index-1].name}！`," ",
                                          "success",{button: false,timer: 2000});
                            }else{
```

```
                        rtc.sendMessage(this.config.macList.mac_602,`{CD1=${
                            this.data602[index-1].val},D1=?}`);     // 發送開關命令
                        swal(` 辨識到手勢 ${index}, 關閉 ${
                            this.data602[index-1].name} ! `," ",
                                "success",{button: false,timer: 2000});
                    }
                }
                setTimeout(() => {
                    gestureThrottle = true
                }, 3000);
            }
        }else{
            count[gesture] = 1
        }
    }
    setTimeout(() => {
        resultThrottle = true
    }, 500);
  }
 }
}
```

(2) 語音辨識演算法。智慧家居系統透過 Ajax 介面將音訊資料傳遞給百度語音辨識介面進行語音辨識，百度語音辨識介面是透過單次推理介面呼叫的。百度語音辨識介面的參數請參考表 4.11。前端應用處理範例如下：

```
/*****************************************************************************************
* 名稱：speechRecognition()
* 功能：按一下「開始錄音」按鈕開始錄製音訊，再次按一下該按鈕可將錄製的音訊發送到後端進行辨識，
        辨識結果將顯示在頁面中，系統根據辨識結果對相應的家居裝置進行開關操作
*****************************************************************************************/
speechRecognition() {
    if (this.recordVal == ' 開始錄音 ') {
        this.recordVal = ' 結束錄音 '
        // 開始錄音
        recorder.start().then(() => {
        }, (error) => {
            // 出錯了
```

```
            console.log(`${error.name} : ${error.message}`);
        });

    } else {
        this.recordVal = '開始錄音'
        //結束錄音
        recorder.stop();
        let formData = new FormData();
        formData.set('file_name',recorder.getWAVBlob(),'audio.wav');
        formData.append('param_data', JSON.stringify({
            "APP_ID": this.config.user.baidu_id,
            "API_KEY": this.config.user.baidu_apikey,
            "SECRET_KEY": this.config.user.baidu_secretkey
        }));
        $.ajax({
            url: this.config.user.edge_addr + this.linkData[2],
            method: 'POST',
            processData: false,                    //必需的
            contentType: false,                    //必需的
            dataType: 'json',
            data: formData,
            headers: { 'X-CSRFToken': this.getCookie('csrftoken') },
            success: result => {
                if(result.code == 200){
                    this.recordRsult = result.result_data
                    if(result.result_data.indexOf('開啟窗簾') > -1){
                        swal(`辨識到開啟窗簾！窗簾開啟中...`," ",
                            "success",{button: false,timer: 2000});
                        rtc.sendMessage(this.config.macList.mac_602,'{OD1=4,D1=?}');
                    }
                    if(result.result_data.indexOf('關閉窗簾') > -1){
                        swal(`辨識到關閉窗簾！窗簾關閉中...`," ",
                            "success",{button: false,timer: 2000});
                        rtc.sendMessage(this.config.macList.mac_602,'{CD1=4,D1=?}');
                    }
                    if(result.result_data.indexOf('開啟風扇') > -1){
                        swal(`辨識到開啟風扇！風扇開啟中...`," ",
                            "success",{button: false,timer: 2000});
                        rtc.sendMessage(this.config.macList.mac_602,'{OD1=8,D1=?}');
```

```
            }
            if(result.result_data.indexOf('關閉風扇') > -1){
                swal(`辨識到關閉風扇！風扇關閉中...`," ",
                    "success",{button: false,timer: 2000});
                rtc.sendMessage(this.config.macList.mac_602,'{CD1=8,D1=?}');
            }
            if(result.result_data.indexOf('開啟燈光') > -1 &&
                                result.result_data.indexOf('燈') > -1){
                swal(`辨識到開啟燈光！LED燈開啟中...`," ",
                    "success",{button: false,timer: 2000});
                rtc.sendMessage(this.config.macList.mac_602,'{OD1=48,D1=?}');
            }
            if(result.result_data.indexOf('關閉燈光') > -1 &&
                                result.result_data.indexOf('燈') > -1){
                swal(`辨識到關閉燈光！LED燈關閉中...`," ",
                    "success",{button: false,timer: 2000});
                rtc.sendMessage(this.config.macList.mac_602,'{CD1=48,D1=?}');
            }
            if(result.result_data.indexOf('開啟空調') > -1){
                swal(`辨識到開啟空調！空調開啟中...`," ",
                    "success",{button: false,timer: 2000});
                rtc.sendMessage(this.config.macList.mac_602,'{OD1=64,D1=?}');
            }
            if(result.result_data.indexOf('關閉空調') > -1){
                swal(`辨識到關閉空調！空調關閉中...`," ",
                    "success",{button: false,timer: 2000});
                rtc.sendMessage(this.config.macList.mac_602,'{CD1=64,D1=?}');
            }
            if(result.result_data.indexOf('開啟加濕器') > -1){
                swal(`辨識到開啟加濕器！加濕器開啟中...`," ",
                    "success",{button: false,timer: 2000});
                rtc.sendMessage(this.config.macList.mac_602,'{OD1=128,D1=?}');
            }
            if(result.result_data.indexOf('關閉加濕器') > -1){
                swal(`辨識到關閉加濕器！加濕器關閉中...`," ",
                    "success",{button: false,timer: 2000});
                rtc.sendMessage(this.config.macList.mac_602,'{CD1=128,D1=?}');
            }
        }else{
```

```
                swal(`控制命令錯誤！`," ","error",{button: false,timer: 2000});
            }
        },
        error: function(error){
            console.log(error);
            swal(`語音辨識失敗！`," ","error",{button: false,timer: 2000});
        }
    });
  }
}
```

（3）人體檢測演算法。當光柵感測器警告時，智慧家居系統透過 EventSource 介面獲取處理後的視訊流，透過即時推理介面呼叫人體檢測演算法來辨識視訊流中的人體，傳回 base64 編碼的結果影像和結果資料。前端應用處理範例如下：

```
/*************************************************************************************
* 名稱：getVisitor
* 功能：當光柵感測器警告後呼叫人體檢測演算法進行檢測，並拍照儲存，每隔 15 s 拍照一次
*************************************************************************************/
getVisitor(){
    imgData && imgData.close()
    let resultThrottle = true                               //設定節流結果
    //請求視訊資源
    imgData = new EventSource(this.config.user.edge_addr + this.linkData[3])
    console.log(this.config.user.edge_addr + this.linkData[3]);
    //對視訊資源傳回的資料進行處理
    imgData.onmessage = res => {
        let {result_image} = JSON.parse(res.data)
        this.homeVideoSrc = `data:image/jpeg;base64,${result_image}`
        let {result_data} = JSON.parse(res.data)
        //對辨識到的人體進行判斷，每隔 15 s 拍照一次
        if(result_data.obj_num > 0 && resultThrottle){
            console.log(result_data);
            resultThrottle = false
            this.visitorRecord.unshift({          //將檢測到的人體影像增加到記錄清單中
                image: this.homeVideoSrc,
                time: new Date().toLocaleTimeString()
```

```
            })
            cocoMessage.error(` 檢測到人體，請前往監控記錄頁面查看！`)
            setTimeout(() => {
                resultThrottle = true
            }, 15000);
        }
    }
}
```

4）人體檢測演算法介面設計

```
##########################################################################################
# 檔案：person_detection.py
# 說明：人體檢測
##########################################################################################
from PIL import Image,ImageDraw,ImageFont
import numpy as np
import cv2 as cv
import os
import json
import base64
c_dir = os.path.split(os.path.realpath(__file__))[0]

class PersonDetection(object):
    def __init__(self, model_path="models/person_detection"):
        self.model_path = model_path
        self.person_model = PersonDet()
        self.person_model.init(self.model_path)

    def image_to_base64(self, img):
        image = cv.imencode('.jpg', img, [cv.IMWRITE_JPEG_QUALITY, 60])[1]
        image_encode = base64.b64encode(image).decode()
        return image_encode

    def base64_to_image(self, b64):
        img = base64.b64decode(b64.encode('utf-8'))
        img = np.asarray(bytearray(img), dtype="uint8")
        img = cv.imdecode(img, cv.IMREAD_COLOR)
        return img
```

```
def draw_pos(self, img, objs):
    img_rgb = cv.cvtColor(img, cv.COLOR_BGR2RGB)
    pilimg = Image.fromarray(img_rgb)
    # 建立 ImageDraw 繪圖類別
    draw = ImageDraw.Draw(pilimg)
    # 設定字型
    font_size = 20
    font_path = c_dir+"/../../font/wqy-microhei.ttc"
    font_hei = ImageFont.truetype(font_path, font_size, encoding="utf-8")

    for obj in objs:
        loc = obj["location"]
        draw.rectangle((loc["left"], loc["top"], loc["left"]+loc["width"],
                    loc["top"]+loc["height"]), outline='green',width=2)
        msg =  "%.2f"%obj["score"]
        draw.text((loc["left"], loc["top"]-font_size*1), msg, (0, 255, 0), font=font_hei)
    result = cv.cvtColor(np.array(pilimg), cv.COLOR_RGB2BGR)
    return result
def inference(self, image, param_data):
    #code：辨識成功傳回 200
    #msg：相關提示訊息
    #origin_image：原始影像
    #result_image：處理之後的影像
    #result_data：結果資料
    return_result = {'code': 200, 'msg': None, 'origin_image':
                None, 'result_image': None, 'result_data': None}

    # 即時視頻界面：@__app.route('/stream/<action>')
    #image：攝影機即時傳遞過來的影像
    #param_data：必須為 None
    result = self.person_model.detect(image)
    result = json.loads(result)
    if result["code"] == 200 and result["result"]["obj_num"] > 0:
        r_image = self.draw_pos(image, result["result"]["obj_list"])
    else:
        r_image = image
    return_result["code"] = result["code"]
    return_result["msg"] = result["msg"]
    return_result["result_image"] = self.image_to_base64(r_image)
```

```
        return_result["result_data"] = result["result"]
        return return_result

#單元測試，如果處理類別中引用了檔案，則在單元測試中要修改檔案路徑
if __name__=='__main__':

    from persondet import PersonDet
    #建立視訊捕捉物件
    cap=cv.VideoCapture(0)
    if cap.isOpened()!=1:
        pass
    #迴圈獲取影像、處理影像、顯示影像
    while True:
        ret,img=cap.read()
        if ret==False:
            break
        #建立影像處理物件
        img_object=PersonDetection(c_dir+'/../../models/person_detection')
        #呼叫影像處理函式對影像進行加工處理
        result=img_object.inference(img,None)
        frame = img_object.base64_to_image(result["result_image"])

        #影像顯示
        cv.imshow('frame',frame)
        key=cv.waitKey(1)
        if key==ord('q'):
            break
    cap.release()
    cv.destroyAllWindows()
else :
    from .persondet import PersonDet
```

5）語音辨識演算法介面設計

```
##########################################################################################
#檔案：baidu_speech_recognition.py
#說明：語音辨識
##########################################################################################
import os
```

```
import wave
import numpy as np
from aip import AipSpeech
import ffmpeg
import tempfile

class BaiduSpeechRecognition(object):
    def __init__(self):
        pass

    def __check_wav_file(self,filePath):
        # 讀取 wav 檔案
        wave_file = wave.open(filePath, 'r')
        # 獲取檔案的每秒顯示畫面和通道
        frame_rate = wave_file.getframerate()
        channels = wave_file.getnchannels()
        wave_file.close()
        if frame_rate == 16000 and channels == 1:
            return True
            #feature_path=filePath
        else:
            return False

    def inference(self, wave_data, param_data):
        #code：辨識成功傳回 200
        #msg：相關提示訊息
        #origin_image：原始影像
        #result_image：處理之後的影像
        #result_data：結果資料
        return_result = {'code': 200, 'msg': None, 'origin_image': None,
                         'result_image': None, 'result_data': None}

        # 應用請求介面：@__app.route('/file/<action>', methods=["POST"])
        #wave_data：應用傳遞過來的資料（根據實際應用可能為影像、音訊、視訊、文字）：
        # 語音資料，格式為 wav，每秒顯示畫面為 16000
        #param_data：應用傳遞過來的參數，不能為空
        if param_data != None:
            fd, path = tempfile.mkstemp()
            try:
```

```
                with os.fdopen(fd, 'wb') as tmp:
                    tmp.write(wave_data)
                if not self.__check_wav_file(path):
                    fd2, path2 = tempfile.mkstemp()
                    ffmpeg.input(path).output(path2, ar=16000).run()
                    os.remove(path)
                    path = path2
                f = open(path, "rb")
                f.seek(4096)
                pcm_data = f.read()
                f.close()
            finally:
                os.remove(path)

        # 呼叫百度語音辨識介面，透過以下使用者金鑰連接百度伺服器
        #APP_ID：百度應用 ID
        #API_KEY：百度 API_KEY
        #SECRET_KEY：百度使用者金鑰
        client = AipSpeech(param_data['APP_ID'], param_data['API_KEY'], param_data['SECRET_
KEY'])
        # 語音檔案的格式為 pcm，每秒顯示畫面為 16000，dev_pid 為普通話（純中文辨識）
        response = client.asr(pcm_data,'pcm', 16000, {'dev_pid': 1537,})

        # 處理伺服器傳回結果
        if response['err_msg']=='success.':
            return_result["code"] = 200
            return_result["msg"] = " 語音辨識成功！ "
            return_result["result_data"] = response['result'][0]
        else:
            return_result["code"] = 500
            return_result["msg"] = response['err_msg']
        return return_result

# 單元測試，如果處理類別中引用了檔案，則在單元測試中要修改檔案路徑
if __name__=='__main__':
    # 建立音訊處理物件
    test = BaiduSpeechRecognition()
    param_data = {"APP_ID":"12345678", "API_KEY":"12345678", "SECRET_KEY":"12345678"}
    with open("./test.wav", "rb") as f:
```

```
        wdat = f.read()
        result = test.inference(wdat, param_data)
    print(result["result_data"])
```

6）手勢辨識演算法介面設計

```
###################################################################################
# 檔案：handpose_detection.py
# 說明：手勢辨識
###################################################################################
from PIL import Image,ImageDraw,ImageFont
import numpy as np
import cv2 as cv
import os
import json
import base64
c_dir = os.path.split(os.path.realpath(__file__))[0]

class HandposeDetection(object):
    def __init__(self, model_path="models/handpose_detection"):
        self.model_path = model_path
        self.handpose_model = HandDetector()
        self.handpose_model.init(self.model_path)

    def image_to_base64(self, img):
        image = cv.imencode('.jpg', img, [cv.IMWRITE_JPEG_QUALITY, 60])[1]
        image_encode = base64.b64encode(image).decode()
        return image_encode

    def base64_to_image(self, b64):
        img = base64.b64decode(b64.encode('utf-8'))
        img = np.asarray(bytearray(img), dtype="uint8")
        img = cv.imdecode(img, cv.IMREAD_COLOR)
        return img

    def draw_pos(self, img, objs):
        img_rgb = cv.cvtColor(img, cv.COLOR_BGR2RGB)
        pilimg = Image.fromarray(img_rgb)
        # 建立 ImageDraw 繪圖類別
```

```
        draw = ImageDraw.Draw(pilimg)
        #設定字型
        font_size = 20
        font_path = c_dir+"/../../font/wqy-microhei.ttc"
        font_hei = ImageFont.truetype(font_path, font_size, encoding="utf-8")

        for obj in objs:
            loc = obj["location"]
            draw.rectangle((loc["left"], loc["top"], loc["left"]+loc["width"],
                        loc["top"]+loc["height"]), outline='green',width=2)
            msg = obj["name"]+": %.2f"%obj["score"]
            draw.text((loc["left"], loc["top"]-font_size*1), msg, (0, 255, 0),
font=font_hei)

            color1 = (10, 215, 255)
            color2 = (255, 115, 55)
            color3 = (5, 255, 55)
            color4 = (25, 15, 255)
            color5 = (225, 15, 55)
            marks = obj["mark"]
            for j in range(len(marks)):
                kp = obj["mark"][j]

                draw.ellipse(((kp["x"]-4, kp["y"]-4), (kp["x"]+4,kp["y"]+4)),
                         fill=None,outline=(255,0,0),width=2)
                color = (color1,color2,color3,color4,color5)
                ii = j //4
                if  j==0 or j / 4 != ii:
                    draw.line(((marks[j]["x"],marks[j]["y"]),(marks[j+1]["x"],
                           marks[j+1]["y"])), fill=color[ii],width=2)

            draw.line(((marks[0]["x"],marks[0]["y"]),(marks[5]["x"],
                    marks[5]["y"])), fill=color[1],width=2)
            draw.line(((marks[0]["x"],marks[0]["y"]),(marks[9]["x"],
                    marks[9]["y"])), fill=color[2],width=2)
            draw.line(((marks[0]["x"],marks[0]["y"]),(marks[13]["x"],
                    marks[13]["y"])), fill=color[3],width=2)
            draw.line(((marks[0]["x"],marks[0]["y"]),(marks[17]["x"],
                    marks[17]["y"])), fill=color[4],width=2)
```

```
        result = cv.cvtColor(np.array(pilimg), cv.COLOR_RGB2BGR)
        return result

    def inference(self, image, param_data):
        #code：辨識成功傳回 200
        #msg：相關提示訊息
        #origin_image：原始影像
        #result_image：處理之後的影像
        #result_data：結果資料
        return_result = {'code': 200, 'msg': None, 'origin_image': None,
                         'result_image': None, 'result_data': None}

        # 即時視頻界面：@__app.route('/stream/<action>')
        #image：攝影機即時傳遞過來的影像
        #param_data：必須為 None
        result = self.handpose_model.detect(image)
        result = json.loads(result)
        if result["code"] == 200 and result["result"]["obj_num"] > 0:
            r_image = self.draw_pos(image, result["result"]["obj_list"])
        else:
            r_image = image
        return_result["code"] = result["code"]
        return_result["msg"] = result["msg"]
        return_result["result_image"] = self.image_to_base64(r_image)
        return_result["result_data"] = result["result"]
        return return_result

# 單元測試，如果處理類別中引用了檔案，則在單元測試中要修改檔案路徑
if __name__=='__main__':

    from handpose import HandDetector
    # 建立視訊捕捉物件
    cap=cv.VideoCapture(0)
    if cap.isOpened()!=1:
        pass
    # 迴圈獲取影像、處理影像、顯示影像
    while True:
        ret,img=cap.read()
```

```
        if ret==False:
            break
        # 建立影像處理物件
        img_object=HandposeDetection(c_dir+'/../../models/handpose_detection')
        # 呼叫影像處理函式對影像進行加工處理
        result=img_object.inference(img,None)
        frame = img_object.base64_to_image(result["result_image"])

        # 影像顯示
        cv.imshow('frame',frame)
        key=cv.waitKey(1)
        if key==ord('q'):
            break
    cap.release()
    cv.destroyAllWindows()
else :
    from .handpose import HandDetector
```

5.1.2 開發步驟與驗證

5.1.2.1 專案部署

1）硬體部署

見 4.1.2.1 節。

2）專案部署

（1）執行 MobaXterm 工具，透過 SSH 登入到邊緣計算閘道。

（2）在 SSH 終端執行以下命令，建立開發專案目錄。

```
$ mkdir -p ~/aiedge-exp
```

（3）透過 SSH 將本專案開發專案程式和 aicam 專案套件上傳到 ~/aiedge-exp 目錄下，並採用 unzip 命令進行解壓縮。

```
$ unzip edge_smarthome.zip
$ unzip aicam.zip -d edge_smarthome
```

（4）修改專案設定檔 static/edge_smarthome/js/config.js 內的智雲帳號、百度帳號、硬體位址、邊緣服務位址等資訊，範例如下：

```
user: {
    id: '12345678',                                   // 智雲帳號
    key: '12345678',                                  // 智雲金鑰
    addr: 'wss://api.zhiyun360.com:28090',            // 智雲端服務位址（呼叫錄音需要 HTTPS 連結）
    edge_addr: 'https://192.168.100.200:1446',        // 邊緣服務位址（呼叫錄音需要 HTTPS 連結）
    baidu_id: '12345678',                             // 百度應用 ID
    baidu_apikey: '12345678',                         // 百度應用 APIKEY
    baidu_secretkey: '12345678',                      // 百度應用 SECREKEY
},

// 定義本機存放區參數（MAC 位址）
    macList: {
    mac_601: '01:12:4B:00:E3:7D:D6:64',               //Sensor-A 擷取類感測器
    mac_602: '01:12:4B:00:27:22:AC:4E',               //Sensor-B 控制類感測器
    mac_603: '01:12:4B:00:E5:24:1F:F1',               //Sensor-C 保全類感測器
},
```

（5）透過 SSH 將修改好的檔案上傳到邊緣計算閘道。

3）專案執行

（1）在 SSH 終端輸入以下命令執行開發專案。

```
$ cd ~/aiedge-exp/edge_smarthome
$ chmod 755 start_aicam.sh
$ conda activate py36_tf114_torch15_cpu_cv345     //Ubuntu 20.04 作業系統下需要切換環境
$ ./start_aicam.sh
// 開始執行指令稿
* Serving Flask app "start_aicam" (lazy loading)
* Environment: production
   WARNING: Do not use the development server in a production environment.
   Use a production WSGI server instead.
* Debug mode: off
* Running on http://0.0.0.0:4001/ (Press CTRL+C to quit)
```

（2）在使用者端或邊緣計算閘道端開啟 Chrome 瀏覽器，輸入專案頁面位址 https:// 192.168.100.200:1446/static/edge_smarthome/index.html，即可查看專案內容。

5.1.2.2 智慧家居系統開發驗證

1）智慧物聯

（1）應用設定。第一次登入智慧家居系統後，按一下主頁右上方的「設定」按鈕可設定相應的參數（預設情況下會讀取 static/edge_smarthome/js/config.js 內的初始配置），如智雲帳號、節點位址、百度 AI 帳號等。設定參數後，在智雲帳號介面按一下「連接」按鈕即可連接到資料服務，連接成功後會彈出訊息提示。智慧家居系統的應用設定頁面如圖 5.6 所示。

▲ 圖 5.6　智慧家居系統的應用設定介面

（2）應用互動。在智慧家居系統的首頁可以看到智慧家居硬體的資料，包括 Sensor-A 擷取類感測器的資料（如溫度、濕度、光照度、空氣品質、大氣壓力），Sensor-B 控制類感測器的資料（窗簾、風扇、LED 燈、空調、加濕器），以及 Sensor-C 保全類感測器的資料〔如人體、霍爾（門磁）、火焰、瓦斯、光柵〕，如圖 5.7 所示。

▲ 圖 5.7 智慧家居系統的首頁（手動模式）

Sensor-A 擷取類感測器在預設情況下每 30 s 更新一次資料，Sensor-C 保全類感測器在發生警示事件時每隔 3 s 更新一次資料，首頁的資料會相應地進行更新。Sensor-B 控制類感測器可透過按一下首頁中的相應圖示來進行開關操作。

（3）模式控制。按一下首頁右下角的模式按鈕可切換手動模式和自動模式。當設定為自動模式時，AI 辨識功能將不可用，系統可根據感測器設定的設定值自動控制相關的硬體裝置，如圖 5.8 所示

▲ 圖 5.8 智慧家居系統的首頁（自動模式）

在自動模式下，智慧家居系統可以根據擷取的環境資訊自動調節家居裝置。舉例來說，當實際溫度高於設定的溫度設定值時，就會開啟空調（用於降溫）；當溫度低於設定的溫度設定值時，就會關閉空調（停止降溫）。對於家居環境的光照度、空氣品質，智慧家居系統也可以根據實際的光照度、濕度、空氣品質，來自動調節相應的窗簾、加濕器和風扇等家居裝置。

2）手勢互動

（1）在攝影機的視線範圍中擺好手勢進行手勢辨識，當成功辨識到手勢後，可對相應的家居裝置操作。舉例來說，當辨識到手勢 5 時，則對加濕器操作，如圖 5.9 所示。

（2）在成功辨識手勢後，等待 3 s 後才會進行下一次的手勢辨識，如果之前的某個家居裝置是開啟狀態，再次辨識到相應的手勢時，則會關閉該家居裝置。舉例來說，再次辨識到手勢 5 時，就關閉加濕器，如圖 5.10 所示。

3）語音互動

（1）按一下智慧家居系統首頁右下角的麥克風圖表即可開始錄製語音，並透過語音辨識來控制家居裝置。舉例來說，可透過語音「開啟窗簾」來控制步進馬達反轉，從而開啟窗簾，如圖 5.11 所示。

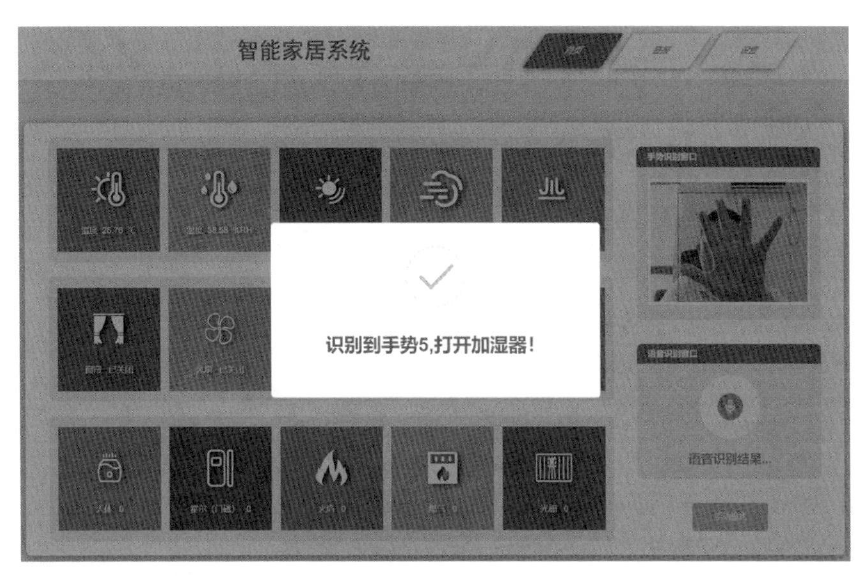

▲ 圖 5.9 根據手勢對加濕器操作（開啟加濕器）

▲ 圖 5.10 根據手勢對加濕器操作（關閉加濕器）

4）入侵監測

當使用不透光的卡片穿過光柵感測器時，光柵感測器會警告（上報狀態 1，光柵感測器會每 3 s 上報一次狀態 1），此時智慧家居系統會通過人體檢測算法進行人體檢測。當辨識到人體時，智慧家居系統會彈出警示訊息，並將影像儲存到警告頁面。當光柵感測器一直處於警告狀態時，智慧家居系統會每隔 15 s 檢測一次人體，並儲存相應的影像。智慧家居系統的警告介面如圖 5.12 所示，智慧家居系統在檢測到人體入侵時抓拍的畫面如圖 5.13 所示。

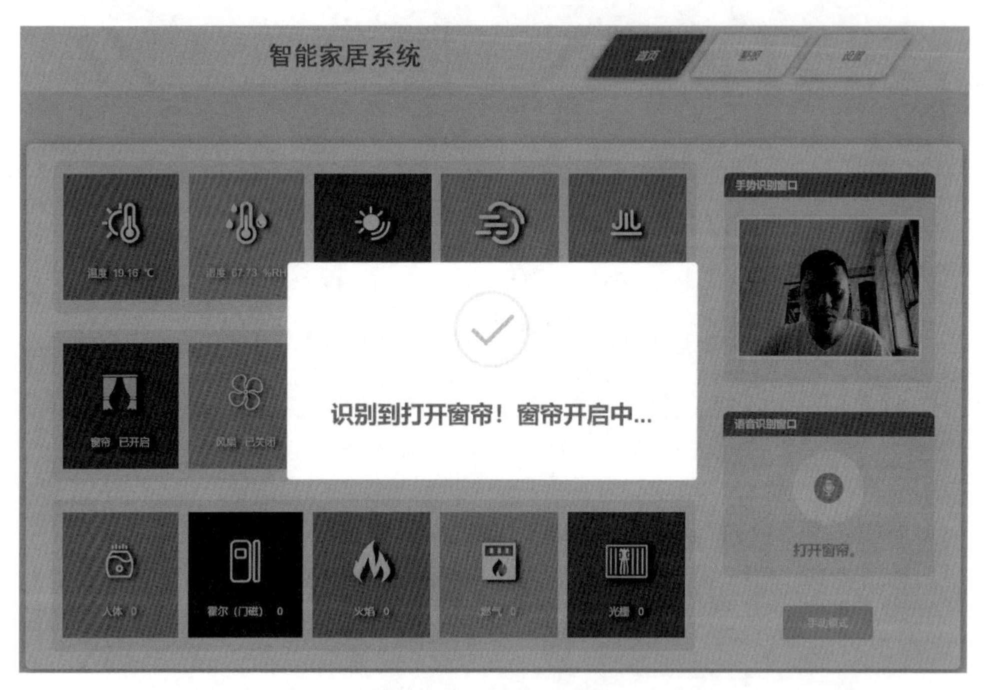

▲ 圖 5.11 開啟窗簾

▲ 圖 5.12 智慧家居系統的警告介面

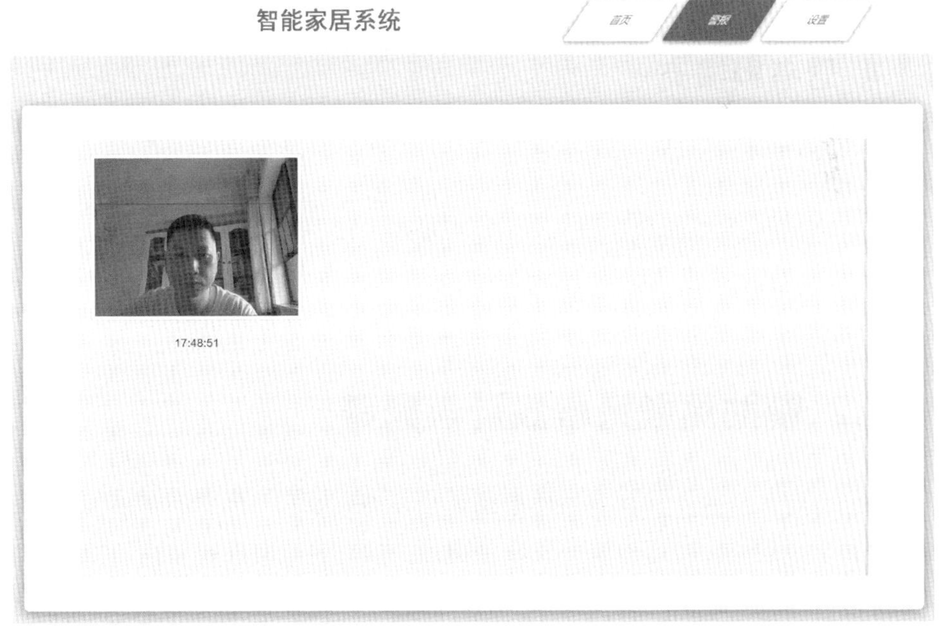

▲ 圖 5.13 智慧家居系統在檢測到人體入侵時抓拍的畫面

如果採用的是利用虛擬模擬建立的光柵感測器，光柵感測器的狀態設定為 1，表示當前處於光柵警告狀態，如圖 5.14 所示。

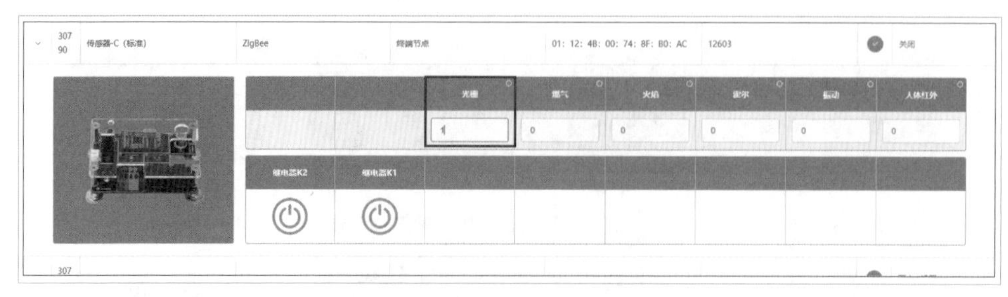

▲ 圖 5.14 將虛擬模擬建立的光柵感測器的狀態設定為 1

5.1.3 本節小結

本節首先分析了智慧家居系統的需求和發展趨勢；然後介紹了智慧家居系統的系統框架和系統功能，在此基礎上完成了硬體設計和通訊設計；接著借助 AiCam 平臺舉出了智慧家居系統的開發流程、演算法互動、人體檢測演算法、語音辨識介面設計；最後對智慧家居系統進行了驗證。

5.1.4 思考與擴充

（1）簡述智慧家居系統的發展趨勢。

（2）機器視覺技術在智慧家居系統中是如何應用的？

（3）簡述智慧家居系統的開發過程。

5.2 輔助駕駛系統設計與開發

輔助駕駛系統旨在為駕駛員提供額外的安全和便利，該系統透過感測器、電腦視覺、控制系統和資料處理等技術來輔助駕駛員駕駛車輛。輔助駕駛系統可在不同的操控等級下執行，從簡單的輔助駕駛到更高級的自動駕駛。

本節的基礎知識如下：

- 了解輔助駕駛系統的應用場景、需求分析。
- 掌握輔助駕駛系統的框架設計、功能設計、硬體設計。
- 掌握基於手勢辨識、語音辨識和駕駛行為檢測的輔助駕駛系統開發。

5.2.1 原理分析與開發設計

5.2.1.1 輔助駕駛系統框架分析

1）需求分析

輔助駕駛系統是一種整合了感測器、電腦視覺、深度學習和控制系統等技術的汽車安全技術，旨在提高駕駛的安全性、舒適性和效率。輔助駕駛系統涉及的主要技術如下：

（1）感測器技術。輔助駕駛系統需要透過多種感測器來感知車輛周圍的環境，主要的感測器包括：

- 雷達：用於檢測與其他車輛、行人和障礙物的相對位置和速度。
- 攝影機：透過電腦視覺技術來辨識道路標識、車道線和交通訊號等。
- 雷射雷達：提供更精確的距離測量。
- 超音波感測器：用於對近距離障礙物進行檢測，如停車輔助系統。
- GPS：用於車輛定位和導航。

（2）資料處理和電腦視覺。輔助駕駛系統使用電腦視覺技術和深度學習演算法來處理感測器資料，用於辨識道路標識、車輛、行人和其他交通參與者，建立車輛周圍的環境模型。這些演算法可以執行在車輛上的高性能計算平臺，如 GPU 或 FPGA。

（3）控制系統。輔助駕駛系統的控制部分負責根據感知資料採取行動，如轉向、煞車和加速，這些控制系統通常包括自動駕駛控制演算法，以及與車輛操控相關的系統（如轉向、煞車、油門）。

（4）高精度地圖。輔助駕駛系統需要使用高精度地圖來輔助定位和路徑規劃，高精度地圖包含了道路的幾何資訊、交通訊號和其他重要的導航資料。

（5）通訊和雲端連接。有些輔助駕駛系統還連接雲端服務器，以獲取即時交通和路況資訊，同時可以與其他車輛進行通訊，以改善交通流量和安全性。

輔助駕駛系統具有多種用途，旨在提高駕駛的安全性、舒適性和效率，主要包括：

（1）自我調整巡航控制系統。自我調整巡航控制系統使用雷達、雷射雷達或攝影機來監測前方車輛的速度和距離，自動調整車輛的速度以保持安全的車距，可提供更舒適的駕駛體驗，減少駕駛員的疲勞。

（2）車道保持輔助系統。車道保持輔助系統使用攝影機來檢測車輛在車道內的位置，並在必要時自動調整方向，以防止車輛意外偏離車道，有助減少事故，特別是在長時間駕駛或疲勞駕駛時。

（3）自動停車系統。自動停車系統利用感測器和攝影機來幫助駕駛員進行平行或垂直停車，可避免停車過程中的刮擦和碰撞，使停車更加方便。

（4）交通擁堵輔助系統。交通擁堵輔助系統可以在交通擁堵時自動控制車輛的加速和煞車，減少駕駛員的壓力，提高交通流暢性。

（5）盲點監測系統。盲點監測系統使用感測器來檢測車輛周圍的盲點區域，當有其他車輛進入盲點時，系統會發出警告，幫助駕駛員避免變道時的事故。

（6）自動緊急煞車系統。自動緊急煞車系統在檢測到潛在碰撞威脅時可以自動剎車，以減少事故的嚴重程度或避免事故。

（7）高速公路輔助駕駛系統。高速公路輔助駕駛系統允許駕駛員在高速公路上使用更高級別的協助工具，如自動駕駛、車道變更輔助等。

（8）輔助停車系統。除了自動停車，一些車型還配備了輔助停車系統，可幫助駕駛員找到合適的停車位，並輔助駕駛員完成停車。

（9）駕駛員監控系統。駕駛員監控系統可監測駕駛員的注意力和疲勞程度，並在必要時提醒駕駛員休息或警告駕駛員注意安全。

（10）交通號誌辨識系統。交通號誌辨識系統使用攝影機辨識道路上的交通號誌，如限速標識、停車標識和禁止標識，並向駕駛員顯示相關資訊。

2）系統框架

從邊緣計算的角度看，輔助駕駛系統可分為硬體層、邊緣層、應用層，如圖 5.15 所示。

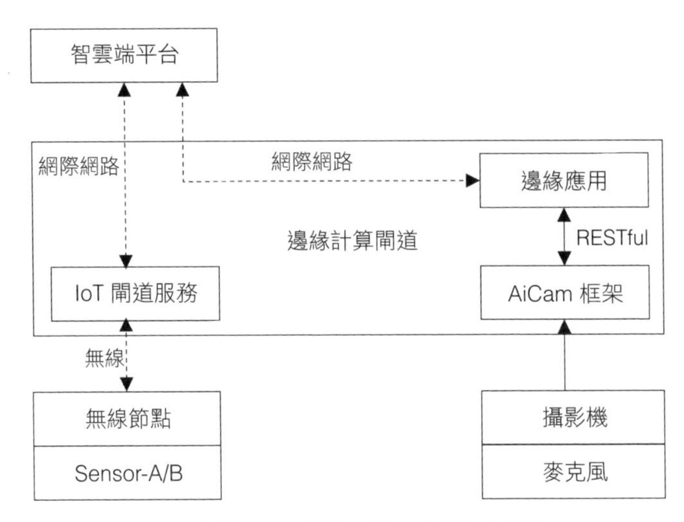

▲ 圖 5.15 輔助駕駛系統的結構

（1）硬體層：無線節點和 Sensor-A 擷取類感測器、Sensor-B 控制類感測器組成了輔助駕駛系統的硬體層。

（2）邊緣層：包括邊緣計算閘道內建 IoT 閘道服務和 AiCam 框架。IoT 閘道服務負責接收和下發無線節點的資料，發送給應用端或將資料發給雲端的智雲端平台。AiCam 框架內建了演算法、模型、視訊串流推送等服務，支援應用層的邊緣計算推理任務。

（3）應用層：透過智雲介面與 IoT 的硬體層互動（預設與雲端的智雲端平台的介面互動），透過 AiCam 的 RESTful 介面與演算法層互動。

5.2.1.2 功能設計

1）功能模組設計

本專案的智慧輔助系統可分為以下功能模組：

（1）智慧物聯：即時顯示車內環境、裝置的資料。

（2）手勢互動：呼叫攝影機即時進行手勢開關車內裝置的操作，在本專案中，手勢 1 表示對車窗進行開關操作、手勢 2 表示對風扇進行開關操作、手勢 3 表示對空調進行開關操作。

（3）語音互動：按一下輔助駕駛系統介面的「語音辨識」按鈕可錄製語音，錄製結束後進行語音辨識，從而開啟 / 關閉車窗、風扇、空調。

（4）駕駛行為檢測：利用攝影機對駕駛員的頭部、面部及眼部動作等進行檢測，針對駕駛員疲勞及注意力分散等危險狀態進行即時預警。

（5）應用設定：對專案的智雲帳號、裝置位址、百度帳號進行設定。

2）功能框架設計

輔助駕駛系統的功能框架如圖 5.16 所示。

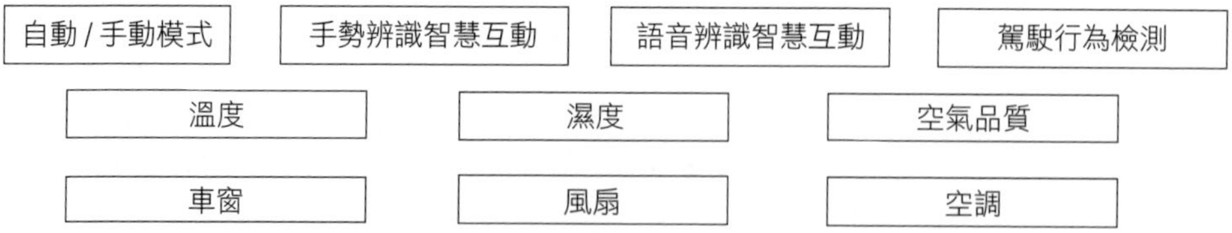

自動 / 手動模式	手勢辨識智慧互動	語音辨識智慧互動	駕駛行為檢測
溫度	濕度	空氣品質	
車窗	風扇	空調	

▲ 圖 5.16 輔助駕駛系統的功能框架

5.2.1.3 硬體與通訊設計

1）硬體設計

輔助駕駛系統採用 LiteB 無線節點、Sensor-A 擷取類感測器、Sensor-B 控制類感測器來完成一套輔助駕駛系統硬體的架設，也可以透過虛擬模擬軟體來建立一個輔助駕駛系統專案，並增加對應的感測器。

（1）Sensor-A 擷取類感測器：主要用於檢測溫濕度、大氣壓力，其虛擬控制平臺如圖 5.3 所示。

（2）Sensor-B 控制類感測器：主要用於控制窗簾（步進馬達）、風扇、LED 燈、空調（繼電器 K1），其虛擬控制平臺如圖 5.4 所示。

2）通訊協定設計

輔助駕駛系統的通訊協定如表 5.2 所示。

▼ 表 5.2 輔助駕駛系統的通訊協定

節點名稱	TYPE	參 數	屬 性	許可權	說 明
Sensor-A 擷取類感測器	601	A0	溫度	R	溫度值，浮點型，精度為 0.1，範圍為 − 40.0 ～ 105.0，單位為℃
		A1	濕度	R	濕度值，浮點型，精度為 0.1，範圍為 0 ～ 100.0，單位為 %RH
		A2	光照度	R	光照度值，浮點型，精度為 0.1，範圍為 0 ～ 65535.0，單位為 lx
		A3	空氣品質	R	空氣品質值，表徵空氣污染程度，整數，範圍為 0 ～ 20000，單位為 ppm
		A4	大氣壓力	R	大氣壓力值，浮點型，精度為 0.1，範圍為 800.0 ～ 1200.0，單位為 hPa

（續表）

節點名稱	TYPE	參 數	屬 性	許可權	說 明
Sensor-A 擷取類感測器	601	A5	跌倒狀態	R	透過三軸感測器計算出跌倒狀態，0 表示未跌倒，1 表示跌倒
		A6	距離	R	距離值，浮點型，精度為 0.1，範圍為 10.0 ～ 80.0，單位為 cm
		D0(OD0/CD0)	上報狀態	R/W	D0 的 Bit0 ～ Bit7 分別代表 A0 ～ A7 的上報狀態，1 表示主動上報，0 表示不上報
		D1(OD1/CD1)	繼電器	R/W	D1 的 Bit6 ～ Bit7 分別代表繼電器 K1、K2 的開關狀態，0 表示斷開，1 表示吸合
		V0	上報時間間隔	R/W	A0 ～ A7 和 D1 的主動上報時間間隔，預設為 30，單位為 s
Sensor-B 控制類感測器	602	D1(OD1/CD1)	RGB	R/W	D1 的 Bit0 ～ Bit1 代表 RGB 三色燈的顏色狀態，00 表示關、01 表示紅色、10 表示綠色、11 表示藍色
		D1(OD1/CD1)	步進馬達	R/W	D1 的 Bit2 表示步進馬達的正反轉動狀態，0 表示正轉、1 表示反轉
		D1(OD1/CD1)	風扇 / 蜂鳴器	R/W	D1 的 Bit3 表示風扇 / 蜂鳴器的開關狀態，0 表示關閉，1 表示開啟
		D1(OD1/CD1)	LED	R/W	D1 的 Bit4 ～ Bit5 表示 LED1、LED2 的開關狀態，0 表示關閉，1 表示開啟
		D1(OD1/CD1)	繼電器	R/W	D1 的 Bit6 ～ Bit7 表示繼電器 K1、K2 的開關狀態，0 表示斷開，1 表示吸合
		V0	上報間隔	R/W	A0 ～ A7 和 D1 的循環上報時間間隔

3）硬體通訊設計

前端應用中的硬體控制部分透過智雲 ZCloud API 連接到硬體系統，前端應用處理範例如下：

```
/********************************************************************************
* 名稱：getConnect()
* 功能：建立即時連接服務，監聽資料並進行處理
********************************************************************************/
getConnect() {                                          // 建立連接服務的函式
    rtc = new WSNRTConnect(this.config.user.id, this.config.user.key)
    rtc.setServerAddr(this.config.user.addr);
    rtc.connect();
    rtc.onConnect = () => {                             // 連接成功回呼函式
        this.onlineBtn = '斷開'
        setTimeout(() => {
            if (this.onlineBtn == '斷開') {
                cocoMessage.success(`資料服務連接成功！查詢資料中...`)
                // 發起資料查詢
                rtc.sendMessage(this.config.macList.mac_601, this.config.sensor.mac_601.
query);
                rtc.sendMessage(this.config.macList.mac_602, this.config.sensor.mac_602.
query);
            }
        }, 200);
    }
    rtc.onConnectLost = () => {                         // 資料服務掉線回呼函式
        this.onlineBtn = '連接'
        cocoMessage.error(`資料服務連接失敗！請檢查網路或 IDKEY...`)
    };

    rtc.onmessageArrive = (mac, dat) => {               // 訊息處理回呼函式
        if (dat[0] == '{' && dat[dat.length - 1] == '}') {
            let its = dat.slice(1, -1).split(',')
            // 截取背景傳回的 JSON 物件（去掉 {} 符號）後，以「,」分割為陣列
            for (let i = 0; i < its.length; i++) {      // 迴圈遍歷陣列的每一個值
                let t = its[i].split("=");              // 將每個值以「=」分割為陣列
                if (t.length != 2) continue;
                //mac_601 擷取類感測器
```

```
                if (mac == this.config.macList.mac_601) {
                    if (t[0] == 'A0') {                                    // 溫度
                        this.data601[0].value = t[1]
                    }
                    if (t[0] == 'A1') {                                    // 濕度
                        this.data601[1].value = t[1]
                    }
                    if (t[0] == 'A3') {                                    //TVOC
                        this.data601[2].value = t[1]
                    }
                }

                //mac_602 控制類感測器
                if (mac == this.config.macList.mac_602) {
                    if (t[0] == 'D1') {
                        if (t[1] & 4) {                                    // 車窗
                            this.data602[0].value = '已開啟'
                        } else {
                            this.data602[0].value = '已關閉'
                        }
                        if (t[1] & 8) {                                    // 風扇
                            this.data602[1].value = '已開啟'
                        } else {
                            this.data602[1].value = '已關閉'
                        }
                        if (t[1] & 64) {                                   // 空調
                            this.data602[2].value = '已開啟'
                        } else {
                            this.data602[2].value = '已關閉'
                        }
                    }
                }

            }
        }
    }
}
*********************************************************************************
* 名稱：control()
```

```
* 功能：根據按一下的模組發送相應的查詢或控制命令
* 參數：index 是對應資料清單的索引
***********************************************************************************
control(index) {
    if (this.ADAS[0].switch) return cocoMessage.error(" 智慧模式下禁止操作 !");
    if (this.onlineBtn == ' 斷開 ') {
        if (this.data602[index].value == ' 已關閉 ') {
            // 發送開關命令
            rtc.sendMessage(this.config.macList.mac_602, `{OD1=${this.data602[index].
val},D1=?}`);
            console.log(this.config.macList.mac_602, `{OD1=${this.data602[index].
val},D1=?}`);
        } else {
            // 發送開關命令
            rtc.sendMessage(this.config.macList.mac_602, `{CD1=${this.data602[index].
val},D1=?}`);
            console.log(this.config.macList.mac_602, `{CD1=${this.data602[index].
val},D1=?}`);
        }
    } else {
        swal(" 資料服務尚未連接！請檢查網路或 IDKEY...", " ", "error", { button: false, timer:
2000 });
    }
},
```

5.2.1.4 專案開發

1）系統框架設計

見 4.1.1.3 節。

2）開發流程

本專案基於 AiCam 平臺開發，開發流程如下：

（1）專案配置見 4.1.1.3 節。

（2）增加模型。輔助駕駛系統用到了手勢辨識深度學習模型，需要在 aicam 專案增加人手檢測模型檔案 models/handpose_detection/handdet.bin 和 models/handpose_detection/handdet.param、手勢辨識模型檔案 models/handpose_detection/handpose.bin 和 models/handpose_detection/handpose.param。

（3）增加演算法。輔助駕駛系統用到了手勢辨識、百度語音辨識介面、百度駕駛行為分析演算法，需要在 aicam 專案增加手勢辨識演算法檔案 algorithm/handpose_detection/handpose_detection.py、百度語音辨識介面檔案 algorithm/baidu_speech_recognition/baidu_speech_ recognition.py、百度駕駛行為分析演算法檔案 algorithm/baidu_driving_behavior_analysis/baidu_ driving_behavior_analysis.py。

（4）增加應用。在 aicam 專案增加演算法專案前端應用 static/edge_autopilot。

3）演算法互動

（1）手勢辨識演算法。輔助駕駛系統採用 EventSource 介面獲取處理後的視訊流，透過即時推理介面呼叫手勢辨識演算法來辨識視訊流中的手勢，傳回 base64 編碼的結果影像和結果資料。前端應用處理範例如下：

```
/*********************************************************************************
* 名稱：toggleVideo
* 功能：切換視訊功能（在普通視訊流和手勢辨識之間切換）
* 註釋：切換到手勢辨識後，對辨識到的手勢次數進行累計，當累計到 5 次時向相應的裝置
        發送開關命令
*********************************************************************************/
toggleVideo() {
    if (this.ADAS[0].switch) return cocoMessage.error("智慧模式下禁止操作！");

    imgData && imgData.close()
    let resultThrottle = true                  // 將節流結果辨識間隔時間設定為 500 ms
    let gestureThrottle = true                 // 將節流閘門開啟間隔時間設定為 3 s
    let count = {}                             // 累計手勢辨識次數

    let url = null
    console.log(this.ADAS[0].switch);
```

```
    if (this.ADAS[1].switch) {
        this.ADAS[1].switch = false
        url = this.config.user.edge_addr + this.linkData[0]
    } else {
        this.ADAS[1].switch = true
        url = this.config.user.edge_addr + this.linkData[1]
    }
    // 請求視訊資源
    imgData = new EventSource(url)
    console.log(url);
    // 對視訊資源傳回的資料進行處理
    imgData.onmessage = res => {
        let { result_image } = JSON.parse(res.data)
        this.videoSrc = `data:image/jpeg;base64,${result_image}`
        let { result_data } = JSON.parse(res.data)
        // 對辨識到的手勢進行判斷（設定增加間隔不得小於 500 ms 一次）
        if (result_data && resultThrottle && this.ADAS[1].switch) {
            resultThrottle = false
            if (result_data.obj_num > 0 && result_data.obj_list[0].score > 0.70 &&
gestureThrottle) {
                console.log(result_data);
                // 設定變數並賦值辨識到的手勢
                let gesture = result_data.obj_list[0].name
                // 為辨識到的每個手勢設定為物件屬性名稱，初始值為 1。當某個手勢的辨識
                // 次數累計達到 5 次時，向相應的裝置發送開關命令，並清除計數
                if (count[gesture]) {
                    count[gesture] += 1
                    if (count[gesture] == 4 && gesture != 'undefined') {
                        count = {}
                        gestureThrottle = false
                        let index
                        if (gesture == 'one') index = 1
                        if (gesture == 'two') index = 2
                        if (gesture == 'three') index = 3
                        //if (gesture == 'four') index = 4
                        //if (gesture == 'five') index = 5
                        console.log(gesture, index);
                        if (index) {
                            if (this.data602[index - 1].value == '已關閉') {
```

```
                                    // 發送開關命令
                                    rtc.sendMessage(this.config.macList.mac_602,
                                            `{OD1=${this.data602[index - 1].val},D1=?}`);
                                    swal(` 辨識到手勢 ${index}, 開啟 ${this.data602[index - 1]
.name} ! `,
                                        " ", "success", { button: false, timer: 2000 });
                                } else {
                                    // 發送開關命令
                                    rtc.sendMessage(this.config.macList.mac_602,
                                            `{CD1=${this.data602[index - 1].val},D1=?}`);
                                    swal(` 辨識到手勢 ${index}, 關閉 ${this.data602[index - 1]
.name} ! `,
                                        " ", "success", { button: false, timer: 2000 });
                                }
                            }
                            setTimeout(() => {
                                gestureThrottle = true
                            }, 3000);
                        }
                    } else {
                        count[gesture] = 1
                    }
                }
                setTimeout(() => {
                    resultThrottle = true
                }, 500);
            }
        }
},
```

手勢辨識演算法的參數如表 5.3 所示。

▼ 表 5.3 手勢辨識演算法的參數

參 數	範 例
url	"/file/baidu_speech_recognition"
method	'POST'
processData	false
contentType	false
dataType	'json'

data	let config = configData
	let blob = recorder.getWAVBlob();
	let formData = new FormData();
	formData.set('file_name',blob,'audio.wav');
	formData.append('param_data', JSON.stringify({"APP_ID":config.user.baidu_id,
	"API_KEY":config.user.baidu_apikey, "SECRET_KEY":config.user.baidu_secretkey}));

success	function(res){} 內容： return_result = {'code': 200, 'msg': None, 'origin_image': None, 'result_image': None, 'result_data': None} 範例： code/msg：200 表示手勢辨識成功、500 表示手勢辨識失敗 result_data：傳回手勢辨識的文字內容

（2）百度語音辨識介面。輔助駕駛系統透過 Ajax 介面將音訊資料傳遞給百度語音辨識介面進行辨識，百度語音辨識介面是透過單次推理介面呼叫的。前端應用處理範例如下：

```
/*********************************************************************************
* 名稱：speechRecognition()
* 功能：按一下「語音辨識」按鈕開始錄音，再次按一下該按鈕可將錄製的音訊發送到後端進行辨識，
*        在系統的介面顯示辨識結果，並根據辨識的結果對相應的硬體操作
```

```
*********************************************************************************/
speechRecognition() {
    if (!this.ADAS[2].switch) {
        if (this.ADAS[0].switch) return cocoMessage.error("智慧模式下禁止操作!");
        this.ADAS[2].switch = true
        //開始錄音
        recorder.start().then(() => {
            cocoMessage.success("錄音中,請發出命令!再次按一下後結束錄音...");
        }, (error) => {
            //出錯了
            console.log(`${error.name} : ${error.message}`);
        });

    } else {
        if (this.ADAS[0].switch) return
        //結束錄音
        recorder.stop();
        this.ADAS[2].switch = false
        cocoMessage.success("錄音完畢,命令辨識中!");

        let formData = new FormData();
        formData.set('file_name', recorder.getWAVBlob(), 'audio.wav');
        formData.append('param_data', JSON.stringify({
            "APP_ID": this.config.user.baidu_id,
            "API_KEY": this.config.user.baidu_apikey,
            "SECRET_KEY": this.config.user.baidu_secretkey
        }));
        $.ajax({
            url: this.config.user.edge_addr + this.linkData[2],
            method: 'POST',
            processData: false,                         //必需的
            contentType: false,                         //必需的
            dataType: 'json',
            data: formData,
            headers: { 'X-CSRFToken': this.getCookie('csrftoken') },
            success: result => {
                if (result.code == 200) {
                    console.log(result.result_data)
                    if (result.result_data.indexOf('開') > -1 && result.result_data.
```

```
indexOf('窗') > -1) {
                    swal(`車窗開啟中...`, " ", "success", { button: false, timer: 2000 });
                    rtc.sendMessage(this.config.macList.mac_602, '{OD1=4,D1=?}');
                }
                if (result.result_data.indexOf('關') > -1 && result.result_data.
indexOf('窗') > -1) {
                    swal(`車窗關閉中...`, " ", "success", { button: false, timer: 2000 });
                    rtc.sendMessage(this.config.macList.mac_602, '{CD1=4,D1=?}');
                }
                if (result.result_data.indexOf('打') > -1 && result.result_data.
indexOf('扇') > -1) {
                    swal(`風扇開啟中...`, " ", "success", { button: false, timer: 2000 });
                    rtc.sendMessage(this.config.macList.mac_602, '{OD1=8,D1=?}');
                }
                if (result.result_data.indexOf('關') > -1 && result.result_data.
indexOf('扇') > -1) {
                    swal(`風扇關閉中...`, " ", "success", { button: false, timer: 2000 });
                    rtc.sendMessage(this.config.macList.mac_602, '{CD1=8,D1=?}');
                }
                if (result.result_data.indexOf('開') > -1 && result.result_data.
indexOf('空調') > -1) {
                    swal(`空調開啟中...`, " ", "success", { button: false, timer: 2000 });
                    rtc.sendMessage(this.config.macList.mac_602, '{OD1=64,D1=?}');
                }
                if (result.result_data.indexOf('關') > -1 && result.result_data.
indexOf('空調') > -1) {
                    swal(`空調關閉中...`, " ", "success", { button: false, timer: 2000 });
                    rtc.sendMessage(this.config.macList.mac_602, '{CD1=64,D1=?}');
                }
            } else {
                swal(`控制命令錯誤！`, " ", "error", { button: false, timer: 2000 });
            }
        },
        error: function (error) {
            console.log(error);
            swal(`語音辨識失敗！`, " ", "error", { button: false, timer: 2000 });
        }
    });
}
```

```
},

/* 切換駕駛模式 */
toggleModel() {
    clearInterval(this.timer)
    if (!this.ADAS[0].switch) {
        //智慧模式下關閉手勢、語音辨識功能
        this.ADAS[1].switch && this.toggleVideo()
        this.ADAS[2].switch && this.speechRecognition()

        this.ADAS[0].name = '智慧模式'
        this.ADAS[0].switch = true
        swal(`已切換為智慧模式`, "智慧模式下將對使用者駕駛行為進行定時監測 1 次 /15 s",
            "success", { button: false, timer: 2000 });
        //開啟駕駛行為監測功能
        this.driverBehavior()
    } else {
        this.ADAS[0].name = '手動模式'
        this.ADAS[0].switch = false
        swal(`已切換為手動模式`, "手動模式下可透過手勢、語音、按一下圖示控制裝置開關！",
            "success", { button: false, timer: 2000 });
    }
},
```

（3）百度駕駛行為分析演算法。輔助駕駛系統透過 Ajax 介面將影像資料傳遞給百度駕駛行為分析演算法進行辨識，百度駕駛行為分析演算法是透過單次推理介面呼叫的。百度駕駛行為分析演算法的參數如表 5.4 所示。

▼ 表 5.4 百度駕駛行為分析演算法的參數

參數	範例
url	"/file/baidu_driving_behavior_analysis"
method	'POST'
processData	false
contentType	false

參數	範例
dataType	'json'

data	let blob = this.dataURItoBlob(this.homeVideoSrc)
	let formData = new FormData();
	formData.append('file_name', blob, 'image.png');
	formData.append('param_data', JSON.stringify({
	"APP_ID": this.config.user.baidu_id,
	"API_KEY": this.config.user.baidu_apikey,
	"SECRET_KEY": this.config.user.baidu_secretkey
	}));

success	function(res){} 內容： return_result = {'code': 200, 'msg': None, 'origin_image': None, 'result_image': None, 'result_data': None} 範例： code/msg：200/ 辨識成功否則就是辨識失敗。 result_data：演算法傳回辨識後的行為結果。

百度駕駛行為分析演算法每 15 s 獲取一次攝影機即時拍攝的影像，並對駕駛行為進行一次分析。前端應用處理範例如下：

```
/*************************************************************************************
* 名稱：driverBehavior
* 功能：在智慧模式下分析駕駛行為
* 註釋：對使用者駕駛行為進行監測，每 15 s 獲取一次攝影機即時拍攝的影像，並對駕駛行為進行
*         一次分析
*************************************************************************************/
 driverBehavior() {
    // 每 15 s 獲取一次攝影機即時拍攝的影像並進行駕駛行為分析
    this.timer = setInterval(() => {
        // 按一下發起專案結果請求、並對傳回的結果進行相應的處理
        let blob = this.dataURItoBlob(this.videoSrc)
        let formData = new FormData();
        formData.append('file_name', blob, 'image.png');
```

```
formData.append('param_data', JSON.stringify({
    "APP_ID": this.config.user.baidu_id,
    "API_KEY": this.config.user.baidu_apikey,
    "SECRET_KEY": this.config.user.baidu_secretkey
}));
$.ajax({
    url: this.config.user.edge_addr + this.linkData[3],
    method: 'POST',
    processData: false,                      // 必需的
    contentType: false,                      // 必需的
    dataType: 'json',
    data: formData,
    success: res => {
        //console.log(res);
        if (res.code == 200) {
            // 增加監測影像清單
            this.imgList.push({
                image: 'data:image/jpeg;base64,' + res.result_image,
                time: new Date().toLocaleTimeString()
            })

            let behavior = ''
            if(res.result_data.person_num > 0){
                let {attributes} = res.result_data.person_info[0]
                console.log(attributes);
                for (let item in attributes) {
                    if(attributes[item].score > 0.7){
                        if(item == 'both_hands_leaving_wheel') behavior +=
                                                 '雙手離開方向盤、';
                        if(item == 'cellphone') behavior += '使用手機、';
                        if(item == 'eyes_closed') behavior += '閉眼、';
                        if(item == 'head_lowered') behavior += '低頭';
                        if(item == 'no_face_mask') behavior += '未正確佩戴口罩、';
                        if(item == 'not_buckling_up') behavior += '未綁安全帶、';
                        if(item == 'not_facing_front') behavior += '角度未朝前方、';
                        if(item == 'smoke') behavior += '吸煙、';
                        if(item == 'yawning') behavior += '打哈欠、';
                    }
                }
```

```
                    if(behavior){
                        swal('違規駕駛！', `監測到使用者：${behavior} 請規範駕駛！`,
                            "error", { button: false, timer: 3000 });
                    }else{
                        behavior = '規範駕駛！'
                    }
                    behavior = behavior.slice(0, -1) + '!'
                    this.resultList.push({
                        behavior,
                        time: new Date().toLocaleTimeString()
                    })
                }
                // 保持捲軸在最下方
                setTimeout(() => {
                    this.$refs.imgList.scrollTop = this.$refs.imgList.scrollHeight
                    this.$refs.textList.scrollTop = this.$refs.textList.scrollHeight
                }, 0);
            } else {
                swal(res.msg, " ", "error", { button: false, timer: 2000 });
            }
        },
        error: function (error) {
            console.log(error);
            swal('請求失敗！', " ", "error", { button: false, timer: 2000 });
        }
    });
  }, 15000);
},
```

4）駕駛行為分析演算法介面設計

```
##########################################################################################
# 檔案：Driving_behavior_analysis.py
# 說明：駕駛行為分析演算法
##########################################################################################
from PIL import Image, ImageDraw, ImageFont
import numpy as np
import cv2 as cv
import os,sys,time
```

```
import json
import base64
from aip import AipBodyAnalysis

class BaiduDrivingBehaviorAnalysis(object):
    def __init__(self, font_path="font/wqy-microhei.ttc"):
        self.font_path = font_path

    def imencode(self,image_np):
        # 將 JPG 格式的影像編碼為資料流程
        data = cv.imencode('.jpg', image_np)[1]
        return data

    def image_to_base64(self, img):
        image = cv.imencode('.jpg', img, [cv.IMWRITE_JPEG_QUALITY, 60])[1]
        image_encode = base64.b64encode(image).decode()
        return image_encode

    def base64_to_image(self, b64):
        img = base64.b64decode(b64.encode('utf-8'))
        img = np.asarray(bytearray(img), dtype="uint8")
        img = cv.imdecode(img, cv.IMREAD_COLOR)
        return img

    def inference(self, image, param_data):
        #code：辨識成功傳回 200
        #msg：相關提示訊息
        #origin_image：原始影像
        #result_image：處理之後的影像
        #result_data：結果資料
        return_result = {'code': 200, 'msg': None, 'origin_image': None, 'result_image':
None, 'result_data': None}

        # 應用請求介面：@__app.route('/file/<action>', methods=["POST"])
        #image：應用傳遞過來的資料（根據實際應用可能為影像、音訊、視訊、文字）
        #param_data：應用傳遞過來的參數，不能為空
        if param_data != None:
            # 讀取應用傳遞過來的影像
            image = np.asarray(bytearray(image), dtype="uint8")
```

```
image = cv.imdecode(image, cv.IMREAD_COLOR)
#對影像資料進行壓縮，以便網路傳輸。
img = self.imencode(image)

#呼叫百度駕駛行為分析介面，透過以下使用者金鑰連接百度伺服器
#APP_ID：百度應用 ID
#API_KEY：百度 API_KEY
#SECRET_KEY：百度使用者金鑰
client = AipBodyAnalysis(param_data['APP_ID'], param_data['API_KEY'],
                    param_data['SECRET_KEY'])

#沒有參數應用百度駕駛行為分析演算法
response=client.driverBehavior(img)
#應用部分
if "error_msg" in response:
    if response['error_msg']!='SUCCESS':
        return_result["code"] = 500
        return_result["msg"] = "駕駛行為分析介面呼叫失敗！"
        return_result["result_data"] = response
        return return_result
if len(response['person_info']) == 0:
    return_result["code"] = 404
    return_result["msg"] = "沒有檢測到駕駛員！"
    return_result["result_data"] = response
    return return_result
if len(response['person_info'])>0:
    #影像輸入
    img_rgb = cv.cvtColor(image, cv.COLOR_BGR2RGB)  #影像色彩格式轉換
    pilimg = Image.fromarray(img_rgb)    #使用 PIL 讀取影像像素陣列
    draw = ImageDraw.Draw(pilimg)
    #設定字型
    font_size = 25
    font_hei = ImageFont.truetype(self.font_path, font_size, encoding="utf-8")
    #獲取駕駛員位置
    loc=response['person_info'][0]['location']
    #獲取資料
    count = 0
    for key,res in response['person_info'][0]['attributes'].items():
        probability=res['score']
```

```
            #若置信值過小，則丟棄
            if probability<0.5:
                continue
            #繪製矩形外框
            draw.rectangle((int(loc["left"]), int(loc["top"]), (int(loc["left"]) +
                        int(loc["width"])), (int(loc["top"]) + int(loc["height"]))),
            outline='green', width=1)
            #給影像增加文字
            #判斷有沒有打哈欠和閉眼，因為只張嘴可能不一定是打哈欠，
            #也可能是說話，故無法判斷是否疲勞駕駛
            if key == "yawning":
                if "eyes_closed" in response['person_info'][0]['attributes'].
keys():
                    draw.text((loc["left"], loc["top"]+count), '行為:'+"打哈欠，
                            閉眼可能在疲勞駕駛",fill= 'red', font=font_hei)
                else:
                    draw.text((loc["left"], loc["top"]+count), '行為:'+"說話聊天未
違規",
                            fill= 'red', font=font_hei)
                count = count + 20
            #判斷是否閉眼且角度朝前
            elif key == "eyes_closed":
                if "head_lowered" in response['person_info'][0]['attributes'].
keys():
                    draw.text((loc["left"], loc["top"]+count), '行為:'+"閉眼，
                            低頭可能在疲勞駕駛",fill= 'red', font=font_hei)
                elif "not_facing_front"in response['person_info'][0]['attributes'].
keys():
                    draw.text((loc["left"], loc["top"]+count), '行為:'+"閉眼，角度
未朝前
                            可能在疲勞駕駛",fill= 'red', font=font_hei)
                count = count + 20
            #判斷雙手是否離開方向盤
            elif key == "both_hands_leaving_wheel":
                draw.text((loc["left"], loc["top"]+count), '行為:'+"雙手離開方向盤
違規",
                        fill= 'red', font=font_hei)
                count = count + 20
            #判斷是否綁安全帶
```

```
            elif key == "not_buckling_up":
                draw.text((loc["left"], loc["top"]+count), '行為：'+"未綁安全帶",
                        fill= 'red', font=font_hei)
                count = count + 20
            # 判斷是否在打電話
            elif key == "cellphone":
                draw.text((loc["left"], loc["top"]+count), '行為：'+"使用手機，
                        玩手機",fill= 'red', font=font_hei)
                count = count + 20
            # 判斷是否在抽煙
            elif key == "smoke":
                draw.text((loc["left"], loc["top"]+count), '行為：'+"吸煙",
                        fill= 'red', font=font_hei)
                count = count + 20

        # 輸出影像
        result = cv.cvtColor(np.array(pilimg), cv.COLOR_RGB2BGR)
        return_result["code"] = 200
        return_result["msg"] = "多主體監測成功！"
        return_result["origin_image"] = self.image_to_base64(image)
        return_result["result_image"] = self.image_to_base64(result)
        return_result["result_data"] = response
    else:
        return_result["code"] = 500
        return_result["msg"] = "百度介面呼叫失敗！"
        return_result["result_data"] = response
    # 即時視頻界面：@__app.route('/stream/<action>')
    #image：攝影機即時傳遞過來的影像
    #param_data：必須為 None
    else:
        return_result["result_image"] = self.image_to_base64(image)

    return return_result

# 單元測試，如果處理類別中引用了檔案，則在單元測試中要修改檔案路徑
if __name__=='__main__':
    # 建立影像處理物件
    img_object = BaiduDrivingBehaviorAnalysis()
```

```
    # 讀取測試影像
    img = cv.imread("./test.jpg")
    # 將影像編碼成資料流程
    img = img_object.imencode(img)

    # 設定參數
    param_data = {"APP_ID":"123456", "API_KEY":"123456", "SECRET_KEY":"123456"}
    img_object.font_path = "../../font/wqy-microhei.ttc"

    # 呼叫介面處理影像並傳回結果
    result = img_object.inference(img, param_data)
    if result["code"] == 200:
        frame = img_object.base64_to_image(result["result_image"])
        print(result["result_data"])

        # 影像顯示
        cv.imshow('frame',frame)
        while True:
            key=cv.waitKey(1)
            if key==ord('q'):
                break
        cv.destroyAllWindows()
    else:
        print(" 辨識失敗！ ")
```

5）百度語音辨識演算法介面設計

```
###########################################################################################
# 檔案：baidu_speech_recognition.py
# 說明：百度語音辨識介面
###########################################################################################
import os
import wave
import numpy as np
from aip import AipSpeech
import ffmpeg
import tempfile

class BaiduSpeechRecognition(object):
```

```
def __init__(self):
    pass

def __check_wav_file(self,filePath):
    # 讀取 wav 檔案
    wave_file = wave.open(filePath, 'r')
    # 獲取檔案的每秒顯示畫面和通道
    frame_rate = wave_file.getframerate()
    channels = wave_file.getnchannels()
    wave_file.close()
    if frame_rate == 16000 and channels == 1:
        return True
        #feature_path=filePath
    else:
        return False

def inference(self, wave_data, param_data):
    #code：辨識成功傳回 200
    #msg：相關提示訊息
    #origin_image：原始影像
    #result_image：處理之後的影像
    #result_data：結果資料
    return_result = {'code': 200, 'msg': None, 'origin_image': None, 'result_image':
None, 'result_data': None}

    # 應用請求介面：@__app.route('/file/<action>', methods=["POST"])
    #wave_data：應用傳遞過來的資料（根據實際應用可能為影像、音訊、視訊、文字）：
    # 語音資料，格式為 wav，每秒顯示畫面為 16000
    #param_data：應用傳遞過來的參數，不能為空
    if param_data != None:
        fd, path = tempfile.mkstemp()
        try:
            with os.fdopen(fd, 'wb') as tmp:
                tmp.write(wave_data)
            if not self.__check_wav_file(path):
                fd2, path2 = tempfile.mkstemp()
                ffmpeg.input(path).output(path2, ar=16000).run()
                os.remove(path)
                path = path2
```

```
                f = open(path, "rb")
                f.seek(4096)
                pcm_data = f.read()
                f.close()
            finally:
                os.remove(path)

        # 呼叫百度語音辨識介面，透過以下使用者金鑰連接百度伺服器
        #APP_ID：百度應用 ID
        #API_KEY：百度 API_KEY
        #SECRET_KEY：百度使用者金鑰
        client = AipSpeech(param_data['APP_ID'], param_data['API_KEY'], param_data['SECRET_
KEY'])
        # 語音檔案的格式為 pcm，每秒顯示畫面為 16000，dev_pid 表示普通話（純中文辨識）
        response = client.asr(pcm_data,'pcm', 16000, {'dev_pid': 1537,})

        # 處理伺服器傳回結果
        if response['err_msg']=='success.':
            return_result["code"] = 200
            return_result["msg"] = " 語音辨識成功！ "
            return_result["result_data"] = response['result'][0]
        else:
            return_result["code"] = 500
            return_result["msg"] = response['err_msg']
        return return_result

# 單元測試，如果處理類別中引用了檔案，則在單元測試中要修改檔案路徑
if __name__=='__main__':
    # 建立音訊處理物件
    test = BaiduSpeechRecognition()
    param_data = {"APP_ID":"12345678", "API_KEY":"12345678", "SECRET_KEY":"12345678"}
    with open("./test.wav", "rb") as f:
        wdat = f.read()
        result = test.inference(wdat, param_data)
    print(result["result_data"])
```

6）手勢辨識演算法介面設計

```
##########################################################################################
# 檔案：handpose_detection.py
# 說明：手勢辨識演算法
##########################################################################################
from PIL import Image,ImageDraw,ImageFont
import numpy as np
import cv2 as cv
import os
import json
import base64
c_dir = os.path.split(os.path.realpath(__file__))[0]

class HandposeDetection(object):
    def __init__(self, model_path="models/handpose_detection"):
        self.model_path = model_path
        self.handpose_model = HandDetector()
        self.handpose_model.init(self.model_path)

    def image_to_base64(self, img):
        image = cv.imencode('.jpg', img, [cv.IMWRITE_JPEG_QUALITY, 60])[1]
        image_encode = base64.b64encode(image).decode()
        return image_encode

    def base64_to_image(self, b64):
        img = base64.b64decode(b64.encode('utf-8'))
        img = np.asarray(bytearray(img), dtype="uint8")
        img = cv.imdecode(img, cv.IMREAD_COLOR)
        return img

    def draw_pos(self, img, objs):
        img_rgb = cv.cvtColor(img, cv.COLOR_BGR2RGB)
        pilimg = Image.fromarray(img_rgb)
        # 建立 ImageDraw 繪圖類別
        draw = ImageDraw.Draw(pilimg)
        # 設定字型
        font_size = 20
        font_path = c_dir+"/../../font/wqy-microhei.ttc"
```

```
        font_hei = ImageFont.truetype(font_path, font_size, encoding="utf-8")

        for obj in objs:
            loc = obj["location"]
            draw.rectangle((loc["left"], loc["top"], loc["left"]+loc["width"],
                        loc["top"]+loc["height"]), outline='green',width=2)
            msg = obj["name"]+": %.2f"%obj["score"]
            draw.text((loc["left"], loc["top"]-font_size*1), msg, (0, 255, 0), font=font_
hei)

            color1 = (10, 215, 255)
            color2 = (255, 115, 55)
            color3 = (5, 255, 55)
            color4 = (25, 15, 255)
            color5 = (225, 15, 55)
            marks = obj["mark"]
            for j in range(len(marks)):
                kp = obj["mark"][j]

                draw.ellipse(((kp["x"]-4, kp["y"]-4), (kp["x"]+4,kp["y"]+4)),
                        fill=None,outline=(255,0,0),width=2)
                color = (color1,color2,color3,color4,color5)
                ii = j //4
                if  j==0 or j / 4 != ii:
                    draw.line(((marks[j]["x"],marks[j]["y"]),(marks[j+1]["x"],marks[j+1]
["y"])),
                            fill=color[ii],width=2)

            draw.line(((marks[0]["x"],marks[0]["y"]),(marks[5]["x"],marks[5]["y"])),
fill=color[1],width=2)
            draw.line(((marks[0]["x"],marks[0]["y"]),(marks[9]["x"],marks[9]["y"])),
fill=color[2],width=2)
            draw.line(((marks[0]["x"],marks[0]["y"]),(marks[13]["x"],marks[13]["y"])),
fill=color[3],width=2)
            draw.line(((marks[0]["x"],marks[0]["y"]),(marks[17]["x"],marks[17]["y"])),
fill=color[4],width=2)
```

```
        result = cv.cvtColor(np.array(pilimg), cv.COLOR_RGB2BGR)
        return result

    def inference(self, image, param_data):
        #code：辨識成功傳回 200
        #msg：相關提示訊息
        #origin_image：原始影像
        #result_image：處理之後的影像
        #result_data：結果資料
        return_result = {'code': 200, 'msg': None, 'origin_image': None,
                         'result_image': None, 'result_data': None}

        # 即時視頻界面：@__app.route('/stream/<action>')
        #image：攝影機即時傳遞過來的影像
        #param_data：必須為 None
        result = self.handpose_model.detect(image)
        result = json.loads(result)
        if result["code"] == 200 and result["result"]["obj_num"] > 0:
            r_image = self.draw_pos(image, result["result"]["obj_list"])
        else:
            r_image = image
        return_result["code"] = result["code"]
        return_result["msg"] = result["msg"]
        return_result["result_image"] = self.image_to_base64(r_image)
        return_result["result_data"] = result["result"]
        return return_result

# 單元測試，如果處理類別中引用了檔案，則在單元測試中要修改檔案路徑
if __name__=='__main__':

    from handpose import HandDetector
    # 建立視訊捕捉物件
    cap=cv.VideoCapture(0)
    if cap.isOpened()!=1:
        pass
    # 迴圈獲取影像、處理影像、顯示影像
    while True:
```

```
        ret,img=cap.read()
        if ret==False:
            break
        #建立影像處理物件
        img_object=HandposeDetection(c_dir+'/../../models/handpose_detection')
        #呼叫影像處理函式對影像進行加工處理
        result=img_object.inference(img,None)
        frame = img_object.base64_to_image(result["result_image"])

        #影像顯示
        cv.imshow('frame',frame)
        key=cv.waitKey(1)
        if key==ord('q'):
            break
    cap.release()
    cv.destroyAllWindows()
else :
    from .handpose import HandDetector
```

5.2.2 開發步驟與驗證

5.2.2.1 專案部署

1）硬體部署

見 4.1.2.1 節。

2）專案部署

（1）執行 MobaXterm 工具，透過 SSH 登入到邊緣計算閘道。

（2）在 SSH 終端建立開發專案目錄。

（3）透過 SSH 將本專案開發專案程式和 aicam 專案套件上傳到 ~/aiedge-exp 目錄下，並採用 unzip 命令解壓。

```
$ unzip edge_autopilot.zip
$ unzip aicam.zip -d edge_autopilot
```

（4）參考 5.1.2.1 節的內容，修改專案設定檔 static/edge_fan/js/config.js 內的智雲帳號、硬體位址、邊緣服務位址等資訊。

（5）透過 SSH 將修改好的檔案上傳到邊緣計算閘道。

3）專案執行

（1）在 SSH 終端輸入以下命令執行開發專案。

```
$ cd ~/aiedge-exp/edge_autopilot
$ chmod 755 start_aicam.sh
$ conda activate py36_tf114_torch15_cpu_cv345     //Ubuntu 20.04 作業系統下需要切換環境
$ ./start_aicam.sh
```

（2）在使用者端或邊緣計算閘道端開啟 Chrome 瀏覽器，輸入專案頁面位址 https: //192.168.100.200:1446/static/edge_autopilot/index.html，即可查看專案內容。

5.2.2.2 輔助駕駛系統的驗證

1）智慧物聯

（1）應用設定：第一次登入輔助駕駛系統後，按一下首頁右上角的「設定」按鈕可以設定相關的參數（在預設情況下，系統會讀取 static/edge_autopilot/js/config.js 內的初始配置），如智雲帳號、節點位址、百度 AI 帳號等。設定參數後，在智雲帳號介面按一下「連接」按鈕即可連接到資料服務，連接成功後會彈出訊息提示。輔助駕駛系統的應用設定頁面如圖 5.17 所示。

▲ 圖 5.17 輔助駕駛系統的應用設定介面

（2）應用互動。在輔助駕駛系統的首頁可以看到車內硬體的資料，包括 Sensor-A 擷取類感測器的資料〔如溫度、濕度、空氣品質（TVOC）〕，Sensor-B 控制類感測器的資料〔車窗、風扇（通風）、空調〕，如圖 5.18 所示。透過觀察虛擬模擬平臺或實際硬體的變化，可判斷應用互動的結果。

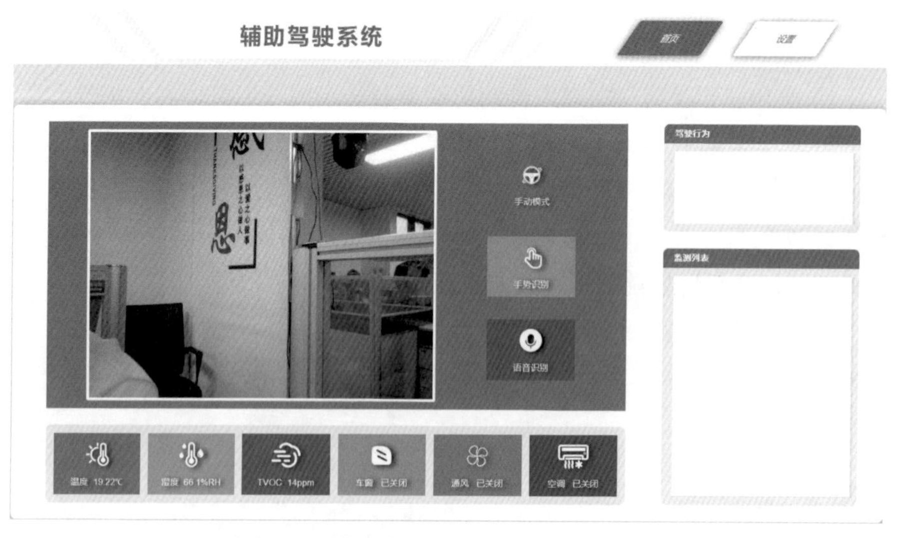

▲ 圖 5.18 輔助駕駛系統控制平臺（手動）

Sensor-A 擷取類感測器在預設情況下每 15 s 更新一次資料，首頁的資料會相應地進行更新。Sensor-B 控制類感測器可透過按一下首頁中的相應圖示來進行開關操作。

（3）模式控制。按一下輔助駕駛系統首頁中的模式按鈕可切換手動模式和智慧模式。當設定為智慧模式時，AI 辨識功能將不可用，系統在預設情況下會每隔 15 s 獲取攝影機即時拍攝的駕駛行為影像，並呼叫百度駕駛行為分析演算法進行分析。

① 未檢測到駕駛員時的介面如圖 5.19 所示。

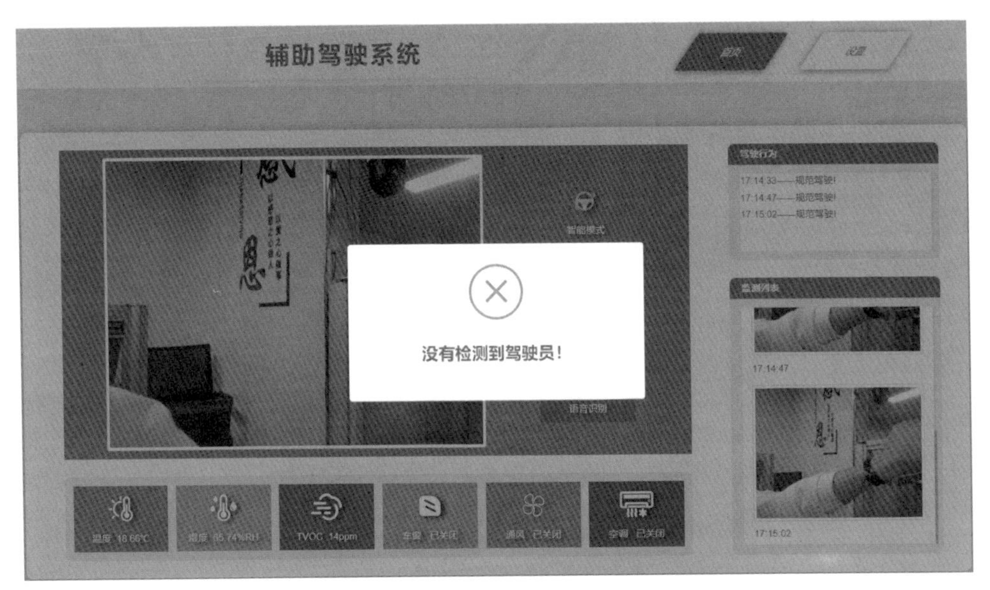

▲ 圖 5.19 未檢測到駕駛員時的介面

② 檢測到駕駛員時的介面如圖 5.20 所示。

▲ 圖 5.20 檢測到駕駛員時的介面

③ 駕駛行為檢測影像如圖 5.21 所示。

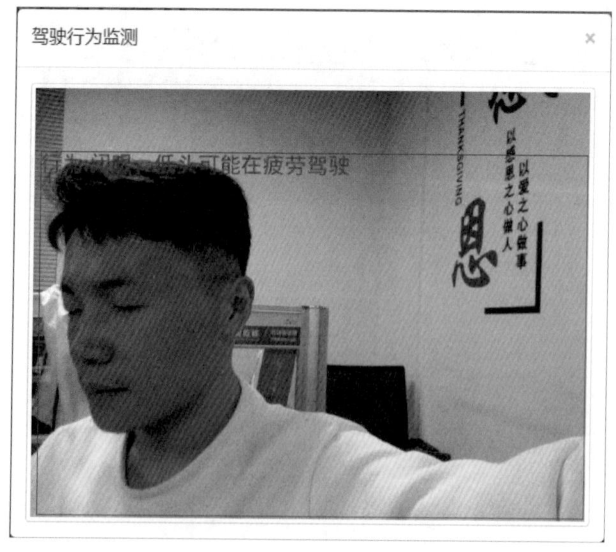

▲ 圖 5.21 駕駛行為檢測影像（不正常行為）

2）手勢互動

（1）在手動模式下，在攝影機的視線內擺好手勢即時進行手勢辨識，成功辨識到手勢後系統可自動對相應的裝置進行開關操作，如圖 5.22 所示。系統連續 5 次辨識同一個手勢則進行相應的操作，手勢 1 表示對車窗進行開關操作，手勢 2 表示對風扇進行開關操作，手勢 3 表示對空調進行開關操作。

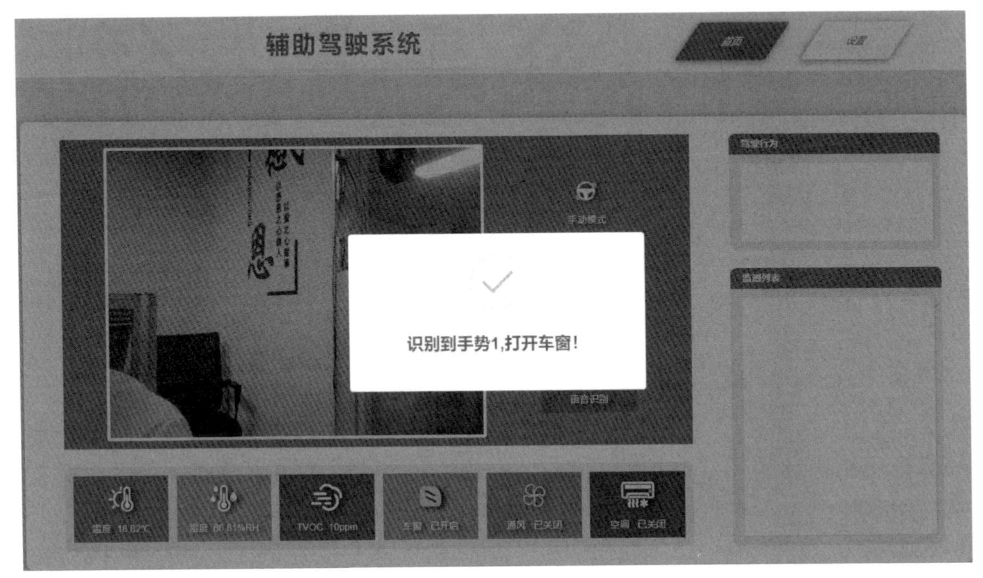

▲ 圖 5.22 手勢辨識成功（開啟車窗）

（2）手勢辨識成功後，再次辨識到同樣的手勢時，如果之前裝置處於開啟狀態，則本次操作將關閉該裝置，如圖 5.23 所示。

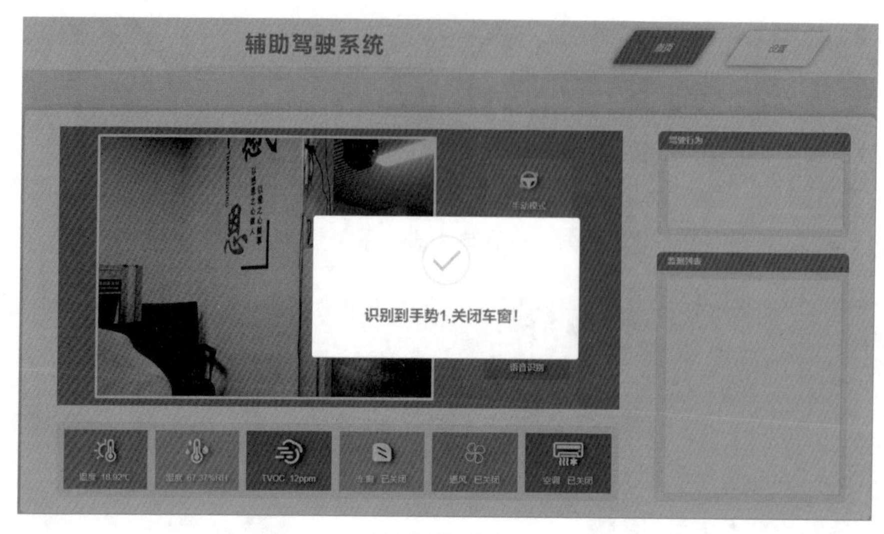

▲ 圖 5.23 手勢辨識成功（關閉車窗）

3）語音互動

（1）在手動模式下，按一下首頁中的「語音辨識」按鈕即可開始錄製語音，再次按一下該按鈕可結束錄製並進行語音辨識，根據語音辨識結果對車內裝置進行開關操作。舉例來說，錄製「開啟車窗」語音後，輔助駕駛系統將開啟車窗，如圖 5.24 所示。

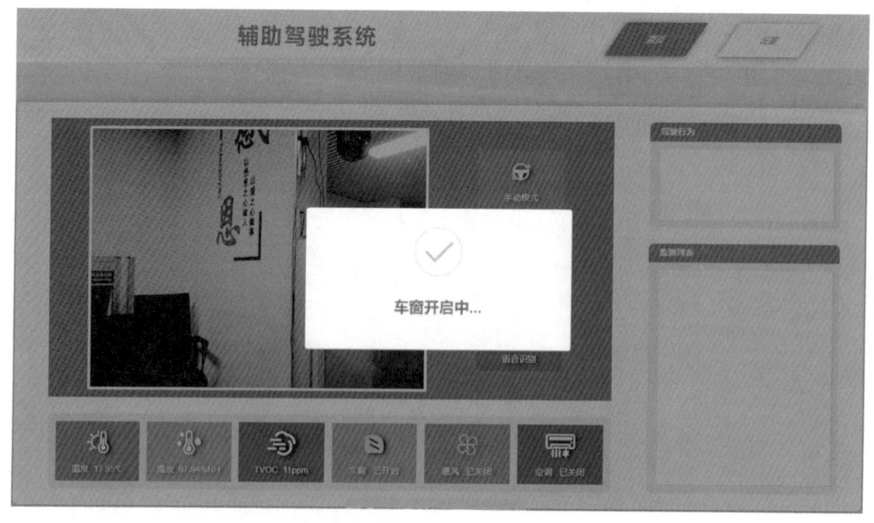

▲ 圖 5.24 語音辨識控制成功的介面（開啟車窗）

5.2.3 本節小結

本節首先分析了智慧輔助駕駛的需求和發展趨勢；然後介紹了輔助駕駛系統的系統框架和系統功能，在此基礎上完成了硬體設計和通訊設計；接著借助AiCam 平臺舉出了輔助駕駛系統的開發流程、演算法互動、語音辨識控制裝置、駕駛行為分析演算法介面設計、百度語音辨識介面設計、手勢辨識演算法介面設計；最後對輔助駕駛系統進行了驗證。

5.2.4 思考與擴充

（1）智慧輔助駕駛需要什麼前提條件？

（2）如何實現駕駛行為分析？

（3）簡述輔助駕駛系統的開發過程。

參考文獻

[1] 英特爾亞太研發有限公司．邊緣計算技術與應用 [M]．北京：電子工業出版社，2021．

[2] 楊術．5G 新時代與邊緣計算 [M]．北京：電子工業出版社，2022．

[3] 蔔天聰．電腦視覺導向的高能效邊緣計算架構關鍵技術研究 [D]．長春：吉林大學，2022．

[4] 雷鑫．基於邊緣計算與深度學習的即時人臉辨識關鍵技術 [D]．南昌：南昌大學，2021．

[5] 冉雪．基於 YOLO 的物件辨識演算法設計與實現 [D]．重慶：重慶大學，2020．

[6] 張明偉．基於邊緣計算的物件辨識與辨識演算法 [D]．福州：福建師範大學，2020．

[7] 顧笛兒，盧華，謝人超，等．邊緣計算開放原始碼平臺整體說明 [J]．網路與資訊安全學報，2021，7（2）：22-34．

[8] 徐坤坤．5G MEC 導向的邊緣計算平臺實現和部署方案研究 [D]．深圳：深圳大學，2020．

[9] 王健．輕量級邊緣計算平臺方案設計與應用研究 [D]．北京：北京郵電大學，2019．

[10] 趙梓銘，劉芳，蔡志平，等．邊緣計算：平臺、應用與挑戰 [J]．電腦研究與發展，2018，55（2）：327-337．

[11] 劉晶宇，楊鵬．基於 YOLOv5 改進的遙感影像物件辨識 [J]．電腦時代，2023（7）：50-55．

[12] 張勇．基於 YOLOv3 的交通物件辨識演算法研究 [D]．淮南：安徽理工大學，2021．

[13] 康莊．基於改進 YOLOv3 的交通樞紐行人檢測與追蹤技術研究 [D]．贛州：江西理工大學，2021．

[14] 李珣，劉瑤，李鵬飛，等．基於 Darknet 框架下 YOLOv2 演算法的車輛多物件辨識方法 [J]．交通運輸工程學報，2018，18（6）：142-158．

[15] 劉瑤．改進 Darknet 框架的多物件辨識與辨識方法研究 [D]．西安：西安工程大學，2019．

[16] 郭濤，郭家，李宗南，等．基於 Darknet 深度學習框架的桃花檢測方法 [J]．中國農業資訊，2021，33（6）：25-33．

[17] 陳旭．基於深度學習的無人機影像物件辨識演算法研究 [D]．杭州：杭州電子科技大學，2022．

[18] 黃煜真，元澤懷，陳嘉瑞，等．PyTorch 框架下基於 CNN 的人臉辨識方法研究 [J]．資訊與電腦（理論版），2022，34（10）：193-195．

[19] 李蔣．基於深度學習 PyTorch 框架下 YOLOv3 的交通號幟燈檢測 [J]．汽車電器，2022（6）：4-7．

[20] 黃玉萍，梁煒萱，肖祖環．基於 TensorFlow 和 PyTorch 的深度學習框架對比分析 [J]．現代資訊科技，2020，4（4）：80-82，87．

[21] 焦利偉，張敏，麻連偉，等．基於 PyTorch 框架架設 U-Net 網路模型的遙感影像建築物提取研究 [J]．河南城建學院學報，2020，29（4）：52-57．

[22] 王甜．PyTorch 至 ONNX 的神經網路格式轉換的研究 [D]．西安：西安電子科技大學，2022．

[23] 李秉濤・基於輕量級 CNN 的即時物件辨識研究 [D]・貴陽：貴州大學，2022・

[24] 張英傑・深度卷積神經網路嵌入式推理框架的設計與實現 [D]・廣州：華南理工大學，2020・

[25] 邊緣計算產業聯盟，機器視覺產業聯盟，智慧視覺產業聯盟・邊緣計算視覺基礎設施白皮 [R]・北京：邊緣計算產業聯盟，2022・

[26] 百度智慧雲・「雲智一體」技術與應用解析系列白皮書（第三期）：智慧物聯網篇 [R]・北京：百度，2021・

[27] 中興通訊股份有限公司・中興通訊 Common Edge 邊緣計算白皮書 [R]・深圳：中興通訊股份有限公司，2021・

[28] 邊緣計算產業聯盟，工業網際網路產業聯盟・邊緣計算參考框架 2.0[R]・北京：邊緣計算產業聯盟，2017・

[29] Zhang J, Yan Y; Lades, M. Face recognition: eigenface, elastic matching, and neural nets[J]. Proceedings of the IEEE,1997, 85 (9): 1423-1435.

[30] 葛宏孔，羅恒利，董佳媛・基於深度學習的非實驗室場景人臉屬性辨識 [J]・電腦科學，2019,46（z2）：246-250・

[31] Ramadhan M V, Muchtar K, Nurdin Y, et al.. Comparative analysis of deep learning models for detecting face mask[J]. Procedia Computer Science, 2023, 216: 48-56.

[32] Li F, Li X, Liu Q, et al.. Occlusion handling and multi-scale pedestrian detection based on deep learning: a review[J]. IEEE Access, 2022,10: 19937-19957.

[33] Kumar A, Kaur A, Kumar M, et al.. Face detection techniques: a review. artificial intelligence review[J]. Multimedia Tools and Applications, 2019, 52: 927-948.

[34] Xu C, Li Z, Tian X, et al.. Vehicle detection based on modified YOLOv3 and deformable convolutional network[J]. IEEE Access, 2019,7: 65763-65772.

[35] Wu Y, Ji S, Wang Y, et al.. Real-time traffic sign recognition based on YOLO and deep residual network[C]. In Proceedings of the 2020 IEEE International Conference on Big Data, Artificial Intelligence and Internet of Things Engineering, 2020.

[36] Chan W Y, Chiu C C, Gales M J, et al.. Listen, attend and spell[C]. In Proceedings of the 2016 IEEE International Conference on Acoustics, Speech and Signal Processing (ICASSP), 2016.

[37] Hu J, Shen L, Sun G. Squeeze-and-excitation networks[C]. Proceedings of theIEEE Conference on Computer Vision and Pattern Recognition, 2018.

[38] Ren S, He K, Girshick R, et al.. Faster R-CNN: towards real-time object detection with region proposal networks[J]. IEEE Transactions on Pattern Analysis and Machine Intelligence, 2017, 39(6): 1137-49.

[39] Xu C, Li Z, Tian X, et al.. Vehicle detection based on modified YOLOv3 and deformable convolutional network[J]. IEEE Access, 2019, 7: 65763-65772.

[40] 唐昊．基於 MobileNet v2 的輕量級人臉辨識神經網路系統的設計與實現[D]．重慶：重慶大學，2022．

[41] Shi B, Bai X, Yao C. An End-to-End trainable neural network for image-based sequence recognition and its application to scene text recognition[J]. IEEE Transactions on Pattern Analysis and Machine Intelligence, 2017, 39(11):2298-2304.

[42] Yang J, Liu D, Zhang J, et al.. Vehicle license plate recognition method based on convolutional neural network[J]. IEEE Access, 2021, 9: 24405-24416.

MEMO

深智數位
股份有限公司

深智數位
股份有限公司